Wayward Women

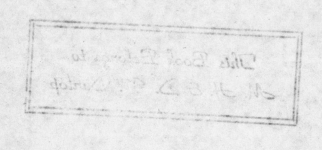

Wayward Women

A Guide to
Women Travellers

JANE ROBINSON

Oxford   New York
OXFORD UNIVERSITY PRESS

Oxford University Press, Walton Street, Oxford OX2 6DP
Oxford New York Toronto
Delhi Bombay Calcutta Madras Karachi
Kuala Lumpur Singapore Hong Kong Tokyo
Nairobi Dar es Salaam Cape Town
Melbourne Auckland Madrid
and associated companies in
Berlin Ibadan

Oxford is a trade mark of Oxford University Press

First published 1990
First issued as an Oxford University Press paperback 1991

British Library Cataloguing in Publication Data
Robinson, Jane
Wayward women: a guide to women travellers.
1. Travel by women—Biographies
I. Title
016.91082
ISBN 0–19–282822–3

Library of Congress Cataloging in Publication Data
Robinson, Jane.
Wayward women: a guide to women travellers/Jane Robinson.
p. cm.
Reprint. Originally published 1990.
Includes bibliographical references and index.
1. Women travellers—Bibliography. I. Title
016.91082—dc20 Z6011.R65 1991 [G200] 90–47140
ISBN 0–19–282822–3

5 7 9 10 8 6

Printed in Great Britain by
Biddles Ltd.,
Guildford and King's Lynn

To Helen; for Neil

Preface

Wayward Women was originally going to have been a plain bibliography: a book about books, not people. I was working in an antiquarian bookshop specializing in travel when the idea of writing it was first suggested to me, and had often bewailed (in a purely academic sense) the lack of a reference guide to what seemed an important but little-documented body of literature. Women travellers had been writing about their journeys for sixteen centuries, after all: why should the traditional image of the species always be that of an intrepid Victorian spinster vigorously prodding the ends of the earth with her parasol? If I could produce a detailed list of all the first-hand travel accounts ever written by women in the English language, not only would an informed new interest in the genre be aroused amongst other booksellers, collectors, librarians, and the like —but my reputation as a bibliographer would be made. Naturally I need not become involved with the women themselves: the books were the thing. Or so I thought.

Snatching all the time I could from my work in the shop I duly began to amass a scholarly array of titles, collations, and issue-points, but of course it did not take long to realize that the exhaustive list I had so airily planned would take a lifetime at least to compile. And anyway, the more information I gathered about these women's books, the less I could resist the temptation to commit the bibliographer's cardinal sin and actually read what they had written. Titles like Kate Marsden's *On Sledge and Horseback to Outcast Siberian Lepers* and Annie Hore's *To Lake Tanganyika in a Bath Chair* quite got the better of me and before I knew what was happening, I had given up my job and embarked on a prodigal journey of my own accompanied by a horde of spirited companions—and all hopes of a bare-bones bibliography were lost.

If I was going to write about the women as well as their

books, I had to decide how to organize them. The female sex has always suffered from retrospective overgeneralization: everyone knows, for example, that medieval women were either peasant witches or amoral aristocrats (and that includes the nuns); all Augustan ones were pampered whores with bulging brown eyes or else vituperative intellectuals; and the Victorian matron was invariably either small and submissive or large and indomitable. A female who travelled abroad was a strange creature in any age—although by the nineteenth century perhaps not quite as shocking as before. Then, if she were a Lady, she could go where she pleased, given suitable male protection, as long as it lay within the bounds of civilization (i.e. the British empire), and even if she were a mere woman she need not be entirely useless. On the contrary, if the American explorer Samuel Hearne was to be believed, 'Women were made for labour: one of them can carry or haul as much as two men do. They also pitch our tents, make and mend our clothing, keep us warm at night . . . they do everything, and are maintained at a trifling expense.' Women did not write books, of course, but if a Lady should take an urge to record her sojourn, the book she produced was bound to be predictable: vastly religious if it was a pilgrimage, lighthearted if a tour, and laughable if it pretended to be an 'expedition'.

But by their very nature, women travel writers were in the past (and still are, to some extent) a nonconformist race, and I soon discovered that their own accounts roundly defy any pigeon-holing. In the fourth century, for instance, in the abbess Etheria's account of a pilgrimage to the Holy Land we read not only of the religious significance of her journey, but of the doltish tour-guide she has been allotted and the difficulties of mountaineering on Mount Sinai. Mary Wollstonecraft is celebrated as one of our pioneer feminists—and yet the secret voyage she made in 1795 to Scandinavia was undertaken all for the love of a cad. Isabella Bird, whose pan-global travels crowded the second half of the nineteenth century, was a meek and dutiful woman at home, but once let loose in what she sublimely called 'the congenial barbarism of the desert' she assumed a most unladylike 'up-to-anything free-legged air', while her contemporary Mary Kingsley canoed herself serenely through the white waters of West African rivers dressed impeccably in black silk and bonnet. Closer to our own time, some of the most glamorous society gals of the 1920s and 1930s were likely to feel just as comfortable in *The Tatler* as in the oily cockpit of a record-breaking Gypsy Moth, or stalking dinner in some God-forsaken central-Asian wasteland. It is no good trying to type-cast them.

In the end I rashly resolved to label these unruly individuals according to the sort of traveller they were on setting forth: pioneer (as in the opening chapter 'Untrodden Peaks and Unfrequented Valleys'), evangelist ('Quite Safe Here with Jesus'), emigrant ('Life in the Bush'), and so on. But even these labels, it must be said, have a tendency to flutter away in the mêlée. One traveller frequently encroaches upon the territory of another, so that missionaries who write about their travels (encouraged to sugar the pill of proselytism with hair-raising tales of adventures in the field) might easily become confused with explorers who discover God on a mountain summit. Governesses and nurses fall prey to wanderlust on their way to an appointment in St Petersburg or the Crimea, and a female foreign correspondent turns anthropologist with as much ease as an army officer's wife turns professional tourist. In fact one of the few things these women have in common is their very originality. All of them have voyaged 'over the straits'—not just, like Louisa Meredith (from whom I borrow the term), across seas from the familiar to a strange land, but across the boundaries of convention and traditional feminine restraints too. And they are stalwart sailors all.

Just to parenthesize a moment, I think it is this singularity that sets a woman's account of her travels apart from a man's. Throughout the centuries spanned by this book, men have been setting out for the world with some definite purpose in mind, with reputations to forge and patrons to please, and their written accounts have been dedicated to tangible results. Women, whether travelling by choice or default, had for the most part no such responsibilities: they left the facts and figures of foreign travel to the men, and dwelt instead on the personal practicalities of getting from A to B, and on impressions. And now that I have myself hurtled headlong into the trap of overgeneralizing, I may as well finish the job properly: men's travel accounts are traditionally concerned with What and Where, and women's with How and Why.

Having explained the difficulties of arranging the authors in *Wayward Women* into some sort of manageable order, I should say a word on how the book was eventually organized. I selected some four hundred writers, all using English as a first language (except for the matchless Etheria whose work was written almost before English existed), mostly of British extraction, and always travelling beyond the frontiers of their native land. A biographical sketch of each lady is headed by the brief first-edition details of her travel accounts, including the books' titles and imprints (transcribed exactly from pre-1850 publications and standardized from later ones), their size (octavo unless otherwise described), their pagination, and the number and nature of their illustrations. I have only listed *first-hand* travel accounts *in book form*—for integrity's sake I have not mentioned articles, fictional or historical works, and so on. The whole company is sorted into chapters, each with a short introduction, and at the end of the book I have included an index of authors, a geographical index, and a list of useful reference books.

Although several publishers have been commendably active on the re-issue front for some time now, many of the books I have described are long out of print and their authors quite forgotten. My final aim in writing *Wayward Women* has been to uncover a wealth of literature—peopled by an astonishing and extravagant array of characters—literature at best of classic quality (and at worst deliciously McGonagallesque) and too long hidden in obscurity. In doing so, perhaps I can repay these women travellers in some small degree for all the courage, wisdom, the sense of discovery and of freedom, and the irrepressible high spirits they have shared with me.

J. H. R.

London,
June 1989

Acknowledgements

IF there were any justice, this list of acknowledgements would be as long as the book: so many people have helped with knowledge, encouragement, and enthusiasm, and I am indebted to them all. Particular thanks must go to the author Peter Hopkirk, whose idea this book was in the first place and whose generous advice has been invaluable; to Caroline Schimmel for her indefatigable support; and to Barbara Grigor-Taylor, my erstwhile employer, for her time and inspiration; also to the Librarians and staff of the following institutions: the Alpine Club, Bodleian Library, British Library, Church Missionary Society, London Library, National Maritime Museum, Reading University Library, Royal Commonwealth Society, Royal Geographical Society, Thomas Cook Ltd.'s Archives, Wednesbury Art Museum, and the Schlesinger Library at Harvard. I owe special thanks too to those authors and publishers who allowed me to search their memories and archives; to my past and present editors at the Oxford University Press; and, for various kind services rendered, to Grace Dibble, Christina Dodwell, Paul Goodwin, Sheila Lochhead, Ella Maillart, Dorothy Middleton, Nell Penfold, Helen Robinson, Paulette Rose, the late Miles Smeeton, Iain and Judith Sproat, Arthur Staples, Edmund Swinglehurst, John Theakstone, Esmé Vizetelly, Carole Walker, David and Carl Warrington, Margaret Washington, Elizabeth Wessels—and Bruce.

Contents

List of Plates

Untrodden Peaks and Unfrequented Valleys

TRADITIONALLY, pioneering has always been a dangerous and male preserve. There was no room for women in the open boats of the pilgrim navigators setting out in the early centuries of Christendom to win new lives (or lose their own) for God, nor in the craft and caravans of those marauding for power and knowledge in their wake. The swashbucklers of the sixteenth and seventeenth centuries, commissioned to sail new seas, map new lands, and cross new continents for their patrons were chiefly the well-favoured gentlemen of the age; and when their empires were built, they were founded on those stout and masculine qualities of patriotism, honour, and a noble lust for challenge and adventure.

Women meanwhile (and just as traditionally) cared more for their family than their country, more for loyalty than integrity, and loved obedience and duty beyond all things. They were too precious and too poorly designed to travel unless they must (although with the growth of colonialism they had to more and more). Only the odd accompanied tourist carried along the elegant routes of the European Grand Tour could ever be expected both to travel and maintain her reputation as a Lady of Quality; for the rest, going where few visitors had gone before was a long, fearful, and vaguely vulgar business. Until the women of this chapter came along, that is.

Perhaps the first to break the mould was Lady Mary WORTLEY MONTAGU, whose decision to join her husband on his appointment as British Ambassador to Turkey in 1716 left the London *beau monde* quite aghast. Like other pioneering women travellers of the eighteenth and early nineteenth centuries—Mary Ann PARKER, who followed the First Fleet to Australia, for example, or Mrs ELWOOD on her epic trek from Eastbourne to Bombay—Lady Mary Wortley Montagu started out merely as an 'accompanist'; once left to herself she became curious, however, and this curiosity developed into a decided appetite for travel. Excessive curiosity was an ill-bred trait in a woman, of course (Ida PFEIFFER scandalized polite society by spinning twice round the world between the ages of fifty and fifty-five just to *see* it), but translated into a book and disguised as literature, such curiosity became quite acceptable. Fashionable, even.

By the time Mrs COLE was writing her enthusiastic (but eminently decorous) guide to the unfeminine art of mountaineering in 1859—and although she herself did not quite have the courage to put her own name to the book—increasing numbers of women were finding a precedent to abandon reputation for the sake of new experience. Leaving their less adventurous sisters to what Louisa JEBB witheringly remembers as a life of 'crochet and drawing rooms', pioneers like Elizabeth LE BLOND (who terrorized her relations by leaping up Alps) and Lady Anne BLUNT (riding deep into the Arabian desert dressed as a bedouin) plunged dauntlessly and headlong into the search, through travel, for freedom.

With freedom came independence: male chaperones were no longer called upon as shields and defenders; they were rather pressed into service as expedition advisers, sponsors, and bureaucratic decoys. With independence, in turn, came confidence. Annie PECK had no doubt that she could achieve her ambition—the first ascent of Mount Huascaran in Mexico—but if she failed, it would not be because she was a spinster travelling alone. And it was precisely because Alexandra DAVID-NEEL was forbidden on pain of death to make her epic journey that she set out in 1923 to become the first European woman to reach the sacred Tibetan city of Lhasa: she had faith in herself, as every pioneer must.

The stalwart spirit is still flourishing: with the flaunted equalities of the twentieth century lady travellers have emerged as stern competitors not just against men (despite Fanny Bullock WORKMAN's protestations), nor even against history (although the exploits of Mrs BRUCE, Amy JOHNSON, and Naomi JAMES are all the stuff of record books)—but against themselves. Nowadays philosophical challenges are as important as physical ones, and each journey has become a testing and very personal voyage of discovery: a new era of travel has begun.

BAKER, Florence Barbara Maria (1841–1916)

Baker, Anne (ed.): *Morning Star: Florence Baker's diary of the expedition to put down the slave trade on the Nile 1870–1873.* London: Wm. Kimber, 1972; pp. 240, with 17 halftones, 4 facsimile leaves, and a map

Hall, Richard: *Lovers on the Nile* [the lives and travels of Samuel and Florence Baker]. London: Collins, 1980; pp. 254, with 27 halftones and a map

Not so long ago, an old diary was found in an English attic. The writing was strong and individual, the language engagingly idiosyncratic, and the adventures it recorded almost incredible. The author turned out to be the lover (and later the wife) of one of Victorian Britain's foremost African explorers, and her subject was a journey she undertook in the early 1870s into the heart of the 'Dark Continent' at his side.

Florence Barbara Maria Finnian von Sass must be one of the best-kept secrets in the annals of African exploration. Her achievements were unprecedented, but Victorian propriety and her own sense of absolute loyalty pushed her stiffly into the background whilst her companion, Sir Samuel Baker, was fêted as champion and darling of the empire. Her history is suitably mysterious. Sportsman Baker had been four years widowed when in 1859 he accepted an invitation to hunt boar by the Danube. Passing through the Hungarian town of Widin he came upon a slave sale in the market-place (so the legend goes) and fell promptly in love with the lot about to be sold to a lascivious Turk. It was later claimed that Florence was an orphan refugee of noble Romanian stock, but Baker neither knew nor cared; he simply outbid the Turk and swept off the beautiful seventeen-year-old on horseback.

To avoid embarrassing the aunts who were looking after his children back home, Baker stayed in Hungary whilst the inseparable couple planned their travelling future. Marriage somehow seemed irrelevant, he said; and if Florence was a slave still, then so, no less, was he. They went everywhere

together. Their two great expeditions—the first to Abyssinia (Ethiopia) in
1861–4 in search of the source of the Nile, and the second (described in her
diary) to report on and if possible suppress the flourishing trade in slaves
along its banks—were riddled with hardships. Not only had Florence to
cope with the usual desperate fevers, lack of food, and constant danger of
wild animals (or men), but with the added complications of camp mutinies,
tribal wars, and treacherous local politics too, and all against a background
of the accumulating fatigue of years' marching in an enervating climate.
Still, in her husband's own words, she possessed 'a share of sang-froid
admirably adapted for African travel. Mrs Baker is not a screamer.' She
never faltered, and would vary her routine by spending one 'rest' day busy
bottling pickles, perhaps, sewing trousers, or making beads for currency,
another preparing the seemly celebration of Afternoon Tea in a jungle
clearing or gamely tackling a 'not very young hippo' for dinner (washed
down with champagne), and the next optimistically writing home for ' 1 pair
best rather short French stays with 6 pr. silk long stay laces' and '12 good
fine handkerchiefs'.

Although Sam and Florence married in London in 1865, there was
always a vague sense of scandal associated with the new Lady Baker. Queen
Victoria refused to accept her at court, for example, and London society,
whilst thrilled to accommodate fashionable Sir Sam, always preferred his
eccentric and rather dubious peccadillo of a wife to be left at home. But
Florence was quite happy to be eclipsed and was satisfied to care for Baker
(increasingly 'not quite the thing', as she put it, in his old age) at their Devon
home until his death in 1893. She lived on for another twenty-three years,
cared for by her stepchildren (the eldest of whom was only six years younger
than she) and proud to remember her private role in Baker's very public
success.

❖

BELL, Gertrude Lowthian (1868–1926)

[Anon.]: *Safar Nameh: Persian Pictures. A Book of Travel.* London: Richard Bentley,
1894; pp. 294

The Desert and the Sown. London: Heinemann, 1907; pp. 347, with colour frontispiece, 161 halftones, and a map

Amurath to Amurath. London: William Heinemann, 1911; pp. 370, with 235
halftones and a map

The Letters of Gertrude Bell. Selected and Edited by Lady Bell. London: E. Benn,
1927; 2 vols., pp. 402/(403)–791, with 39 halftones and 2 maps

The Earlier Letters of Gertrude Bell. Collected and Edited by Elsa Richmond. London:
E. Benn, 1937; pp. 347, with 8 halftones

Burgoyne, Elizabeth: *Gertrude Bell from her Personal Papers.* London: E. Benn, 1958–61; 2 vols., pp. 320/399, with 30 halftones and a facsimile leaf

Gertrude Bell's life was composed of a potent mixture of the traditional and the unconventional. She was born into a wealthy, enlightened family of industrialists in the north of England and although she became in 1888 the first woman ever to take a first-class degree in Modern History at Oxford, and learned languages like Farsi and Arabic in her youth as though they were French or German, she spent the first four years of what might have been a career living quietly at home as her stepmother's companion. And although she fell wildly in love with a young British diplomat on a visit to her relative the Ambassador at Tehran in 1892, when forbidden by her parents to marry him she came straight and meekly home, eschewing the palpable romance of Persia for daughterly obedience. She was nearly twenty-five. She travelled twice round the world with her family, and between 1899 and 1902 became so proficient a scrambler in the Alps that the great Himalayan pioneer Filippo de Filippi offered her a place on his next expedition, and William Coolidge described her as 'the best of all mountaineers'. And yet she still wrote to ask her father's permission before undertaking any journey, and *never* walked in London without a chaperone. She was an ardent anti-suffragette, as vulnerable to the expectations of others as any well-brought-up young lady of her time—and one of the most original travellers and explorers Britain has ever produced.

When Gertrude's father died in 1904, she bolted back to the East, the scene of the one love of her life; it was familiar enough to be comforting (she had translated the Persian *Poems from the Divan of Hafiz* in 1897 and had studied its ancient history since her last visit), and it was anonymous enough to stimulate her precocious intellect. In 1905, then, with all the money she would ever need and an occupation in archaeology to keep her amused, she set out for what she called 'the empty world'.

Up until 1913, Gertrude travelled extensively in Syria (*The Desert and the Sown*), along the Euphrates in what is now Iraq (*The Thousand and One Churches*), into eastern Turkey (*Amurath to Amurath*), and then to Assyria (*The Palace and Mosque of Ukhaidir*). She collected scraps of knowledge like shards along the way, and armed herself peaceably with curiosity and an acute political awareness; soon she had become accepted and trusted by everyone she met (albeit as an unfathomable but benevolent eccentric), whatever their tribe or sympathies.

In 1914 Gertrude set out on her most ambitious journey, an expedition into the uncharted heart of Arabia. And even though the reverberations of the outbreak of war meant she must turn back at Hail, she had still penetrated further into the desert than any other European woman, including her friend Lady BLUNT. This unprecedented trek was not just the usual matter of wanderlust, scholarship, and curiosity. Gertrude had fallen desperately in love—*really* desperately—with a British officer in Arabia, Major Charles Doughty-Wylie; this time it was not her duty to her parents that thwarted her, but Doughty-Wylie's to his wife, whom he would not leave. Gertrude could only endure the affair by travelling when the major could not be with her, and by writing to him almost daily from the depths of the desert ('I have filled all the hollow places in the world with my desire for you'). And even then it was hardly endurable: just before his death at Gallipoli, she issued him an ultimatum—and bitterly regretted it ever afterwards.

To assuage her confusion after Gallipoli Gertrude threw herself into the war by enlisting first as a nurse in France and then, more suitably, as a political officer with the Arab Intelligence Bureau in Cairo. After a brief stint in India in 1916 she settled in Baghdad as the only female member of the British Expeditionary Force in Mesopotamia, and there she stayed—first as political adviser to King Faisal in the new-born country of Iraq and then, as her services were declined and then forgotten, as the director of the National Museum she founded there in 1923. The impulse to travel seems to have atrophied with Doughty-Wylie's death, there being nothing left to return to any more once the travelling was done; as her life became less and less demanding—less and less fulfilling—she grew passive, faded unwontedly into the Iraqi background, and two days before her fifty-eighth birthday took an overdose and quietly died.

◆

BLUNT, Lady Anne, Baroness Wentworth (1837–1917)

Bedouin Tribes of the Euphrates. Edited, with some account of the Arabs and their Horses, by W[ilfrid] S[cawen] B[lunt]. London: John Murray, 1879; 2 vols., pp. 346/283, with 12 wood engravings and a map.

A Pilgrimage to Nejd, The Cradle of the Arab Race. A Visit to the Court of the Arab Emir, and 'Our Persian Campaign'. London: John Murray, 1881; 2 vols., pp. 273/283, with 14 wood engravings, 17 vignettes, and a map

One would expect the granddaughter of Lord Byron and wife of the extravagant Arab scholar and poet Wilfrid Scawen Blunt to be rather an exotic and flamboyant figure. In fact Lady Anne Blunt was one of the coolest and most level headed of all lady travellers. These two works, describing journeys made between 1877 and 1879, are weighty with careful observations—natural, historical, and political—and soon became standard texts on their area as well as classic travel accounts.

Anne Isabella was the only daughter of the first Earl of Lovelace and, as such, in receipt of a hefty annual income. This enhanced her chances both of marriage and of travel, and although she and Blunt separated after thirty years together, they made good pilgrim companions to their own particular promised land of Arabia. Both travelled unsentimentally and with no other motive than pure and learned curiosity. To find out how the bedouin lived, Lady Anne lived like one herself: she became a temporary nomad, riding the two thousand miles from the Mediterranean to the Persian Gulf for the most part in Arab dress, and without guides or the usual caravan. This was quite an innovation, and prompted Blunt to dub his wife 'the first bona-fide tourist who has taken the Euphrates Road'.

Also unprecedented was the way in which these two 'persons of distinction from England' were treated by the bedouin. Because they travelled unprejudiced and unassuming, the Blunts were equally trusted and accepted (even though one of them was only a woman); indeed, the Amir himself presented them with his choicest brood mares to take home for their famous Crabbet Arabian Stud. Despite all this, Lady Anne denied that she was in any way remarkable—she had just had 'the good fortune', she said, 'to see a little more than is generally seen, and to learn a few things more than are generally known'.

❧

BRUCE, The Honourable Mrs Victor (Mildred Mary) (1895–1990)

Nine Thousand Miles in Eight Weeks: Being an Account of an Epic Journey by Motor-Car through Eleven Countries and Two Continents. London: Heath Cranton, 1927; pp. 254, with 51 halftones and endpaper map

The Woman Owner-Driver: Practical Advice on Motoring Manners. London: Iliffe [1928]; pp. 53, illustrated throughout

The Bluebird's Flight. London: Chapman and Hall, 1931; pp. 292, with 46 halftones and 4 maps

Nine Lives Plus: Record-Breaking on Land, Sea and in the Air [etc.]. London: Pelham, 1977; pp. 192, with 62 halftones

The amazing Mrs Bruce first started breaking records at the age of fifteen, when the police caught her doing 67 m.p.h. on her brother's motor cycle and she became the first ever woman to be charged for speeding. From then on she never looked back. Cars were her greatest love; she won the *Coupe des Dames* at the Monte Carlo Rally of 1927, and broke numerous track speed and distance records at Montlhéry before joining the Alvis works team at Brooklands. For relaxation she would take her AC saloon on what were then thoroughly outlandish trips—to the Sahara, perhaps, or Lapland, further north than anyone had ever driven before. And when she was bored with cars, she took to motor boats, making the first crossing of the Yellow Sea and breaking the 24-hour long-distance record on the Atlantic in 1929. Aeroplanes came next: in a shop window one morning in Mayfair she happened upon a chromium-plated Bluebird and bought it, there and then, for £555. Eager to test her hastily-acquired skills as a pilot she promptly flew round the world (folding up the plane and putting it on a liner across the Pacific and Atlantic). She established the women's record for the longest solo flight (London to Japan) in 1930, joined a flying circus, then set up an air courier service at Croydon—and so it goes on. Mrs Bruce was still at it at the age of seventy-eight, test-driving for the Ford Motor Company at Thruxton, and vowed never to decelerate until she had to: 'going slowly always made me tired.'

❧

COLE, Mrs H.W.

[Anon.]: *A Lady's Tour round Monte Rosa: with Visits to the Italian Valleys . . . in a Series of Excursions in the Years 1850–56–58*. London: Longmans, 1859; pp. 402, with hand-coloured vignette, 4 chromolithographs, 8 wood engravings, and a map

Mrs Cole was one of the earliest women to write about her own mountaineering experiences, and her book was responsible for enticing a whole crop of later authoresses to the foothills or upper slopes of the Alps.

It is a very thorough book, setting out clearly and efficiently the preparations and pitfalls of a complete circuit of Monte Rosa (pleasantly feminine name), at 15,217 ft. 'the second-highest mountain in Europe', and not finally conquered until five years after Mrs Cole's first visit, in 1855. Her advice to lady travellers is eminently sensible. Nothing that can possibly be dispensed with should be taken, because extra loads make the porters slow. One should always wear a broad-brimmed hat, so that awkward parasols may be left behind (although it must be said that the odd parasol comes in very useful for frightening over-frisky bullocks away—Zélie COLVILE reckoned she owed her life to one). Small rings should be sewn into the seams of one's skirt and a cord passed through them 'in such a way that the whole dress may be drawn up at a moment's notice' to keep it dry, to stop it catching on one's boots (studded with Link's Glacier Nails), and to avoid brushing stones down with it into the path of the gentlemen following behind. And since it is difficult to find side-saddles in the Italian and Swiss valleys, let alone horses, one should equip oneself with a Whippy's Folding Portable Side-Saddle—they fit most varieties of four-legged creature. Such sterling and richly illustrated advice, given at a time when Switzerland was just beginning to blossom as the 'playground of Europe', made Mrs Cole's book a mountaineering classic.

DAVID-NEEL, Alexandra (1868–1969)

My Journey to Lhasa: The Personal Story of the only White Woman who Succeeded in Entering the Forbidden City. London: Heinemann, 1927 [published simultaneously in New York by Harper and later the same year in Paris as *Voyage d'une Parisienne à Lhassa*]; pp. 310, with 44 halftones

Tibetan Journey. London: John Lane, 1936 [first published in Paris as *Grand Tibet*, 1933]; pp. 276, with 22 halftones and a map

'[I]f "Heaven is the Lord's", the earth is the inheritance of Man, and . . . consequently any honest traveller has the right to walk as he chooses, all over the globe which is his.' This was the Frenchwoman Alexandra David-Neel's justification for stealing through the desolate winter plains and mountains of Tibet towards the sacred capital Lhasa in 1923. The journey had nothing to do with missionary zeal, like Annie TAYLOR's or Susie RIJNHART's, nor with a passion for exploration and discovery; Mrs David-Neel went to Lhasa simply because she was not allowed to—and because it was her spiritual home.

She was an experienced traveller, having run away from her unhappy home several times as a child—once even as far as England. Later, from her base in Paris where she managed to combine a somewhat bizarre career as *première chanteuse* at the Opéra-Comique with the serious study of theosophy and Bhuddism at the Sorbonne, Alexandra visited India, Ceylon, and north Africa. She met her husband in Tunis and although they parted five days after the marriage, Philippe remained her chief correspondent and literary agent all his life.

Alexandra returned to India in 1911 as a journalist commissioned to interview the Dalai Lama at Kalimpong; this 'short assignation' went on to last for fourteen years. From Sikkim, where she met Yongden, the young priest who later became her adopted son and shared all her travels, she ventured to a Lamaist monastery just inside the Tibetan border, and then on through Burma and Bhutan, Japan and Korea, and across the Gobi desert to another monastery at Kumbum on the Sino-Tibetan frontier. There she lived for three years, perfecting her spiritual powers (which she claimed included the art of *thumo reskiang*, or raising the body temperature by enveloping oneself in imaginary flames) and preparing for the pilgrimage to Lhasa.

At last, in her fifty-fifth year, the right time came to set out across the border. She travelled disguised as a mendicant nun with her plaits inked and lengthened with yak's hair and charcoal and cocoa smeared on her face, and surreptitiously laden with maps, watches, thermometers and a compass. Accompanied by trusty Yongden, and after four and a half months' journey through the worst conditions Tibet could offer, Alexandra David-Neel became the first Western woman to enter Lhasa.

She lived undetected in the shadow of the Potala Palace for a further two months and would have stayed longer, had she not been called as a witness in a local dispute: her disguise wouldn't stand scrutiny that close, she thought, and so she and Yongden took their leave and started south for India. They returned to France in 1925, where Mrs David-Neel spent the next few years writing—not just these travel books, but other more scholarly works on Tibetan mysticism and occultism in general. At the age of sixty-nine she took off again for China, this time staying for eight years, after which brief fling she was content finally to settle down in Provence and to die at the sage old age of one hundred.

❖

DAVISON, (Margaret) Ann

Last Voyage. London: Davies, 1951; pp. 248, with 2 halftones

My Ship is So Small. London: Davies, 1956; pp. 286, with a colour frontispiece, 17 halftones, and a map

By Gemini: A Coastwise Cruise from Miami to Miami. London: Davies, 1962; pp. 239, with 31 halftones and a map

Florida Junket: The Story of a Shoestring Cruise. London: Davies, 1964; pp. 151, with 18 halftones and a map

Ann Davison was the first woman to sail alone across the Atlantic. When she set out towards the West Indies in her 23-foot sloop *Felicity Anne* in 1953 she already regarded herself as a middle-aged woman, past her prime. She did have experience, though. Four years earlier she and her husband had begun a spirited but ill-prepared voyage—the 'last' voyage of her first book—from Liverpool to Cuba, partly for the sake of adventure and partly to avoid bankruptcy and the bailiffs back home. It ended in tragedy: their craft was driven on to rocks off the coast of Cornwall and Frank was killed. Anne barely survived herself. But characteristic courage and a quick refresher course in navigation (she had been an air pilot in her youth) fuelled her bleak determination to conquer the Atlantic in memory of Frank, and, after two months, she succeeded.

The crossing was symbolic for Ann: in America, her new-found land, she married again and settled, continuing her love-hate relationship with the sea in a much gentler fashion by driving her motor boat around the comfortingly azure waters of Florida.

❖

DU FAUR, Freda

The Conquest of Mount Cook and other Climbs: An account of four seasons' mountaineering on the Southern Alps of New Zealand. London: Allen and Unwin, 1915; pp. 250, with 41 halftones

The first complete traverse of Mount Cook (12,349 ft.), the first ascent of Mount Dampier (11,287 ft.), the first traverse of Mount Sefton (10,354 ft.), and the christening of Mounts Du Faur and Cadogan: small wonder the world's mountaineering fraternity (and I use the word advisedly) turned their heads when Miss Du Faur appeared over the horizon. She was an Australian girl, who had never even seen snow, let alone mountains, until a holiday visit to New Zealand's South Island in 1906. From Christchurch she could see majestic Mount Cook, the island's highest peak, and was smitten. She vowed then to be the first woman to climb it, which she did after only two seasons' training in 1910. Three years later she accomplished the first ever traverse of the mountain's three peaks.

Mountaineers are born, not made, and if Freda also had the misfortune to have been born a woman, she was not going to let it cramp her style. Her female friends implored her, for reputation's sake, not to go off for days 'alone' with a guide ('Do not spoil your life for so small a thing as climbing a mountain!'). But the reputation she sought and earned was that of a serious mountaineer, and pioneer of the New Zealand Alps. Unlike fellow climbers Fanny Bullock WORKMAN or Elizabeth LE BLOND, Freda was small, dainty, and vivacious; she wore a check shirt, a 'frill' (a shocking sort of skirt which barely reached the knee—in all the photographs of her the guide is beaming roundly), and a huge and extravagant bonnet. Her great dream, she said, was one day to climb in the Himalaya; sadly illness thwarted her plans, and after 1914 she seems never to have climbed again.

◆

DUNSHEATH, Mrs Joyce (Cissie) (1902–1976)

[with REID, Hilda, GREGORY, Eileen, and DELANY, Frances]: *Mountains and Memsahibs.* By the Members of the Abinger Himalayan Expedition, 1956. London: Constable, 1958; pp. 198, with 16 halftones and endpaper maps

Guest of the Soviets: Moscow and the Caucasus, 1957. London: Constable, 1959; pp. 183, with 13 halftones and endpaper map

[with BAILLIE, Eleanor]: *Afghan Quest: The Story of the Abinger Afghanistan Expedition, 1960.* London: Harrap, 1961; pp. 239, with 16 halftones and a map

Joyce Dunsheath had a highly distinguished mountaineering career. The first expedition she led was to the little-explored Kulu–Spiti–Lahul watershed in the north-western Himalaya. She and another member of the four-woman team (all members of the Ladies' Alpine Club) spent six adventurous weeks driving the 8,500 miles from Abinger in Sussex to Manali in the eastern Punjab. After ten days to acclimatize, the expedition severally or together surveyed the Bara Shigri Glacier, made the first ascent of Cathedral Peak (20,000 ft.), Chapter House (19,100 ft.), and the first feminine lead of Deo Tibba (19,687 ft.). Mrs Dunsheath's next trip was to Russia in 1957. Here, courtesy of the Soviet government, she became the first Englishwoman to climb Mount Elbrus—at 18,482 ft., the Caucasus' highest mountain. Two years later she led a two-woman expedition to the Hindu Kush, where she hoped to climb Mir Samir (19,880 ft.). Foul weather and recalcitrant porters thwarted her plans, however, and the Abinger Afghanistan Expedition had to content itself with climbing Mir Samir's smaller sisters before coming home early. Mrs Dunsheath's next important climb was in 1962, back in the Himalaya, as leader of the first Indian Women's Expedition to the Garwhal mountains, the chain guarded by the giant Nanda Devi. Three years later she was occupied by the Peruvian Andes, and in 1973, at seventy-one, she made her first visit to Mounts Kilimanjaro (19,340 ft.) and Kenya (17,058 ft.) before retiring gracefully into low-altitude old age.

❧

EARHART, Amelia (1898–1937)

20 Hrs. 40 Min.: Our Flight in the Friendship: *The American Girl, First Across the Atlantic by Air, Tells her Story.* New York and London: Putnam's, 1928; pp. 314, with 61 halftones

The Fun Of It: Random Records of My Own Flying and of Women in Aviation. New York: Brewer, Warren, and Putnam, 1932; pp. 218, with 31 halftones

Last Flight. Arranged by George Palmer Putnam [her widower]. New York: Harcourt, Brace, 1937 [published in London by Harrap, 1938]; pp. 230, with 37 halftones and an endpaper map

Amelia Earhart, cool and serene, is a twentieth-century heroine. She was twice the first woman to cross the Atlantic by air: initially, in 1928, as a passenger just a year after Lindbergh's pioneering flight and then, in 1932, flying solo. On each occasion she was greeted with rapturous admiration both in her native America and in Europe; like her contemporaries Amy JOHNSON and Beryl MARKHAM she was featured in all the fashionable magazines of the day as a symbol of the new independent woman: an irresistible mixture of action and allure. There was nothing frivolous about her achievements—she had financed her passion for aviation by working as a telephone operator, shipping clerk, anything to pay for lessons, and subordinated her whole life to flying.

The list of records Amelia established reads like a catalogue of aviation history, and includes the first flights from Hawaii to California and from California to Mexico, and the American transcontinental air speed record. Any time begrudgingly spent on the ground was used to further the cause of career-women like her, principally in her capacity as student adviser for women at Purdue University. In 1937, as her supreme achievement, she planned to 'girdle the globe' in a Lockheed Electra, using the aeroplane as a sort of flying laboratory to test the biological and mechanical effects of long-haul flight. All went well until the last leg of the circumnavigation when the plane, its pilot, and its navigator quite simply vanished.

No one has ever satisfactorily proved what happened to Amelia Earhart. She was last heard of over the Pacific, just beyond New Guinea, but despite extensive searches neither wreckage nor bodies were ever found. Some said she had been shot down by the Japanese as a spy, the whole trip being one huge covert surveillance operation for the US Government. There were even reports that she had been taken prisoner and was still alive long after the Second World War. But however it ended, Amelia's career as a first-generation flyer was unique, and endorsed her own opinion that women 'must try to do things as men have tried. When they fail, their failure must be but a challenge to others.'

❧

EDWARDS, Amelia (Ann) Blandford (1831–1892)

Untrodden Peaks and Unfrequented Valleys: A Midsummer Ramble in the Dolomites. London: Longmans, Green, 1873; quarto, pp. 385, with 9 wood-engraved plates, vignettes throughout, and a map

A Thousand Miles Up The Nile. London: Longmans, Green, 1877; quarto, pp. 732, with 17 wood-engraved plates, 62 vignettes, and 2 maps

Pharaohs, Fellahs and Explorers. London: Osgood, McIlvaine, 1891; pp. 325, with halftones and sketches throughout

Amelia Edwards—cousin of Matilda Betham EDWARDS—was a prolific novelist, poet, and children's historian. Her earliest 'travel' book was published in 1862—a children's picture-book of Belgium. Soon Amelia went further afield, to northern Italy with a friend, and published a hugely popular book on the little-frequented Dolomite mountains (reprinted under the title *A Midsummer Ramble*). Since there were not yet any proper roads (nor Cook's tourists, she was relieved to note) the 'ramble' became a sort of voyage of discovery: she was hooked. A year later she arrived in Egypt—almost by accident, to get out of the European rain on a long holiday—and discovered what was to become her life's love.

Travelling by *dahabiah*, a well-appointed sailing craft peculiar to the Nile, and armed with sketch-book and measuring tape, Amelia carefully recorded all she saw of the temples, graves, and monuments—even discovering a buried chapel of her own—and provided in *A Thousand Miles Up The Nile* the first general archaeological survey of Egypt's ruins. The book is stiff with historical footnotes and careful details—a considerable scientific tome which earned her the admiration of Sir Flinders Petrie, amongst other leading archaeologists of the time. In fact Amelia Edwards was responsible for founding the first chair in Egyptology (a science she helped create) at University College London, and by the time her last book was published, had established herself as one of the few authorities on the subject. She was, of course, a marvellous writer too, which is why *A Thousand Miles* has remained one of the most inspiring travel books in the language.

❖

ELWOOD, Mrs Colonel (Anne Katharine)

Narrative of a Journey Overland from England, by the Continent of Europe, Egypt, and the Red Sea, to India; including A Residence there, and Voyage Home, in the years 1825, 26, 27 and 28. London: Colburn and Bentley, 1830; 2 vols., pp. 429/400, with 6 aquatints

Mrs Elwood was fully aware of the pioneering nature of her journey from Eastbourne to India: 'I believe I may safely say, that I am the only lady who ever travelled thither overland', she tells us (which is not strictly true—Mrs LUSHINGTON had already published a first-hand account of the route in 1829): to her, the journey had all the charms (and alarms) of novelty.

The first leg took the Elwoods through France to Italy, whence they embarked for Alexandria (overland?!); then they sauntered up the Nile to 'Kinné' (Qena), marched across to the Red Sea, and sailed at last to their destination of Bombay. Anne Elwood was proud to note that she had tried almost every vehicle there was, from common donkey cart to 'takhutrouan', a sort of camel-carried palanquin, and from steam-packet to Arab dhow. The six-day journey across the desert of the Thebais was the part she enjoyed the most, away from the heat, rats, and cockroaches of Cairo. The caravan travelled by moonlight, and peeping from her palanquin, she found the romantic beauty of the place quite captivating: 'I believe I was born under a roaming star, and I must say, I infinitely preferred this . . . way of life, unshackled as it was, to the artificial stupidity of civilisation.'

After several months' residence in Bombay and Bhooj, the Elwoods returned home, this time by sea via Ceylon and the Cape. Exhaustive appendices to the book give detailed itineraries, with costs, of every stage of the journey, and a comparative table of continental currencies, as well as hints from the expert to the lady traveller in the desert.

◆

HALL, Mary (1857–1912)

A Woman's Trek from the Cape to Cairo. London: Methuen, 1907; pp. 424, with 64 halftones and 2 maps

A Woman in the Antipodes and in the Far East. London: Methuen, 1914; pp. 374, with 46 halftones

Mary Hall was the first woman to cross the continent of Africa from south to north. She did it unaccompanied (except for the native porters and guides hired along the way) and unarmed (her terrier Mafeking was supposed to warn against approaching lions and savages), covering the 7,000-odd miles in seven months.

She was well-travelled before she ever thought of walking across Africa: what started as a search for health had become a lifetime's habit, and Miss Hall was now a self-styled 'pioneer tourist'. In her second book she *was* a tourist, in her mid-fifties, covering old ground in New Zealand, Australia, and the south Pacific before sailing on to China and home via the Trans-Siberian Railway. But the African journey was real pioneering stuff.

Travel was a simple matter, Mary said, for those who took every precaution and abandoned all fear. She did both, spending a year travelling around South Africa to prepare herself and a busy administrative winter at home before leaving via the Cape for Chinde, Mozambique, in April 1905. Her route took her up the Shiré river to Port Herald (Nsanje), by rail and *machila* (a hammock slung on a pole) to Lake Nyasa, along the lake aboard a 'travelling theological college' to Karonga, then on foot or by *machila* again to Lake Tanganyika. She crossed the water (roughly the length of England) by steamer and 'mission dugout', trekked on to Lake Victoria Nyanza and, after a brief detour east to Nairobi by rail and rickshaw, reached Entebbe, Lake Albert, and Gondokoro. By now she was on the relatively familiar route down the Nile to Khartoum and Cairo—more or less home.

It was an astonishing journey, which Miss Hall took entirely in her diminutive stride. In fact she seems to have been more interested in her companions than in the country they crossed. Her bands of porters became little communities of which she, with her small, rounded figure and thoroughly British bearing, became a local Queen Victoria. She managed them with fond tolerance: they were all overgrown children, she said, who responded best to tact and firmness. Where possible, she would stay at mission stations *en route*, but never regretted having to camp in a native village if need be; the chief might meet her with suspicious bewilderment, but they would always part with gales of laughter and 'the best of friends'. There is something of Mary KINGSLEY about Miss Hall—she enjoyed the ridiculous (bounding along for her morning walk in a petticoat to keep her skirt dry, and crossing rivers with nothing on but a mackintosh and a black straw hat) and her only regret in travelling alone was that there was no one else 'with whom to exchange a smile' over the glorious absurdity of it all.

◆

HUBBARD, Mrs Leonidas (Mena Benson) (1870–1956)

A Woman's Way Through Unknown Labrador: An Account of the Exploration of the Nascaupee and George Rivers. New York: McClure, [London: John Murray], 1908; pp. 305, with 51 halftones and a map

On 20 June 1903 three Canadian pioneers left Quebec for the coast of Labrador. They meant to cross the unexplored peninsula from the mouth of the North-West River up to Ungava Bay. For their leader Leonidas Hubbard it would be the fulfilment of a dream. He had planned and financed the expedition, and intended to produce a full survey of the area on his return.

He never did return. A combination of unusually cruel weather and ill fortune stranded the three men; George Elson and Dillon Wallace survived, but Hubbard died of starvation. His journal was recovered some time later, and sent to his widow Mena in Montreal. The story might have ended there, had not Wallace published an account (*The Long Labrador Trail*, 1905), which scarcely mentioned Hubbard and Elson, and claimed the sole hero of the expedition to be Wallace himself. He also declared that he was to lead a second expedition, this time to succeed where Hubbard had failed. This was too much for Mena Hubbard. To shame Wallace, she determined not only to publish her husband's journal, but to add to it her own account: she would finish the journey herself.

Two years were spent preparing for what now became a race to cross Labrador between Mrs Hubbard and Dillon Wallace. Mena learned the rudiments of navigation and surveying, bought herself a battery of scientific instruments, two cameras, and a fishing rod, and set off on what most people regarded as a flamboyantly romantic suicide mission. She took with her George Elson and three local woodsmen armed with canoes, tents, and limited stores of food. She carried a revolver and a hunting knife slung round her waist, and spent the next forty-three days in a wilderness, 'less homeless than I had ever felt anywhere'. For the most part they canoed, shooting rapids and crossing uncharted lakes; sometimes they would be forced to cut tracks and carry the canoes through Indian country previously thought impassable. At one stage they were storm-bound, as Hubbard had been, and survived on what they could kill around the camp. Yet Mena was never uncomfortable, she says, just occasionally weary.

Not surprisingly, she was fêted as a national heroine on her return to Canada. She had done the 576-mile journey in under two months (Wallace took almost four), becoming the first white woman to cross Labrador, the first person to chart the Naskaupee and George river systems, and the first to note the flora and fauna of the area and to witness the great caribou herds there—apart from the Indians, whom she was the first to describe and photograph. The journey was the finest thing she had done, or was likely to do: although she lived until her eighty-seventh year, she never tried to match it.

❖

JACKSON, Monica

[with STARK, Elizabeth]: *Tents in the Clouds: The First Women's Himalayan Expedition.* London: Collins, 1956; pp. 255, with 36 halftones and 5 maps

The Turkish Time Machine. London: Hodder and Stoughton, 1966; pp. 158, with 17 halftones and endpaper map

The First Women's Himalayan Expedition was not conceived to prove that women could do as well as men; its aims were purely topographical and scientific. That Betty STARK, Evelyn Camrass, and Monica Jackson happened to be the first *cordée feminine* to climb alone in the mountains of Nepal was almost incidental. The expedition set out from the Ladies' Scottish Climbing Club in the spring of 1955 with four objectives: to find a way through the lower gorges to the glacier valleys of the Jugal Himal on the Nepali–Tibetan border; to negotiate the glaciers into the heart of the group; to find passes over the mountain ridges separating one glacier from another, and to check and correct the accuracy of the only (rather vague) map there was of the area. In two months they did all this, and made the ascent of an unknown peak of over 22,000 ft., christening it Gyalgen (after their chief Sherpa) and naming its neighbours 'Big White Peak' and 'Ladies' Peak'. It was a thoroughly efficient and successful affair.

More relaxed and romantic was the 'Hakkari Expedition' Mrs Jackson led to the mountains of south-eastern Turkey in 1965. She flew with her 'anti-mountaineer' husband and four friends to Lake Van (where they explored the mythologically sacred island of Aghtamar), and then trekked to the Demirkapu Pass and Cilo Dag mountains north of the Iraqi border.

❖

JAMES, Naomi (b. 1949)

Woman Alone. London: Daily Express, 1978; pp. 128, with photographs, maps, and diagrams throughout

At One With the Sea: Alone Around the World. London: Hutchinson and Stanley Paul, 1979; pp. 192, with 28 halftones, 21 colour photographs, and endpaper maps

At Sea On Land. London: Hutchinson and Stanley Paul, 1981; pp. 176, with 28 halftones and a map

Courage at Sea: Tales of Heroic Voyages [including the author's 1982 Double-Handed Round Britain Race victory]. London: Stanley Paul, 1987; pp. 207, illustrated throughout

Naomi James had only been sailing for two years when she made her historic voyage. She was born in New Zealand, and was working her way round Europe when she met her husband, the British yachtsman Rob James, in 1975. He taught her how to crew, and encouraged her when after only a season's practice she determined to become the first woman to sail alone round the world.

The difficulties of organization and finance were almost crippling, but by September 1977, with a 53-foot yacht lent by Chay Blyth and sponsored by a daily newspaper, and a small black kitten to keep her company, Naomi was ready to start. It was an incredibly feckless venture—she had never even handled a boat by herself before—but it came off. Several times the yacht and its equipment were seriously damaged by storms, and at one stage she almost turned back. Boris the kitten was inevitably lost overboard after six weeks, and Naomi herself nearly drowned in a capsize off Cape Horn; still she managed to finish in the fastest time ever (nine months all but a day) and, what surprised her even more, arrived home more sane than when she left.

Naomi—now Dame Naomi—became a celebrity; it took more discipline to survive the massed attention that followed the voyage, she said, than to make the voyage itself. And the hype went on: the next year she was the only woman to finish in the Observer Single-handed Transatlantic Race, and in 1982 she and Rob won the Double-Handed Round Britain Race in a trimaran. By now she was exhausted; she decided to retire from top-class sailing (a decision endorsed by Rob's death in an accident in 1983) and has since turned her attention to writing about maritime history, rather than making it.

❖

JOHNSON, Amy (1903–1941)

Sky Roads of the World. London: W. & R. Chambers, 1939; pp. 314, with frontispiece portrait

Myself When Young By Famous Women of Today. Edited by the Countess of Oxford
and Asquith. London: F. Muller, 1938; pp. 131–56, with a portrait plate

Amy Johnson really was a legend in her own lifetime. She was born in Hull,
the daughter of a fish merchant, and after taking a degree in economics at
Sheffield University moved down to London. She drifted through a series
of jobs, becoming an accountant's clerk, shop-girl, copy-writer, and legal
secretary before finding her real career in the air. Her digs happened to be
near the Stag Lane Aerodrome in north London; one Sunday, lured by the
low-flying planes, she took a bus up there with a pound in her pocket and
demanded a lesson. This was 1928.

Two years (and just fifty flying-hours) later, Amy had become the darling
of the skies, by flying a tiny open-cockpit Tiger Moth solo from England to
Australia. The 13,000-mile journey, including fifteen stopovers and three
forced landings, lasted for 19½ days. Amy had taken a job in the workshops at
Stag Lane to finance the flight, and whilst working there had become the
first woman to pass the coveted Ground Engineer's Certificate qualifying
her as a commercial pilot. Her successes continued with a flight from
England to Tokyo in 1931 and a record-breaking run from London to Cape
Town and back in 1936: she collected honours, awards, and medals
wherever she went.

The Press followed her every move—this slender little Yorkshire girl,
nicknamed Johnnie by her fellow engineers, was rapturously transformed
into the Queen of the Air, the Lone Girl Flyer, the Empire's Great Little
Woman, and Winged Victory. It was celebrity that broke up her marriage to
fellow aviator Jim Mollison, she said, and not the pressures of flying. Two of
her many biographies were published before she was out of her twenties,
and she was commissioned by popular magazines to write countless
glamorous articles on flying and her new passions of gliding and
motor-racing.

On the outbreak of the war in 1939 Amy immediately volunteered for the
Air Transport Auxiliary Force, and it was on one of its missions that she
died. Although advised not to, she had insisted on flying through packed,
freezing fog to deliver a plane from Prestwick to Kidlington, near Oxford.
For security reasons, she was flying without radio; she lost her course, ran
out of fuel and ditched in the Thames Estuary on 5 January 1941. A naval
patrol saw her bail out and went to her rescue, but Amy was too numb with
cold to reach for the lifeline they threw her. She was swept under the bow of
the ship, and drowned.

LE BLOND, Mrs Aubrey (Elizabeth, formerly Mrs MAIN and Mrs BURNABY) (1861–1934)

The High Alps in Winter: Or, Mountaineering in Search of Health. London: Sampson Low *et al.*, 1883; pp. 204, with frontispiece portrait, 4 wood engravings, and 2 maps

High Life and Towers of Silence. By the author of *The High Alps* [etc.]. London: Sampson Low *et al.*, 1886; pp. 195, with 8 halftones, and vignettes

My Home in the Alps. London: Sampson Low *et al.*, 1892; pp. 131

True Tales of Mountain Adventure for Non-Climbers Young and Old. London: T. Fisher Unwin, 1903; pp. 299, with 63 halftones

Adventures on the Roof of the World. London: T. Fisher Unwin, 1904; pp. 333, with 103 halftones

Mountaineering in the Land of the Midnight Sun [north Norway]. London: T. Fisher Unwin, 1908; pp. 304, with 71 halftones and a map

Day In, Day Out [autobiography]. London: John Lane, 1928; pp. 264, with 29 halftones

Elizabeth Hawkins-Whitshed was brought up in Ireland, the only daughter of aristocratic parents, and used to all the comforts of life. During her first 'Season' in London, at the age of seventeen, she met and married the gallant Captain Fred Burnaby. He was the soldier, correspondent, and traveller— twenty years her senior—whose books *Ride to Khiva* (1876) and *On Horseback through Asia Minor* (1876) had made him a hero of the Empire: the young Elizabeth looked all set to embark on a dazzling drawing-room career as a Famous Man's Wife.

A visit to Chamonix in 1881 changed all that. Elizabeth had been ill, and it was thought a breath of Alpine air would do her good; at this stage, knowing nothing about mountains and caring less, she confined herself to taking mule rides amongst the lower slopes of the Mont Blanc chain. Most unexpectedly the lure of the ice and snow proved irresistible, however, and by the end of the summer she had an astonishing double ascent of Mont Blanc itself to her credit, as well as a string of lower peaks. Although climbing turned her high-heeled button-boots to pulp, burned her face to copper, and scandalized her family, Elizabeth was eager to show that it was possible to be a Lady Mountaineer. The two were not mutually exclusive, and with good breeding and a sense of adventure, one could make a success of both.

After Burnaby's death in 1885, Elizabeth married an English doctor, J. F. Main, and settled in San Moritz in Switzerland. Here her burgeoning accomplishments included learning to skate (she was the first woman to pass the coveted Men's Skating Test), tobogganing (she helped plan and construct the fabulous Cresta Run), and the singular art of Alpine Bicycling (once spinning all the way to Rome which, she modestly admits, 'caused a bit of a flutter'!). She also delighted in the thoroughly modern and competitive sport of driving motor cars up mountains. She became an authority on snow photography (producing a textbook on the technique in 1895), the author of tourist guidebooks on Spain and Italy, and one of the most celebrated and fashionable hostesses in Switzerland. Meanwhile her reputation as a pioneer of female mountaineering grew; she chose to specialize in 'mountains hitherto unclimbed in winter'—the most gruelling of all—and even the august *Alpine Club Journal* was forced to admit, as early as 1883, that her record must form 'one of the most brilliant chapters in the history of winter mountaineering'. During the summer months she would take her Swiss guides with her to the Lyngen peninsula, north of Tromsö in Norway, where over twenty virgin peaks succumbed to her determined expertise.

Elizabeth travelled further afield in 1912, accompanying her third husband on a tour of Russia, China, and Korea which she summarizes in *Day In, Day Out*. She went on during the Great War to work as a nurse in Dieppe, and lectured to British troops on her career in climbing.

Although that career had practically come to a close around the turn of the century, Elizabeth remained active in the cause of Lady Mountaineers to the end of her days. In 1907 she was elected the first ever President of the Ladies' Alpine Club in London, an association she was instrumental in founding, and she held the post until her death. The appointment was 'one of the nicest things that has ever happened to me', she said, and a fitting honour for the greatest mountaineering Lady of her age.

❦

MILNES WALKER, Nicolette (b. 1943)

When I Put Out to Sea. London: Collins, 1972; pp. 158, with 9 colour photographs and a map

Ann DAVISON was the first woman to sail solo across the Atlantic, in 1953. But Nicolette Milnes Walker was the first to do it non-stop—Ann had gone by way of Spain and the Canary Islands. Miss Milnes Walker organized the voyage as an experiment (she was an industrial psychologist by profession), to see how the human spirit can cope with days of unbroken physical and mental challenge. Very well, it seems: although she had scarcely sailed single-handed before (like Naomi JAMES before her circumnavigation), and although she caught flu a few days out at sea, she managed not only to survive but to relish the 4,000-mile crossing, completing it in just over six weeks. In fact so much did she relish it that she left psychology for the sea and, through various publications on yacht and dinghy sailing, has been faithful ever since.

❖

MOODIE (née STRICKLAND), Susanna (1803–1884)

Roughing it in the Bush: Or, Life in Canada. London: Richard Bentley, 1852; 2 vols., pp. 293/294

Life in the Clearings versus the Bush. London: Richard Bentley, 1853; pp. 384

Susanna Moodie is one of my favourites. She was a real pioneer (and a more reluctant one you could hardly find), she led a raw and adventurous life, had a healthy sense of humour, and, most of all, she could write. These two books are craftsman's pieces, and I think she was the first—man or woman—to treat the travel book with real literary respect.

Writing attracted Susanna far more than travelling: in 1831 she published her first collection of poems and with her sisters Agnes (author of *Lives of the Queens of England*) and Catherine (later Mrs TRAILL, and fellow emigrant), she looked set for an illustrious career. But then she married a younger son, and with fifty thousand other hopefuls in 1832 succumbed to the advertisements and sailed for Canada. Government grants of land were available for the asking and fortunes were just waiting to be made by anyone with a taste for grand scenery and an Irish labourer or two. *Roughing it in the Bush* is the story of the Moodies' first seven years there.

They were terrible years: 'If these sketches should prove the means of deterring one family from sinking their property, and shipwrecking all their hopes, by going to reside in the backwoods of Canada', wrote Susanna, 'I shall . . . feel that I have not toiled and suffered in the wilderness in vain.' The Moodies' first home, waiting for them on the banks of Lake Ontario at Cobourg, was at first mistaken by Susanna for a disused pigsty. Their emigrant neighbours considered them snobbish and soft; try as she would, Susanna could not help weeping at the searing cold of winter and the billowing mosquitoes in summer, while the forest glowered at their backs, guarded by wolves and bears and refusing to be cleared. After three years the Moodies moved up to Indian country, to sixty-six acres at the upper reaches of the Kutchawanook Lake—a mile and a half from Susanna's sister Catherine's homestead. They built themselves a log cabin (which twice almost burned down) and settled down to eke a living out of the land for the next seven years. Susanna learned to survive on milk, bread, fish from the lake, and dandelion tea; when times were really hard one of the children's pets would have to be slaughtered, or a joint of bear begged from a local hunter.

To bring in extra money, Moodie volunteered for the local militia at the outbreak of the Toronto Rebellion in 1837 (leaving his wife to manage the farm alone) and Susanna began writing again. The sketches published in Montreal's *Literary Garland* formed the bases of her two travel books. Meanwhile her hair had turned grey, her skin had coarsened, she looked double her age, clung to her solitude, and was 'no longer fit for the world'. Then, in 1840, the family fortunes changed. Largely thanks to Susanna's efforts, Moodie was appointed Sheriff of the County of Hastings, based at the town of Belleville back on Lake Ontario. This meant leaving the woods at last, being together, and having a regular income: *Life in the Clearings* began.

The Moodies settled at Belleville for the rest of their lives. Ten years after moving there Susanna fell ill, and determined to visit the Falls before she expired, she made a difficult journey to Niagara and back. But she survived—old habits die hard—and carried on writing, to become a prolific and respected novelist before her eventual death at the age of eighty-one.

❦

PARKER, Mary Ann

A Voyage round the World, in the Gorgon Man of War: Captain John Parker. Performed and Written by his Widow [dedication signed 'Mary Ann Parker'], *for the advantage of a numerous family* [etc.]. London: John Nichols, 1795; pp. 149

This is the earliest first-hand account of an Englishwoman's voyage to Australia, and round the world. It was published by subscription three years after the author's return, to raise the money to pay her late husband's debts. Parker had been commissioned in January 1791 to transport a corps of soldiers to New South Wales and bring home some of the marines and their families who had been at Botany Bay since its founding as a penal settlement just three years earlier. It was the Captain's privilege to take his lady, and, although it would mean leaving behind her two children and her mother ('from whom I had never been separated a fortnight at one time during the whole course of my life'), Mary Ann was pleased to oblige. After all, says she proudly, she had already 'travelled into France, Italy, and Spain', so was not exactly a novice.

The voyage out took six months, including breaks at Tenerife and Cape Town, and once the *Gorgon* had arrived at 'New Holland', no efforts were spared to make her welcome. Governor Phillip came on board at Port Jackson, and the Captain's lady was taken by the Governor's to sightsee Botany Bay, 'Sidney Cove', and Paramatta, and to try the local delicacies of emu and kangaroo. Using the notes she kept during her stay, Mary Ann was the first lady traveller to write of the aborigines, or 'inhabitants of New

South Wales', naked and filthy with fish-fat in their hair, and to describe the arrival of a convict ship from England on which, as usual, only a third of the inmates had survived.

After three months, the *Gorgon* set sail for home, laden with kangaroos, opossums, plants, shrubs, animal skins, and birds, via New Zealand, Cape Horn, and—completing the circumnavigation—the Cape of Good Hope again. The return took only half the time of the voyage out; and a week after arriving in London, a son was born to Mary Ann, to replace (she said) the one that had died whilst they were away. How easy she made it all seem.

❦

PECK, Annie S(mith) (1850–1935)

A Search for the Apex of America: High Mountain Climbing in Peru and Bolivia [etc.]. New York: Dodd Mead, 1911 [published the following year in London as *High Mountain Climbing* [etc.] by T. Fisher Unwin]; pp. 370, with 124 halftones and a map

Flying Over South America: Twenty Thousand Miles by Air. Boston and New York: Houghton Mifflin, 1932; pp. 256, with 118 halftones and endpaper map

Annie Peck would have been a remarkable woman even if she had never embarked on the mountaineering career that made her famous. Always a staunch feminist, she won a degree from the University of Michigan, became the first woman to be admitted to the American School of Classical Studies in Athens (all this well before the turn of the century), and taught Greek archaeology at the prestigious Purdue University and Smith College back in the States.

She did not begin climbing until her mid-forties, after glimpsing the Matterhorn (14,782 ft.) on a Swiss holiday. That alluring mountain became one of her first conquests in 1895, and Popocatapetl (17,887 ft.) and Orizaba (18,700 ft.), both in Mexico, soon followed. From now on, everything took second place to the mountains in her life. She chose South America for her special hunting-ground (writing a definitive guidebook in 1913 and an oft-reprinted *Commercial Handbook* in 1922) and made nine expeditions in all to Mexico, Bolivia, and Peru, listing amongst her first ascents the north peak of Huascaran (21,834 ft.), which she managed at the fifth attempt (having imported some Swiss guides) in 1908.

Poor Miss Peck never had the money and equipment with which her countrywoman and rival Fanny Bullock WORKMAN was blessed. Although eager to contribute as much as she could to the advancement of science, she could usually only afford a mercurial barometer to take her triangulations, and more than once she was embarrassingly wide of the mark (embarrassing only because jealous Fanny sent hordes of scientists to double-check the measurements of *anyone* who dared challenge her prowess). Even Annie's outfit was borrowed. It was an elderly eskimo suit from the American Museum of Natural History; perhaps this, and the rather alarming knitted mask she wore (adorned with a fashionable Poirot-like moustache) accounted for the helpless behaviour of most of her porters. Still, she persevered—and succeeded again and again, becoming one of her country's foremost mountaineers (man or woman), a founder-member of the American Alpine Club, and Fellow of the Royal Geographical Society in 1917. She was still at it well into her seventies, climbing regularly in the White Mountains, and in 1931 she arranged for herself an unprecedented tour round the coast of South America by aeroplane. It was almost her last journey: she was a pioneer to the end.

❖

PFEIFFER, Ida (1797–1858)

Visit to the Holy Land, Egypt, and Italy [etc.]. Translated from the German by H. W. Dulcken. London: Ingram, Cooke (National Illustrated Library series), 1852 [first published in Vienna, 1846]; pp. 336, with 8 tinted engravings

A Lady's Voyage round the World: A Selected Translation from the German of Ida Pfeiffer. By Mrs Percy Sinnett. London: Longman, Brown, Green and Longmans, 1851 [first published in Vienna, 1850]; 2 parts, pp. 134/272

Journey to Iceland: And Travels in Sweden and Norway [etc.]. From the German, by Charlotte Fenimore Cooper. London: Richard Bentley, 1852 [first published in Pest, 1846]; pp. 363

A Lady's Second Journey round the World: From London to the Cape of Good Hope, Borneo, Java, Sumatra, Celebes . . . California, Panama, Peru, Ecuador, and the United States. London: Longman, Brown, et al., 1855; 2 vols., pp. 451/423

The Last Travels of Ida Pfeiffer: Inclusive of a Visit to Madagascar. With a Biographical Memoir of the Author. Translated by H. W. Dulcken. London: Routledge, Warne, and Routledge, 1861 [first published in Vienna, 1861]; pp. 338, with mezzotint portrait

Even if she had never written a word of English, Ida Pfeiffer could not possibly be left out of a book like this. She was the first full-time woman traveller of all, and one of the very few who never felt the need to qualify her impulse: she travelled because she wanted to see the world, and saw no reason why she shouldn't.

Mrs Pfeiffer was born in Vienna, the only daughter in a family of eight. She was a mistake: according to her misogynist father she should have been a boy, and until his death in 1806 to all intents and purposes she was. Soon after her mother had eventually got her into petticoats the confused Ida fell violently in love, but the man was thought unsuitable, so that was that. A more eligible candidate was chosen for her in Dr Pfeiffer, a widower a quarter of a century her senior whom she married in 1820. Ida listlessly carried out her wifely duties for him until 1842: by then her two sons were growing up, her husband was living away from home, and her mother was dead. She was forty-five, bored, and free: it was time to indulge her childhood dreams of seeing what lay over the horizon.

Fully expecting to die on the way, Ida chose the Holy Land as an appropriate destination, and having made her will and settled her affairs, she set out down the Danube to Constantinople and beyond. Nine months later she was back. She had been swindled by sea captains, cheated by camel-drivers and exhausted by guides and companions—and could not have enjoyed herself more.

The rest of her life was spent either travelling or preparing to travel. She raised money in the early days by writing unexciting books, selling everything she owned, and begging subscriptions and commissions. Later on her exploits were so well known that shipping companies and railways were begging *her* to accept free passages. Her journeys became increasingly adventurous, as she wandered further and further from the tourist track (into the jungles of Celebes, for example, or mountains of Peru), and nearly petered out altogether in a conspirators' prison in Madagascar in 1857. Her books were sensationally popular and translated and retranslated all over the world. Yet they were only 'simple narratives', she said, and she herself just a plain and ordinary woman who happened to be possessed of 'an insatiable desire for travel'—nothing special at all.

❖

SCOTT, Sheila (1927–1988)

I Must Fly: Adventures of a Woman Pilot. London: Hodder and Stoughton, 1968; pp. 222, with 50 halftones and 4 maps (including 1 on endpapers)

On Top of the World. London: Hodder and Stoughton, 1973 [published in USA as *Barefoot in the Sky*, 1974]; pp. 281, with 43 halftones and 7 maps (including 1 on endpapers)

Elegant, *petite* Sheila Scott was one of aviation's pioneers, and did just about all it's possible for a pilot to do. She learned to fly in 1959, after a short and desultory career as actress and clothes-designer, and was soon a familiar face at races and rallies around the country in her hire-purchased Tiger Moth. Every possible moment was spent in the air—either training for her commercial licence, qualifying as a helicopter pilot, or simply floating in a hot-air balloon—and by the mid-1960s Sheila was Britain's most experienced aviatrix. In 1966 she flew solo round the world in a Piper Comanche, setting just one of her one hundred world-class speed and distance records, and in 1971 became the first person to fly solo from equator to equator over the North Pole. And towards the end of a well-stocked life, just to crown her career, she achieved one of her greatest personal triumphs ever: she learned to drive a car.

❖

SHELDON, May French (1848–1936)

Sultan to Sultan: Adventures among the Masai and other Tribes of East Africa. London: Saxon, 1892 [published the same year in Boston, USA by Arena under the same title, and by Dana Estes as *Adventures in East Africa*]; pp. 435, with 25 halftones, vignettes throughout, and a map

'Ho for East Africa!' bawls May French Sheldon lustily at the beginning of this exhausting book. It is an account of the pioneering safari she led in 1891 from Mombasa to the Masai lands beyond Mount Kilimanjaro and back, full of adventures and enormously (if unconsciously) entertaining.

Its author, an American who settled in London after 1892 and eventually died there at the age of eighty-eight, must be one of the most colourful characters in the history of African exploration. She was every bit a lady, travelling in a variety of feminine flounces (and occasionally a blonde wig) in a great titanic palanquin of wickerwork and cushions, insisting on gallantry from her retinue of 153 porters (whom she was liable to whip if they didn't comply), prudently displaying a pennant attached to her alpenstock proclaiming the deathless legend 'Noli Me Tangere', and always missing her husband Eli Lemon Sheldon who waited patiently at home in Boston for his errant little May. She was a scientist too, with a medical training and a keen interest in geography and ethnology. Her book is brimming with facts and observations on local tribes, and her work on previously unexplored Lake Chala led to her being among the first ever women elected to Fellowship of the Royal Geographical Society in 1892. She had great panache—only May, for example, could possibly have thought of taking a thousand rings inscribed with her name into the Interior to distribute to bedazzled natives—and she had the sort of courage that kept her going even when a thorn had speared itself into her eyeball and one of her men was eaten by lions.

May's past life had prepared her admirably for her 'Adventures amongst the Masai'. She was born of wealthy and liberal American parents who sent her to schools in Europe and several times round the world for her education; her business-man husband was also liberal enough to allow his wife to pursue a writing and publishing career independently of him (her popular novel *Herbert Severance* was published to great acclaim in 1889). When the time came for her to realize her dearest ambition to travel within

Africa, she was in no doubt at all that given her strength of character and resolution, she would do just as well as any man. And so she did, becoming in her own rather indulgent words 'the first lady to attempt to lead a caravan that history had ever known'. May returned to Africa in 1894, to the Belgian Congo this time; it was her last visit. Even though we know she intended to publish an account of this second safari, no such book ever appeared—more's the pity.

❖

STARK, Freya Madeline (1893–1993)

Baghdad Sketches. London: John Murray, 1937 [first published by the *Baghdad Times*, 1932]; pp. 269, with 57 halftones, 10 sketches, and an endpaper map

The Valleys of the Assassins and Other Persian Travels. London: John Murray, 1934; pp. 365, with 33 halftones and 6 maps (1 on endpapers)

The Southern Gates of Arabia: A Journey in the Hadhramaut. London: John Murray, 1936; pp. 328, with 116 halftones and 2 maps

Seen in the Hadhramaut. London: John Murray, 1938; 130 halftones, with pp. 23 letterpress and a map

A Winter in Arabia. London: John Murray, 1940; pp. 328, with 88 halftones and 3 maps

Letters from Syria. London: John Murray, 1942; pp. 194, with 25 halftones and a map

East is West [the Yemen, Egypt, Syria, and Persia]. London: John Murray, 1945; pp. 218, with 84 halftones and a map

Perseus in the Wind [essays, many based on travels]. London: John Murray, 1948; pp. 169

Traveller's Prelude: Autobiography 1893–1927. London: John Murray, 1950; pp. 346, with 24 halftones and a map

Beyond Euphrates: Autobiography 1928–1933. London: John Murray, 1951; pp. 341, with 52 halftones and a map

The Coast of Incense: Autobiography 1933–1939. London: John Murray, 1953; pp. 287, with 51 halftones, line-drawn vignettes, and a map

Ionia: A Quest. London: John Murray, 1954; pp. 263, with 62 halftones and a map

The Lycian Shore. London: John Murray, 1956; pp. 204, with 66 halftones, line-drawn vignettes, and a map

Alexander's Path from Caria to Cilicia. London: John Murray, 1958; pp. 283, with 76 halftones, line-drawn vignettes, and 5 maps

Riding to the Tigris. London: John Murray, 1959; pp. 114, with 39 halftones

Dust in the Lion's Paw: Autobiography 1939–1946. London: John Murray, 1961; pp. 297, with 22 halftones and a map

Rome on the Euphrates: The Story of a Frontier. London: John Murray, 1966; pp. 481, with 56 halftones and 2 maps

The Zodiac Arch [essays]. London: John Murray, 1968; pp. 230

Space, Time and Movement in Landscape [photographs]. London: Her Godson, 1969 [limited edition of 500 copies]; oblong folio, pp. 167 with 120 halftones

The Minaret of Djam: An Excursion in Afghanistan. London: John Murray, 1970; pp. 99, with 51 halftones, line-drawn vignettes, and a map

A Peak in Darien [essays]. London: John Murray, 1976; pp. 109

Rivers of Time [photographs]. Edinburgh: Wm. Blackwood, 1982; quarto, 202 halftones, with pp. 22 Introduction by Alexander Maitland

[Also published: eight volumes of Freya Stark's letters (John Murray, 1974–82); an anthology (*Journey's Echo*, 1963), and several biographical studies of her work]

Dame Freya Stark was the Grand Old Lady of travellers. Historian, philosopher, explorer, and artist, she indulged a devoted audience for nearly sixty years: the perfect successor to such as Lady Mary WORTLEY MONTAGU and Isabella BIRD. She was born in Paris and grew up in Devon and in Asolo, northern Italy, where she died. In the true tradition of the Middle-Eastern explorer, she was inspired to travel by the copy of *The Arabian Nights* given to her as a child, and justified her first visit to Syria and Persia in 1927 by calling it a restorative (her health had never been strong) and a chance to practise the Arabic she had learned when last imprisoned in hospital. It was both of these, and also the start of a love affair with travel that never faltered.

After two years as a journalist in Baghdad, during which time she managed to 'disentangle the absolute wrongness of the map' of part of Persia, Dame Freya turned to the Hadhramaut, south Yemen. She started the Second World War as southern-Arabian expert to the Ministry of Information in London, and later moved out to Aden, Cairo, and Baghdad before lecturing on the Middle East in America and India. Her brief marriage to diplomat Stewart Perowne ended in 1951 and a year later she 'discovered' Turkey, which remained a favoured country ever after. She visited China for the first time in her seventies and drove a jeep across Afghanistan at seventy-six; later she took a film crew on a voyage down the Euphrates and a pony trek in Nepal, and hardly a season passed without some trip somewhere. Her figure became the image for an archetypal British eccentric abroad—comfortable and regal, colourfully draped, and invariably topped with an elaborate titfer, perhaps astride a camel ('always so obliging') or bobbing up-river on an inflated goatskin waving serenely to the passers-by—but there is a far more serious side to her career. In all her journeys she was able to distil and communicate a rich philosophy of travel and to illustrate the art of travelling in time as well as place. She carried the past with her, whether discovering long-buried fortresses in the Valley of the Assassins in Luristan or tracing the footsteps of the ancient incense traders of Arabia, always teaching and learning at the same time. She was, quite simply, a classic.

❧

WORKMAN, Fanny Bullock (1859–1925)

Algerian Memories: A Bicycle Tour over the Atlas to the Sahara. London: T. Fisher Unwin [New York: A. D. F. Randolph] 1895; pp. 215, with 12 halftones and 11 vignettes

Sketches Awheel in Fin de Siècle Iberia. London: T. Fisher Unwin, [New York: Putnam's,] 1897; pp. 232, with 30 halftones and a map

In the Ice World of Himálaya . . . Among the Peaks and Passes of Ladakh, Nubra, Suru, and Baltistan. London: T. Fisher Unwin, [New York: Cassell,] 1900; pp. 204, with 2 photogravures, a sketch, 64 halftones, and 3 maps

Through Town and Jungle: Fourteen Thousand Miles A-Wheel Among the Temples and People of the Indian Plain [etc.]. London: T. Fisher Unwin, 1904; pp. 380, with 202 halftones and a map

Ice-Bound Heights of the Mustagh: An Account of Two Seasons of Pioneer Exploration and High Climbing in the Baltistan Himálaya. London: Constable, [New York: Scribners,] 1908; pp. 444, with 3 colour plates, 3 photogravures, 97 halftones, 2 maps, and vignettes throughout

Peaks and Glaciers of Nun Kun . . . A Record of Pioneer Exploration and Mountaineering in the Punjab Himalaya. London: Constable, [New York: Scribners,] 1909; pp. 204, with 4 colour plates, 88 halftones, and a map

The Call of the Snowy Hispar: A Narrative of Exploration and Mountaineering on the Northern Frontier of India. London: Constable, 1910; pp. 299, with 3 photogravures, 110 halftones, and 2 maps

Two Summers in the Ice-Wilds of Eastern Karakoram: The Exploration of Nineteen Hundred Square Miles of Mountain and Glacier [etc.]. London: T. Fisher Unwin, [New York: Duttons,] 1917; pp. 296, with 4 photogravures, 137 halftones, and 3 maps

Fanny is one of the honorary Americans in this book—and I would not fancy my chances in the afterlife had I dared to leave her out. She was the most determined, uncompromising, and aggressive of the whole tribe, intensely jealous of her steely success as a lady traveller and a self-styled pioneer of pioneers.

Fanny was blessed with a well-heeled and liberal upbringing in New England, and after her education had been 'finished' in Europe she settled down to a cultured married life with William Hunter Workman, a doctor and fellow-American. They started travelling together in 1889 because of William's failing health and, give or take the odd break to visit their daughter Rachel at home, they carried on until the outbreak of the Great War in 1914.

First of all they travelled by bicycle. The Rover Safety Cycle had just arrived and the Workmans, being a stout-hearted pair, soon subdued this newfangled machine into thousands of miles' service in north Africa, Spain, India, Burma, Sri Lanka, and Java: 'a-wheel' was simply the easiest way to travel. Then, in 1902, they turned their eyes to the hills, and a remarkable series of expeditions to the peaks and glaciers of the Himalaya began. They aimed to combine sport with science: Fanny was especially eager to pioneer both as a mountaineer and as a surveyor. With the support of the familiar guides she had shipped in from the Alps, and fired by the missionary zeal of a true feminist, she practically battered the Karakorams into submission, first treading them over with her squat, hob-nailed figure and then pinning them down on virgin maps and charts to take home for the various Geographical Societies of England and America to fight over. For twenty-eight years after her conquest of Pinnacle Peak (22,737 ft.) she held the women's altitude record (in fierce competition with her countrywoman Annie PECK); and she collected a battery of medals and awards for her prowess—although it must be said that in their enthusiasm for results, the Workmans were not always as careful of measurement and topographical accuracy as they might have been.

There is a wonderful photograph of Fanny (taken by William, no doubt) at the summit of some snowy wasteland mountain. She is standing squarely in front of her jammed-in ice-axe, swathed in the thick woolly trappings of a Victorian lady mountaineer (including a short skirt and an elaborate bonnet) and proudly holding a placard aloft with the words writ large for all to see: 'VOTES FOR WOMEN'. She was a staunch supporter of the suffragettes, and eager that her own glory should reflect the more general worthiness of her sex. Although she wrote the above travel books in collaboration with her husband, she took the opportunity in them all to wave her feminist banner, stressing that she and William had taken turns in organizing their eight Karakoram expeditions and that she, as a woman, was equal to any man in everything but his habitual small-mindedness.

After the excitement of their travels and the flurry of lecturing tours that always followed them were brought to a close by the First World War, the Workmans settled into an unwontedly gentle retirement in the south of France, where Fanny died in 1925. She had been a figure even larger than life, one of the women's movement's most vigorous votaries, and a forceful (if slightly farcical) influence on the history of Himalayan exploration.

WORTLEY MONTAGU, Lady Mary (1689–1762)

Letters of the Right Honourable Lady M—y W——y M——e. Written, during her Travels in Europe, Asia and Africa, To Persons of Distinction, Men of Letters, &c. in different Parts of Europe. Which contain, . . . Accounts of the Policy and Manners of the Turks; Drawn from Sources that have been inaccessible to other Travellers [edited by John Cleland]. London: for T. Becket and P. A. De Hondt, 1763; 3 vols., pp. 165/167/134

An Additional [spurious] *Volume to the Letters of the Right Honourable Lady M—y W——y M——e . . .* [forged probably by John Cleland]. London: for T. Becket and P. A. De Hondt, 1767; pp. 142

The Works of the Right Honourable Lady Mary Wortley Montagu. Including her Correspondence, Poems and Essays [edited by James Dallaway; including the letters from France and Italy, 1739–61]. London: R. Phillips, 1803; 5 vols., pp. 309/339/288/326/292(+ xxxv: index), with 2 copper-engraved portraits and 10 sheets of facsimile manuscript

The Complete Letters of Lady Mary Wortley Montagu. Edited by Robert Halsband [including all previously unpublished letters from abroad]. Oxford: Clarendon Press, 1965–7; 3 vols., pp. 468/530/408, with 18 halftones

Sometimes I see this collection of travel writers as a 'monstrous Regiment of women', a tribe of eccentrics forging their way through the years further and further away from the familiar. If this is so, Lady Mary Wortley Montagu must surely be one of its chiefs. The unprecedented adventurousness of her journey to Turkey and the skill with which she described its strangeness, set her well apart from the rest. She was the first authoress to travel abroad for mere curiosity's sake, and call herself, with considerable pride, 'a traveller'.

She was born Mary Pierrepont, eldest daughter of the Duke of Kingston, at the dawn of the Age of Reason. Suitably precocious as a child, she taught herself Greek and Latin and, no doubt encouraged by her cousin Henry Fielding and such friends of the Duke as Congreve, Addison, and Steele, soon became the darling of literary London. She was a mercurial, strong-willed woman, always ready to defy convention for the sake of independence. In 1712 she eloped with young Edward Wortley Montagu (the first of many conquests) and settled with him, against her father's wishes, in a gloomy stately home near Sheffield—with frequent and necessary visits to the *beau monde* in London. Then, in 1716, Montagu was appointed Ambassador to Turkey. Lady Mary had been ill with smallpox (nothing unusual then) and felt stale after the birth of her first child—so she decided to join him. This was a scandalous decision. No woman had done such a thing before; that a fashionable Lady, with a Reputation and a brilliant literary future ahead of her should actually *volunteer* to go to the barbarous East, risking death by a thousand dangers, was unthinkable. She thoroughly enjoyed herself, of course. The ambassadorial party, with its retinue, sailed to Rotterdam and then progressed overland through Holland to Germany and down the Danube to Vienna; after two months at the Austrian court they carried on overland to Adrianople (Edirne) instead of continuing down the Danube, which was frozen. It was an untried route, a journey (noted Lady Mary proudly) 'that has not been undertaken by any Christian, since the time of the Greek Emperors'. They arrived in Constantinople almost a year after leaving London, and settled into a hilltop palace at Pera, their home for the next eighteen months.

Lady Mary's *Letters* (written after her return to England, and circulated in manuscript until their publication after her death) revealed to their readers for the first time what lay behind the jewelled curtain of the East: and it was not at all what they might have expected. The slaves of the seraglio, she said, were far less put-upon than servants back home, and Turkish women had far more liberty than English women. Disguised and unrecognizable in their capes and veils, they lived delicious lives of 'perpetual masquerade': the perfect front for intrigue and adventurism. And far from swooning, as might have been expected, at the fabled excesses of Turkish life, Lady Mary merely concluded that 'gallantry and good-breeding are as different, in different climates, as morality and religion. Who have the rightest notions of both, we shall never know'. She was unashamedly open-minded—even to the extent of allowing her son and baby daughter to be 'engrafted' with the smallpox by a Turkish medicine-woman. This 'engrafting'—essentially innoculation—involved daubing 'matter of the best sort of smallpox' on to a needle, which was then pricked into any vein of the patient's choice. A mild bout of fever would follow, and then guaranteed immunization. She brought the idea back with her to England—seventy years before Jenner's smallpox vaccination—and thus became a medical pioneer, too.

Lady Mary was sad to leave Turkey, but the voyage home via Greece and North Africa offered some compensation and it was not long before she was back into the swing of London political and literary life. She was just beginning to settle down into a dignified old age when, in 1738, she met the Italian poet and dandy Francesco Algarotti and fell utterly in love. Six months later she left for Venice, ostensibly in search of health but really to follow her paramour—although she never admitted it to anyone but her rival in love, Lord Hervey (Algarotti had catholic tastes). She found France and Italy to be full of 'English families of fashion', all of whom were eager to entertain the daring and wonderful Lady Mary. Algarotti managed to avoid her for two years, during which time she rented lodgings on the Grand Canal and toured to Rome and Naples; after their meeting, which was an embarrassing anticlimax, Lady Mary left for Geneva and then Avignon, where she lived in a converted mill.

She did not go home to England again for twenty-two years, settling instead into a life of 'rural retirement' in a succession of cottages in Brescia, Venice, and Padua with a variety of colourful escorts. Her daughter sent her books and pamphlets from home and there were constant guests, a visit to Lady Mary being one of the highlights of any mid-eighteenth-century Grand Tour. She continued to write, publishing poems and essays at home, and was heartily welcomed on her return to London after Montagu's death in 1762. She died of cancer, aged seventy-three, that same year, and a year after that her Turkish Letters were published for the first time. I should not think they have been out of print since.

2

The Glamour of the Back of Beyond

PICTURE a party. There is a large room full of people—all women—chattering animatedly to one another. They are a bizarre collection: one, entitled Beatrix BULSTRODE Manico Gull, is standing four-square in the middle of the throng regaling her companion (a neat, *petite*, and rather twittery-looking young lady named Nina) with a booming story you cannot quite catch—something about Mongolian wolves. Nina readily replies with a fondly remembered anecdote to match, about her Himalayan exploits in a Bath chair. Next to Nina a serene and well-dressed woman with an aristocratic voice is comparing notes on how to steal into sacred Muslim cities with what looks like a young boy but who gives his name as Sarah. Sitting in a corner, with a cluster of avid listeners at her feet, a spreading, exotic-looking figure with slightly peculiar eyes and gamy perfume tells of her reign as Queen Hester of the Desert while an Assam goddess, born in England, looks on. In another corner, stretched out voluptuously on a *chaise-longue*, Lola MONTEZ is discussing sex with the compact, weather-beaten figure of Sarah HECKFORD, and pale little Emily HORNBY in bustle and bonnet is talking mountaineering technique with Julie TULLIS, bronzed and sinewy in shorts and a T-shirt.

A more unlikely group of guests it would be hard to imagine—yet all appear to be utterly engrossed in one another's conversation. What they have in common, firing their enthusiasm beyond the secondary excitements of mere travel and authorship, is an overwhelming spirit of adventure.

This flagrant impulse embarrassed some of them, so they tried to disguise it by assuming some more acceptable excuse for travelling. Louisa JEBB, for example, pretended that she was simply accompanying an invalid friend, and in search of nothing less seemly than a little philosophy when she went to Baghdad in 1907. But it was

Adventure she was really after, and before long she was to be found sitting round camp-fires at night amongst Mesopotamian nomads, baying for rebellion, madness, and savagery under the desert moon. Ada CHESTERTON travelled disguised as a journalist, Stella COURT TREATT and Osa JOHNSON as spunky but obedient little wives, and Eliza FAY, of all things, as a hat salesman.

Others cared too little about what society thought or expected of them to plead any ulterior motive. They were usually the women for whom the most extraordinary undertakings somehow became quite commonplace and calm. It was the most natural thing in the world for Helen CADDICK to leave her spinsterish Midland life for the unknown depths of what is now Malawi, or for Gwen RICHARDSON to mount her own diamond-hunting expedition to the far upper reaches of the Essiquebo River in South America. Or so they make it seem.

The journeys described by these women beguiled by what Ella SYKES called 'the glamour of the Back of Beyond' are amongst the most hair-raising in this book. That is not just because of the abundant traditional dangers, such as enraged lions and natives, swampy fevers or Cape Horn seas, sandstorms and white-outs, and so on. The more insidious treacheries of depression and demoralization, and the fear of rape or exploitation by fellow travellers rather than foreign savages, are just as chilling—and rarely admitted elsewhere.

But whatever the dangers, they were never enough to discourage these restless and questing ladies from their travels. Perhaps Miss Sykes summed up their attitude to travel best of all: it quickened the blood, she said, and made the pulse beat high. Once tried, the experience was not easily forgotten.

BROOK, Elaine (b. 1949)

[with DONNELLY, Julie]: *The Windhorse.* London: Cape, 1986; pp. 223, with 23 colour photographs, 9 halftones, 2 sketches, and a map

Land of the Snow Lion. London: Cape, 1987; pp. 238, with 23 colour photographs, 4 vignettes, and an endpaper map

Elaine Brook first started moving round the world—rather than travelling—to find mountains. By the time she arrived in Tibet, the 'land of the snow lion', with an all-male expedition to Shishapangma (26,291 ft.) in 1982, she had already had several seasons' experience elsewhere in the Himalaya, and in the Alps, the Andes, and the Rockies, and had chosen as nomadic a life-style for herself as possible, to allow for sudden departures to the mountains should an expedition leader ring up—as in this case—at impossibly short notice, one 'man' short. But this Tibetan trip was her last such journey. Being the sole woman on a serious expedition like this one put Elaine in an invidious position: her role was unspecified (she seemed to have been brought along as the expedition scapegoat) and eventually, worn down by the strain and petty jealousies surrounding camp life, she left the men behind and set off for the heartland of Tibet alone. The next few weeks she spent living with Tibetan families (whilst being pursued by the suspicious Chinese authorities and a rabid dog): she had discovered the art of travel for the first time.

Three years later, having become by now a guide on treks through the Himalayan foothills of Nepal, Elaine made one of the most remarkable journeys of her career. She accompanied a friend of hers, Julie DONNELLY, from London to the summit of Kala Patthar, two-thirds the way up Everest, at 18,000 ft. It was a 'sponsored walk', in aid of Guide Dogs for the Blind, frustrating and rewarding in almost equal degree: Julie herself is blind. From her, Elaine learned how to appreciate the sounds and scents of travel, the exquisite feel of a cool stone or the terror of an unseen waterfall: it was a rich few weeks for them both.

Elaine still visits the Himalaya regularly: she and her Nepalese Sherpa husband are based in the Peak District, but work back in the Everest region (and elsewhere) escorting trekkers along the familiar routes Elaine knew first as a mountaineer and then, like Julie, as a pure enthusiast.

⬥

BULSTRODE, Beatrix
A Tour in Mongolia. London: Methuen, 1920; pp. 238, with 28 halftones and a map

Mrs Bulstrode's book opens in 1913 in Peking, and promises from the outset to be gripping stuff, with a strong political flavour (unavoidable at a time when Sino-Mongolian relations were teetering on the brink of war) and plenty of Asian derring-do.

Brisk and bustling Beatrix, game for anything, sets off towards the Gobi desert for no other reason, she says, than 'to revert awhile to the primitive', travelling first by train—second class, to meet people—then aboard the camel thoughtfully supplied by the local mandarin; at one stage she joins a bemused Finnish missionary on his mule-cart insisting he stop every afternoon at tiffin time, and at last, surrounded by the vast and echoing stretches of north-west China, she homes in on a small and unsuspecting Englishman getting away from it all in search of sport.

He is a 'peppery little man', admits Beatrix, but as he seems to be going in the right direction and has boasted that he knows how to milk passing cows, shoot wolves, and confound Russian officials, she elects him her travelling escort. For her part, Beatrix initiates Mr Edward Manico Gull into the mysteries of her favourite hobby, the study and collection of human skulls—until the day she scoops one up with half its filling still inside, thus putting them both off for good. Beatrix's insatiable appetite for the bizarre, the most unexpected and outlandish elements of travel, must have been singularly exhausting for her companion. Or was it? For on their return to London, boisterous Mrs Bulstrode, we are told, became meek Mrs Manico Gull . . .

❧

CADDICK, Helen (*c.* 1842–1926)

A White Woman in Central Africa. London: T. Fisher Unwin, 1900; pp. 242, with 16 halftones

Bravery, curiosity, horse-like strength (at least when on the move), and a gently mocking sense of humour: Helen Caddick had all the right ingredients for the popular Victorian lady traveller. And she was almost as inveterate a diarist as she was a globe-trotter—twelve volumes remain of her meticulously-typed and illustrated journals, describing vast holiday jaunts made all over the world from 1889 until the outbreak of the Great War. So why only this one got in to print I cannot imagine. Miss Caddick was a well respected pillar of her local society—a member of the West Bromwich Education Committee and, more enterprisingly, the Governing Body of Birmingham University; she was of a 'good' family, ample means, and a staunchly independent spirit, and she travelled quite simply, she said, for the love of it. The journey she made to Central Africa in 1898 is typical of her refreshingly liberal attitude to tourism: because so many people urged her to abandon the idea of visiting Lake Tanganyika alone, she determined to do so. She was already touring South Africa and Rhodesia (Zimbabwe) so it would be no trouble (to her) to catch a steamer up the Zambesi and Shiré rivers to the Great Lakes of what is now Malawi.

Miss Caddick distinguished herself from other Victorian travellers to the tropics by scorning that sacred essential quinine during her six-month expedition (and consequently completely avoiding fever), and by roundly enjoying every moment of the trip, flinching neither when the rats who frequented her bed 'in a thoroughly happy manner' got their feet stuck in her ear, nor when she had to prise tropical maggots from beneath her toenails: that kind of thing was part and parcel of The Experience. Her porters—'my boys'—were fun and well-disposed (like the coolies she employed in 1909 to row her a thousand miles up a Chinese river), and she found her *machila*, a pole-strung hammock, quite 'the most delightful mode of travelling' overland. Her relationship with the natives was uncommonly uncluttered: like Mary KINGSLEY she appreciated the harm some mission-aries were already doing by 'civilizing' them, and dryly notes that her fellow Europeans were 'greatly amused and interested to meet someone who has come merely to see the land and the people'. But that is all she ever wanted to do.

Miss Caddick spent her post-travelling years carefully documenting a rich collection of native art and artefacts which she bequeathed to the local museum, and showing people her photographs and telling them the odd tale of her foreign adventures—probably surprised that they were interested, but no doubt happy to oblige.

❖

CHESTERTON, Mrs Cecil (Ada Elizabeth) (*c.*1888–1962)

My Russian Venture. London: Harrap, 1931; pp. 283

Young China and New Japan. London: Harrap, 1933; pp. 311, with 45 halftones and a map

Ada and 'Bunny', two middle-aged friends, decided to go to China. They had never seen a Native before, and were thrilled at the prospect. One evening, skittishly deciding to explore their steamer, they came upon a member of the crew:

There was not a sign of human habitation; we wandered from deck to deck, through phantom alleyways that echoed to our feet, up ladders stark as scaffold poles. And then, in the spectral gleam of a solitary electric lamp, a figure flitted across my vision, his face for a moment centred by the light. 'Bunny,' I said, 'that is a Chinese.'

One might be forgiven for inferring that Ada Chesterton—sister-in-law of G.K. and founder of the 'Cecil Homes' for fallen women—was one of the most embarrassing sort of lady travellers: those who caricature all they see and find Foreigners quaint and amusing. In fact she was just the opposite (and a gifted parodist!). She was a serious and incisive commentator on social conditions—special correspondent of the *Daily Express*—whose travels both at home and abroad were dedicated to the 'new and resurgent force' of Communism.

Her Russian Venture was a daring one, involving a clandestine crossing from north-east Poland into Stalinist Ukraine and a tour by train, farm cart, or on foot in search of the political ideal. She returned to England (with some unbidden Russian lice) neither disillusioned nor cheered by her journey: the Russian peasants she spoke to and travelled with lived miserable and ill-afforded lives, but there was no doubt that the Soviet system was full of 'dynamic potentialities'. Eager to visit that other crucible of collectivism, Ada and her friend 'Bunny' set off for China the following year. Again she travelled amongst villages and peasants, from Shanghai up the Yangtze to Hankow and then on to Japan, proving to her own satisfaction the 'consolation and protection' of Communism. These were dangerous journeys as well as physically demanding, but Ada in her zeal ignored discomfort. Or else she courted it, as in London in 1925 when she lived the life of a 'down-and-out' to research her books *In Darkest London* (1926) and *Women of the Underworld* (1930), as the closest means of identifying with her subject.

◆

CHRISTIE, Ella R. (1861–1949)

Through Khiva to Golden Samarkand: The Remarkable Story of a Woman's Adventurous Journey Alone through the Deserts of Central Asia to the Heart of Turkestan. London: Seeley Service, 1925; pp. 280, with 55 halftones and a map

In fact Miss Christie made *two* adventurous journeys into Turkestan in the years leading up to the First World War: one from Moscow and the Aral Sea to Kokand and the second farther south from Constantinople (Istanbul) and the Caspian to Andizhan. She travelled to see 'what lay on that comparatively bare spot on the map' and because she found the lure of the magic names of Bokhara and Samarkand quite simply irresistible. At first the Russian authorities were rather taken aback by this suspiciously ambitious Englishwoman's plans, but she withstood all their interrogation and disincentives: Miss Christie was determined to go, and go she would. Her narrative is objective, articulate, and scholarly; remarkable as her story is, she keeps her own personality very much at bay and it is the richness of the country rather than the virtuosity of its visitor that impresses us: the fact that she made this unprecedented journey as a woman alone, in treacherous terrain at a treacherous time, is taken almost for granted.

◆

COBBOLD, Lady Evelyn (1867–1963)

Wayfarers in the Libyan Desert. London: Humphreys, 1912; quarto, pp. 128, with 15 halftones

Pilgrimage to Mecca. London: John Murray, 1934; pp. 260, with 20 halftones and a map

Kenya—The Land of Illusion. London: John Murray, 1935; pp. 236, with 23 halftones and a map

Lady Evelyn Cobbold was the first Englishwoman to make the pilgrimage to Mecca as a Muslim. Her first book is rather stylized, a wafting account of the country where she spent her childhood holidays and which inspired her to Islam, and her last a pedestrian description of a series of safaris and sightseeing trips in East Africa. But *Pilgrimage to Mecca* is quite different. It is a valuable record of the hadj: for once, a woman's view from the inside out. We do not forget that the author is a Lady—she stays with the distinguished St John Philbys in Jeddah and travels to Mecca in a large limousine with chilled chicken and soda-water in a hamper in the back—but the picture she gives of the experience is unelaborate and revealing, and detailed enough to serve as a guidebook as well as a travel account.

COURT TREATT, Stella

Cape to Cairo: The Record of a Historic Motor Journey. London: Harrap, 1927; pp. 251, with 64 photogravures and a map

Sudan Sand: Filming the Baggara Arabs. London: Harrap, 1930; pp. 252, with 63 halftones and a map

Stella Court Treatt was the perfect 1920s mixture of bright young thing and glamorous adventurer. She was a wealthy and sultry-looking tomboy, too tiny to wear anything but boys' khaki shorts and shirts on her epic drive from one end of Africa to the other, and as handy with a rifle as she was with a cocktail-shaker.

Africa was in her blood: generations of her family had fought, explored, and hunted there. Major Chaplin Court Treatt (a clipped 'C.T.' for short) was also a bit of an 'African' chap: he was considerably older than his wife, a tall, manly figure with a pipe and an insouciant smile, who had worked for several years on the Trans-African Air Route and probably regarded the continent (like many other Englishmen at the time) rather as his own private estate, with its native inhabitants the family retainers. It was C.T. who suggested this 7,000-mile jaunt in a Crossley motor car—it would cheer Stella up, he said, after her recent bout in hospital. It was to be an 'all red' journey—the road (or swamp or desert tract) would never leave those lands so proudly coloured Empire-pink on the globe, and since there would be four other fellows to look after her besides C.T., Stella need never worry. And for the most part, she did not. Only occasionally, perhaps after a particularly bad run of breakdowns or a long floody delay or days and days of sand in one's pink gin did C.T.'s little woman give in: 'I hate self-pity but, hang it all, there are limits, and it isn't easy to be a woman and go through some of the things that I seem continually to have to endure. But it's dashed weak of me even to think of these things'.

Stella stuck the journey out to the end, of course, and went back to Africa two years later as C.T.'s helpmeet during the shooting of his film *Stampede*. She wrote up the story-line in a novel of the same name, and produced her last travel book about the location in Sudan. But by then, C.T.'s ardour was cooling ('I hate self-pity') and soon after their return to England Stella filed for a divorce on the fashionable grounds of his adultery ('but, hang it all, there are limits'). She disappeared to India, married again, and retired into relieved obscurity.

❧

CRESSY-MARCKS, Violet Olivia (d. 1970)

Up the Amazon and Over the Andes. London: Hodder and Stoughton, 1932; pp. 336, with 37 halftones and 3 maps

Journey into China. London: Hodder and Stoughton, 1940; pp. 324, with 79 halftones and 4 maps

Mrs Cressy-Marcks's *Times* obituary was eulogistic: no contemporary woman had equalled the distances she travelled, it said, nor even approached them. Although that is not strictly true, she was one of the most aggressive and questing of her peers. In 1925 she travelled the length of Africa from the Cape to Cairo, and then tackled the frozen journey through Scandinavia to Murmansk by sleigh; after the 1929 trip across South America she took on Afghanistan, Turkey, and Siberia. Then followed conquests of Ethiopia and Kenya (during the Italian invasion), and of war-torn China in 1938.

They were punishing journeys, all of them, but Violet Cressy-Marcks confessed a fascination with boundaries and seems always to have been attracted to the edge of things. She stretched herself to meet self-imposed targets and particular dangers, and the two expeditions she chronicled make exhausting reading. During the Second World War, she became a military correspondent for the *Daily Express* in the Far East (based in China, where she managed to interview Mao Tse-tung at the Red Army HQ and stowed away on a troop train to the forbidden borders of Tibet) and in the quarter century before her death she made countless gentler expeditions. A Fellow of the Royal Geographical and the Zoological Societies (although primarily an archaeologist), she strenuously collected as much scientific data as she could, and confronted any little local difficulty—a head porter committing suicide in the Amazon, for example, or gouging the occasional snake-bite from her foot—with stern courage. She was one of the most committed and thorough of women travellers, constantly trying to prove and improve herself, and, it seems, quite fearless.

❖

DUNCAN, Jane Ellen (1848–1909)

A Summer Ride through Western Tibet. London: Smith, Elder, 1906; pp. 341, with 93 halftones, vignettes, and a map

Jane Duncan's travels are full of delight. Finding herself in Srinagar in April 1904 with the summer approaching, she decided to take to the hills. She rode to Leh, full of the joys of travelling alone, and then on to explore Ladakh and Baltistan or 'Little Tibet'. She relished all she saw—weddings, funerals, even the odd attack by bandits ('I am afraid I am rather proud of that!')—and regularly arranged picnics for her bemused porters. She was so exhilarated by the clear air and the abandonment of her side-saddle (she chose most daringly to ride astride, like men) that when she chanced upon Ella CHRISTIE one day—both treating their meeting in the emptiest corner of Asia as the most natural thing in the world—she chose not to accompany her but to carry on alone. For half a year her home was an eighty-pound 'Cabul' tent, with brass bedstead and separate 'bathroom', guarded by a favourite Indian servant, an ex-batman to Sir Francis Younghusband, who slept under the flysheet; it saw her through the high passes of the frontier with Tibet proper, along the harsh and towering Indus valley, and eventually back to Leh after the finest six months this slight Scots spinster had ever known.

❖

FAY, Mrs Eliza (1756–1816)

Original Letters from India; containing a Narrative of a Journey through Egypt, and the Author's Imprisonment at Calicut By Hyder Ally. To which is added, An Abstract of Three subsequent Voyages to India. Calcutta: [no publisher given], 1817; pp. 404, with an engraved frontispiece portrait

Most accounts of women travellers to India are written by memsahibs, by those privileged in one way or another by rank or family standing. Mrs Fay was no such gentlewoman. She was the young wife of a lowly advocate sent to the Calcutta Supreme Court in 1779, just six years after Warren Hastings had become Governor-General of 'British India' in Bengal, and at a time when upper-class imperialism was only just beginning to emerge from the mercantile British society there. Her account is highly original, not just because of her social position and the historical circumstances, but because of her inimitably vivid style. Such niceties as grammar and felicity of phrase don't worry Eliza Fay: she has too good a story to tell for that.

Her book is divided into two parts: one covering the Fays' journey out to India, which involved a shocking crossing from France to Italy (principally because Eliza had never realized that 'the Alp' involved more than one mountain) and a near fatal attack in the Egyptian desert on the way from Cairo to Suez, along with their stay there together and sensational imprisonment by the anti-British dictator Hyder Ali in Calicut; the second summarizes Eliza's later voyages to India and on to America. The Fays' marriage had not lasted long (E. M. Forster in the introduction to his 1925 edition says they were both 'quarrelsome and underbred'): in any case Fay soon lost his job, his reputation, and his wife, and they sailed home separately in 1784. Eliza returned to India several times as an entrepreneur (more enthusiastic than successful) to set up a hat shop and an export business of fine muslins to America—she took them there herself. And although the letters in the second part of the book were retrospectively written from London, Eliza actually died back in Calcutta.

No doubt the publication of the book was her latest project, but she did not live to see it published. Perhaps no one *would* publish it (certainly the writing has little literary merit); when it did eventually appear it was printed by the administrators of her meagre estate in the hope of realizing something for the creditors. A censorious and bowdlerized edition was reprinted in Calcutta in 1908, but even that cannot suppress Eliza Fay's pungent power of observation and irresistible optimism.

GRAHAM, Maria (afterwards CALLCOTT, Lady) (1785–1842)

Journal of a Residence in India. Edinburgh: Geo. Ramsay for Archibald Constable, 1812; quarto, pp. 211, with 15 engravings (1 hand-coloured)

Letters on India [etc.]. London: Longman, Hurst, *et al.*, 1814; pp. 382, with 9 etchings and a map

Three Months passed in the Mountains East of Rome, During the year 1819. London: For Longman, Hurst, *et al.*, 1820; pp. 305, with 6 aquatints

Journal of a Voyage to Brazil and Residence there During Part of the Years 1821, 1822, 1823. London: Longman, Hurst, *et al.*, and John Murray, 1824; quarto, pp. 335, with 11 aquatints and 11 wood-engraved vignettes

Journal of a Residence in Chile, during the year 1822. And a voyage from Chile to Brazil in 1823. London: Longman, Hurst, *et al.*, and John Murray, 1824; quarto, pp. 512, with 14 aquatints and 10 wood-engraved vignettes

Maria Graham's life as a traveller was remarkable—rather more so than the travel books she wrote, which although beautifully illustrated make unfortunately heavy going. She was a rear-admiral's daughter, who fell in love on her first long voyage with a sea captain, and married him within a month of landing on the exotic shores of India. The Grahams' wedding journey involved another voyage round the coast of the subcontinent, frequently interrupted by forays to temples and ruins inland. They visited Ceylon together, and on the way home to England stopped off at Cape Town and St Helena. As soon as Captain Graham's next leave was due they went to Italy; two years after that their first real voyage together began, with Graham in command of his own ship bound for Brazil.

The Atlantic crossing was successful; not so the subsequent voyage round the Horn to Valparaiso in Chile. It was not storms but fever that beset the ship, and the Captain himself was one of the first victims. He died off Cape Horn in April 1822, and Maria's journal fell silent. She took it up again on her arrival in Chile, where she was lent a cottage by friends and in time settled down to some sort of life amongst the British there.

She spent over a year in Chile, surviving the Great Earthquake of 1822, and the voyage back through the roaring forties to Brazil in 1823. From Rio de Janeiro she secured a passage home, but not before accepting a post as governess to Donna Maria, daughter of the Prince of Portugal (and Emperor of Brazil). The voyage home allowed her just enough time to see her books on Brazil and Chile through publication, and within months she was back in Rio, installed at the Imperial Palace.

By 1827 she was home again, and married to the artist Augustus Callcott. With him she travelled through Europe, and he illustrated several of her books on Continental art and architecture. Maria began at this stage to reel off children's moral tales, too, culminating in the celebrated *Little Arthur's History of England*. She died something of a literary lioness in London in 1842, aged fifty-seven.

Maria Graham's travel books are stiff with history and politics—Brazil was at the time of her visits torn with the insurrections of the northern States against the European Portuguese in the south—and they are all rather impersonal. Such reliance on facts and figures made them easy targets for criticism, and there was a fiery exchange in the 1830s between Maria and the Geological Society over her 'fanciful' descriptions of the 1822 Chilean earthquake, resulting in a volley of indignant pamphlets. She was a unique traveller—a born sailor and one of the first Englishwomen to write of life in

South America; it is a shame that her talents for travel and for writing were not more happily married.

❖

GRAHAM BOWER, Ursula (1914–1988)

Naga Path. London: John Murray, 1950; pp. 260, with 22 halftones and 2 maps

The Hidden Land. London: John Murray, 1953; pp. 244, with 26 halftones and a map

Ursula Graham Bower made her first visit to Assam in 1937, when a friend in Manipur invited her to stay for a while. Ten years later she knew more about the area and its people than any other European. She had fallen in love with the Naga tribes on the hills behind Manipur, and contrived to be sent out there as an anthropological photographer by the Royal Geographical and Central Asian Societies soon after her return to Britain. This time she travelled alone, and in a remarkably short time became accepted by the 'primitive' Naga people not just as their medicine-woman and sage, but as their saviour. She was the reincarnation, they said, of a traditional Naga goddess, and Miss Graham Bower bore the responsibility well. Trips home became shorter and fewer, and by the time the war broke out, she was to all intents and purposes a resident of the Assam hills, known by native tribesmen and British Government alike as the 'Naga Queen'. During the war she organized a guerilla force amongst her neighbours against the Japanese, and ran a refugee canteen on the road from nearby Burma. She also trekked to Tibet: that was a wedding journey. The man she married was a British colonel who had heard of the legendary Naga Queen and spent his month's leave finding, wooing, and winning her. Together they travelled after the war to the Api Tani valley on the Tibetan border (the 'Hidden Land' of her second book), where Colonel Betts had been appointed Political Officer. Ursula continued the lone treks that had been part of her life in the Naga hills until her husband's illness forced them to return to London. The further she travelled from India the more her spirit seems to have dwindled; the second book closes on a grey London pavement in utter depression. Not surprisingly, the Betts were off again as soon as they could manage, and after a six-year stint in Kenya from 1948, they settled at last on the suitably remote and primitive Isle of Mull.

❖

HECKFORD, Mrs (Sarah) (1839–1903)

A Lady Trader in the Transvaal. London: Sampson Low, Marston, *et al.*, 1882; pp. 412

Sarah Heckford's story is extraordinary. She was born in Dublin, orphaned at ten and sent to relatives in London where she lived an indolent life, dreaming vaguely of becoming a doctor. Then the cholera epidemic of 1866 broke out, and much to the alarm of her family Sarah volunteered to become a nurse. She was sent straight to Wapping District Cholera Hospital, and stayed there until the plague had run its course. In 1867 she married one of the doctors at Wapping, and together they planned to set up the first children's hospital in London's poverty-stricken East End. When Dr Heckford died in 1871 Sarah carried on alone, and by 1876 the hospital was built.

The next two years were spent in India, where the exhausted widow had gone for a change of air. Her work with the Zenana Mission in Calcutta led to a most exotic appointment as resident medical adviser to the fabulous Shah Jahan Begam of Bhopal. The experience stirred a rather disconcerting spirit of adventure in Sarah's stout and middle-aged breast, and by 1878 she could contain herself no longer. The Empire had taken a new jewel to its crown that year: the Transvaal. A Farming, Mining and Trading Association was formed to tempt adventurers to the new promised land; Sarah was amongst its first shareholders, and before the year was out had decided to emigrate.

She arrived in Durban in December, and immediately proceeded on a 450-mile journey by bullock cart and horseback through Natal and the Orange Free State to the village of Rustenburg, north of Pretoria. After a year as governess to an Afrikaner family there, she launched into the highly unorthodox career that was to occupy the rest of her life. During the summer she was a farmer, working land, sheep, and cattle. After harvest she became what was locally known as a *smous*. She would trek to Pretoria, pick up various goods ordered from home or bought in the local markets, and travel up to the veld in her cart to trade with the Boers and Afrikaners at their winter camps. She would swap anything—cotton for eggs, boots for pumpkins, brandy for corn—and come summer would have enough money to buy more land and engage enough Kaffirs to work it with her. Thus she earned her livelihood, all through the unquiet prelude to the Boer War.

When times were bad (during the Siege of Pretoria when her cattle and corn were requisitioned without compensation, for example) Sarah would turn her hand to other things: selling shares in gold-mines, perhaps, or running a bullock-drawn transport fleet through the uplands. During the war itself she became a self-appointed spy, reporting distant enemy movements to the British. She died a year after peace, and according to *The Times* she was 'one of the most extraordinary women to whom the British nation has given birth'. To her neighbours in South Africa—where she is remembered better today than in Britain—she was simply the strangest *smous* you ever saw.

HOBSON, Sarah (b. 1947)

Through Persia in Disguise. London: John Murray, 1973; pp. 175, with 13 halftones and a map

Family Web: A Story of India. London: John Murray, 1978; pp. 284, with 16 halftones, 3 plans, and a family tree

Two-Way Ticket [Peru, Norway, Bangladesh, Borneo, Hebrides, and Mauretania]. London: MacDonald, 1982; pp. 61, with colour photographs throughout

Sarah Hobson is a particular kind of explorer: a discoverer (in the truest sense of the word) of culture and the way people live. She made her first journey at the age of twenty-three. Wanting to explore Islamic faith, to study Islamic design, and to taste some good old-fashioned adventure, she decided to visit the forbidden shrine of Qum, in what used to be northern Persia.

Forbidden, that is, to infidel females. So she became a man. She cropped her hair, wore an elastic girdle round her chest under baggy khaki clothes, kitted herself out with pipe, razor, and a pair of large suede boots, and set off. John, as she was now called, hitch-hiked to Istanbul, and then caught a bus to Tehran. Once in Iran, she used unreliable public transport, cadged the odd lift, and bought an elderly motor bike to get around the country. She did penetrate the shrine at Qum, and lived amongst students there who, if not convinced by her disguise, at least respected her need for it.

Sarah's next journey was to India, where her husband Tony Mayer was making a film based on a peasant family in a village near Bangalore. Sarah lived with the family both during and after the filming, and *Family Web* concerns her gradual absorption into their culture. This experience with film drew her into a television series for children, *Two-Way Ticket*, which involved her visiting six different families around the world and inviting a child from each of them back to her own Yorkshire farmhouse; more films and various documentaries have followed, based on her recent travels to the Gambia and Senegal.

❧

HORNBY, Emily (d. 1906)

A Nile Journal. By E.H. Liverpool: J. A. Thompson [for private circulation], 1906; pp. 192, with a frontispiece portrait

Sinai and Petra: The Journals of Emily Hornby, in 1899 and 1901. London: James Nisbet [1907]; pp. 244, with 10 colour plates by F. M. Hornby

Mountaineering Records. By E.H. Liverpool: J. A. Thompson, 1907; pp. 352

Emily Hornby's sisters had her journal of a holiday voyage down the Nile privately printed 'for those who would have heard about the journey from her . . . [and] who knew how much she loved travelling'. It was published as a memorial to the author, who died soon after the 89-day cruise of pneumonia, contracted during the chill nights of desert excursions. A year later her private accounts of treks in Sinai and Petra were also published. The Middle East seems to have been swarming with Hornby sisters and cousins during the last years of the nineteenth century: die-hard female tourists of the old school, who took pride in travelling independently and regarded the increasing herds of Cook's Excursionists with disdain. And yet the journals are strangely spiritless and dull: Emily was used to sterner stuff, as her astonishing *Mountaineering Records* show.

The records were compiled by her sisters from the scant notes Emily kept between 1873 and 1896 during seasons in the Alps, Dolomites, and Tyrol. With her money tucked inside her stays and a flask of cold tea, she conquered an awesome fifty-nine peaks, including all the Alpine classics (Mont Blanc, the Matterhorn (twice!), Monte Rosa, Dente Blanche . . .) and, if her own testimony is to be believed, was a mountaineer of the first order.

Emily's extraordinary achievements have gone largely unnoticed: she climbed for private pleasure rather than public attention and had none of the flamboyance of, say, Mrs LE BLOND or Fanny Bullock WORKMAN. One wonders how many more there were like her, unrecorded and forgotten.

❧

JEBB, Louisa (Mrs Roland Wilkins) (d. 1929)

By Desert Ways to Baghdad. London: T. Fisher Unwin, 1908; pp. 318, with 74 halftones and a map

Neither Miss Jebb nor her friend liked the sea. Deserts were far more romantic, and when the friend's father offered to finance a tour for them, they decided to become nomads for a season and explore Asia Minor. Friend went in search of health, and Louisa, probably dazed with the heady success of her famous work *The Smallholdings of England* (1906), went in search of 'the fundamental'. They travelled by train (second class) as far as the Anatolian plateau in Turkey, and then on horseback over the Taurus mountains and across the Euphrates to Mesopotamia. To reach Baghdad, they floated down the Tigris on a raft made from 260 inflated goatskins. They camped, of course, lived on rice and chocolate, and hardly ever washed—all delightfully primitive.

Louisa's account of the journey is detailed enough to be of topographical value, and nicely witty; in fact the only time she loses her sense of humour is when she actually finds what she was looking for. The search for the 'fundamental' was consummated one evening on the banks of the Tigris in Kurdistan. The locals were celebrating some festival, laughing, singing, and madly dancing in the moonlight. Louisa wistfully looked on, bewailing her native primness: 'I . . . poor sane, fool, can only remember that I once did crochet work in drawing rooms', until all of a sudden: 'a feeling of wild rebellion took hold of me; I sprang into the circle. "Make me mad!" I cried out, "I want to be mad too!" . . . And soon I too was a savage, a glorious, free savage under the white moon.' The ecstasy did not last long. The two girls were confined to the Babylon Hotel on their arrival in Baghdad, and after crossing the Syrian desert to Palmyra (by now a well-hoofed tourist route), they were sent home to horrid, civilized England.

❖

JOHNSON, Osa (Mrs Martin) (1894–1953)

I Married Adventure: The Lives and Adventures of Martin and Osa Johnson. Philadelphia: J. B. Lippincott, [London: Hutchinson,] 1940; pp. 376, with 83 halftones

Four Years in Paradise [British East Africa]. Philadelphia: J. B. Lippincott, 1941; pp. 345, with 74 halftones

Bride in the Solomons. Boston: Houghton Mifflin, 1944 [London: Harrap, 1945]; pp. 251, with 64 halftones

Last Adventure: The Martin Johnsons in Borneo. New York: W. Morrow, 1966 [London: Jarrolds, 1967]; pp. 233, with 31 halftones and a map

'We're going round the world, Osa!' Heady stuff for a dainty little Kansas girl, who had married at sixteen a handsome film photographer just back from the exotic South Seas. Osa Johnson would have been quite happy to keep home and settle down—she had never been further than thirty miles from the house where she was born—but Martin Johnson and their whirlwind romance so bowled her over that all she could answer was 'Well—all right, dear.'

The Johnsons never did settle down: immediately after their wedding in 1910, they toured the Wild West showing a film Martin had made on his voyage to the South Pacific. Osa sang Hawaiian songs at each performance, and by 1912 they had earned enough to finance an expedition to the Solomon Islands and New Hebrides. Its objective was to preserve a primitive lifestyle on film before it evaporated into civilization. Osa proved a natural helpmeet to her husband: 'the best pal a man ever had'. She learned to shoot and fish, keeping the expedition in food; with her long fair hair and smiling prettiness she diverted the natives whilst Martin filmed them.

Much of their time was spent on the notorious island of Malekula; although most previous white visitors there had been killed (the remains of one of them was cooking on the spit when the Johnsons arrived), Osa and Martin managed not only to survive but to return later and give the cannibals an al fresco film-show of themselves. They visited Borneo next, where they spent a year filming the head-hunters of the unexplored Kinabatangan River. It was here that Martin first began photographing wildlife, and it soon became his speciality.

During the years the Johnsons lived on safari in 'Paradise' (an unrevealed site on the Kenyan border with Ethiopia), on the Serengeti plains, and in the Belgian Congo, they were commissioned by the Museum of Natural History in New York to produce a complete film record of the game of Africa. Osa had by now become an expert markswoman; her job was to cover Martin whilst he filmed. Some of the most exciting passages in their commercial films were taken as elephants charged or lions sprang at the camera: Osa might be weeping the while but she always brought them down, often literally at Martin's feet. For relaxation they learned to fly, bought an aircraft each, and flew side by side on joy-rides to Mounts Kilimanjaro or Kenya, meanwhile producing some of the earliest aerial film of Africa.

The Johnsons' last journey together was back to north Borneo. They took one of the planes with them, and spent two years living and filming amongst the Tenggara tribe there. When Martin was killed in a plane crash back in America, after twenty-seven years of blissful marriage Osa carried on their work alone. She produced more films from material they had already collected, and as well as these volumes of autobiography (the last of which was not discovered until ten years after her death) wrote a series of children's books about the menagerie of pets she had collected on her travels. She led Twentieth Century Fox's expedition to Africa for the filming of their epic *Stanley and Livingstone* in 1938 and was planning to return to 'Paradise' when she died in New York, in her sixtieth year.

❧

MAILLART, Ella K. (b. 1903)

Turkestan Solo: One Woman's Expedition from the Tien Shan to the Kizil Kum. London: Putnam's, 1934 [first published in Paris the same year as *Des Montes célestes* [etc.]]; pp. 307, with 64 halftones and 2 maps (1 on endpapers)

Forbidden Journey: From Peking to Kashmir. London: Heinemann, 1937 [first published in Paris the same year as *Oasis interdites* [etc.]]; pp. 312, with 64 halftones and 3 maps

Gypsy Afloat [North Sea, Channel, and Mediterranean]. London: Heinemann, 1942; pp. 240, with 33 halftones, line-drawn vignettes, and endpaper maps

Cruises and Caravans [autobiography]. London: J. M. Dent, 1942; pp. 161, with 27 halftones and endpaper map

The Cruel Way [Geneva to Peshawar]. London: Heinemann, 1947; pp. 217, with 73 halftones and endpaper maps

'Ti Puss [India]. London: Heinemann, 1951; pp. 213, with 19 halftones and a map

The Land of the Sherpas. London: Hodder and Stoughton, 1955; pp. 61, with 78 halftones and a map

For Ella Maillart, a journey is not just a matter of travelling from here to there. Although she is quite capable of playing the fearless and intrepid explorer, whom the critics dub 'a dauntless lady', 'of outstanding courage and endurance', 'among the great travellers of the world', the landscapes and boundaries of the spirit have grown to be her most important discoveries. A real journey must be a personal quest for what she calls, quite simply, 'the truth'.

Mlle Maillart began her travels in 1914, when she was taught to sail and ski on the lakes and alps of her native Switzerland. She ski'd with a passion, and went on to represent her country at single-handed sailing in the Paris Olympic Games of 1924. As soon as she could leave school she went to Britain to learn English (the language in which she wrote all but her first two books) and then—aged twenty—sailed with three girl-friends from Marseilles to Athens, where they sold their 10-ton yawl. It was all very exciting, but not satisfying enough for Ella. So she went to Moscow, ostensibly to write about the film industry there, but really to observe Russian life and Communism at first hand.

Her first book was an account of a remarkable trek in 1932 from Moscow to the easternmost borders of Russian Turkestan, where she lived amongst the dignified Kirghiz and Kazakh tribesmen: this was more like it. Commissioned in 1934 as a special correspondent with the French newspaper *Petit Parisien* she went next to Manchuria, and met the young English journalist and explorer Peter Fleming. Independently they had planned to return to Europe overland, through the impossible terrain of Chinese Turkestan (impossible for political reasons as well as physical ones), but when each discovered the other's plans, they decided to join forces. This was not for any romantic reason—Peter was engaged to the actress Celia Johnson and in any case, Ella was not the sentimental type. It was simply that they had more chance of succeeding together than alone. They made strange travelling companions: the handsome Old Etonian with his 'affected manner and languid Oxford accent', and the fresh-faced and obstinate Swiss sportswoman; their respective accounts (*Forbidden Journey* and Fleming's *News from Tartary*) of the eight-month march through the Gobi and Takla Makan deserts to the gorges of the Hindu Kush make intriguing reading.

Ella's next chronicled journey was just as arduous in its way; it involved guiding a drug-addicted friend in an 18-h.p. Ford from Geneva (and through a nervous breakdown) to Kabul and Peshawar in 1939. There in India, at the bewildering outbreak of the Second World War, Ella began to find what she was looking for. She settled in Tiruvannamalai, south of Madras, and spent the next five years living as simply and prayerfully as possible. Occasional forays were made as far as Tibet or Nepal (often with her beloved cat 'Ti-puss), but for the most part Ella was content to sit at the feet of her guru Ramana Maharishi, exploring the 'unmapped territory of my own mind'.

Although based in Switzerland now (where she regularly ski'd until a few years ago) Ella still visits Asia. But these days, perhaps she doesn't *need* to travel: she has found 'the truth' within herself.

❧

MAZUCHELLI, Nina Elizabeth (1832–1914)

The Indian Alps and How We Crossed Them: Being a Narrative of Two Years' Residence in the Eastern Himalaya and Two Months' Tour into the Interior. By a Lady Pioneer. London: Longmans, Green, 1876; quarto, pp. 612, with 10 chromolithographs, vignettes throughout, and a map

'Magyarland': Being the Narrative of our Travels through the Highlands and Lowlands of Hungary. By A Fellow of the Carpathian Society. London: Sampson Low, Marston, *et al.*, 1881; 2 vols., pp. 379/311, with 2 wood-engraved plates, and vignettes throughout

Nina Mazuchelli is either thoroughly exasperating or thoroughly admirable—it is difficult to decide which. Her book on Hungary is an excellent one: enthusiastic, informative, and written with an irrepressible sense of fun. She chose Hungary because of the Carpathians and the Tatra: mountains were her heart's desire.

The love affair began in India, when Nina and Francis Mazuchelli, a British army chaplain, were posted to Darjeeling after eleven years on the plains. As soon as she saw the summit of Kanchenjunga shimmering in the distance, she was lost. She managed to contain herself for eighteen months, and then, as the time to return to England drew nearer, she made a 'startling proposition'. She, Francis, and their friend 'C' (the District Officer), would become pioneers, and explore the glaciers of the eastern Himalaya. Francis was obviously used to indulging his pretty wife's rhapsodic whims, and before long they had engaged countless coolies to carry their equipment (which ranged from a china dinner-service to a collection of iron bedsteads) to the 'nearest point of earth to heaven'. Nina, with sketch-book in hand, was tucked daintily into a Barielly dandy (a sort of Bath chair slung on poles instead of wheels)—Mrs Syntax in search of the picturesque, C called her—whilst he and Francis clopped along behind, wearing holiday hats and smoking their pipes. And off they went: 'How romantic, how sweetly Arcadian', sighed Mrs Mazuchelli.

The *ingénue*—they were nearly killed, of course. As soon as the first blizzard arrived, half the porters fled and the other half burst into tears. The Dandy-wallahs became snow-blind and started staggering and lurching their precious load from precipice to crevasse; the guide declared himself lost (they were well off the map by now); and soon all that was left of the food was a box of damp biscuits. As luck would have it, the coolies' feet had been bleeding for most of the journey and they were able to retrace their own gory footsteps through the snow until they arrived at a village, which they promptly sacked—with C's official permission, of course. So the Mazuchellis survived.

It was a farcical expedition—and Nina was the first to admit it, making her account an affectionate burlesque, with herself as a heroine of the old Impulsia GUSHINGTON school. But Nina's account has become by default a classic of mountaineering literature—she was, after all, the first Englishwoman to have travelled so far into the eastern Himalaya (they all but reached the Tibetan border before turning back)—a real 'Lady Pioneer'. Is the joke on her, or us?

❖

MILLER, Christian (b. 1920)

Daisy, Daisy: A Journey across America on a Bicycle. London: Routledge and Kegan Paul, 1980; pp. 180, with 5 maps

Mrs Miller was a well established granny before she even had the idea of cycling across America. She had worked in an aircraft factory during the war, published a novel and several short stories, and travelled in what she shamefacedly calls 'a high degree of luxury' all over the world, settling safely back into family life again afterwards. But then there came a day, she says, when she suddenly realized that she was no longer domestically indispensable, and so she bought a folding bike, a tent, and a saddlebag, and flew to Washington DC. Her route took her over the Appalachians to St Louis, across the Great Plain and weaving round the Rockies to Denver, Salt Lake City, Yellowstone, Missoula, and Eugene. She finally fetched up on the Pacific coast, 4,500 miles from her starting-point, with molten brake-pads, frozen toes, toothache—and a sizeable sigh of relief.

❖

MONTEZ, Lola (i.e. GILBERT, Maria Dolores Eliza Rosanna)
(1818–1861)

Autobiography and Lectures of Lola Montez (Countess of Landsfeld). [Edited by the Revd
Charles Chauncy Burr.] London: Jones and Blackwood [1858: published the same
year in New York and Philadelphia, and in London again by Ward Lock 'for Gilbert',
pp. 192, without the portrait]; pp. 202, with a portrait plate

Lola Montez was a real, ripe adventuress. Although biographers have
clustered around her like bees round a honeypot, the facts of her prodigious
life are still not completely clear. Her own autobiography was 'ghosted' by
an American clergyman (posing, in the book, as a childhood girl-friend) and
somewhat romantic, to say the least.

The general opinion is that Lola (a diminutive of Dolores) was born in
Limerick, Ireland, during her parents' honeymoon. Ensign Gilbert, her
father, was posted to India when she was five, and for the next four years she
lived in Calcutta and Dinapore. A wayward child, she was sent home in
1826 to the sobering care of Gilbert's Scottish relations at Montrose and
later to friends of the family at Bath. There she met Thomas James, a young
army captain, and eloped with him (to escape a marriage her mother had
arranged for her in India) in 1837. A year later he too was ordered to India,
and Lola—already celebrated at home for her coquettish beauty—soon
became the darling of the Raj: a 'good little thing', Emily EDEN called her (a
sentiment no doubt echoed by James's thigh-slapping messmates), and 'a
star' amongst the dun young wives around her.

But Lola was unhappy in India—her runaway marriage to the captain had
been a mistake, and in 1842 she left for England without him. Not long
afterwards he divorced her, on the grounds of her adultery on the voyage
home (meanwhile, he had eloped again—the habit was obviously catching).
Lola needed money. No doubt inspired by the likes of Sarah Siddons and
Miss KEMBLE, she enrolled at Frances Kelly's Dramatic School for
Actresses. To complete her education she joined a dancing master in
Madrid, and made her first appearance on the London stage in 1843 as
'Lola Montez, the Spanish Dancer'.

She was an instant success. Glamorous and provocative, she claimed the
hearts of all the roués of the town—including Lord Ranelagh who, when
Lola rebuffed him, arranged for the theatre to close her down. None
daunted, she moved on to Brussels, and spent the next five years touring
Europe and making fools of kings and courtiers. She eventually settled in
Bavaria, the mistress and 'best friend' of elderly King Ludwig. He gave her
gifts almost as lavish as those old Ranjit Singh, or Tsar Nicholas I, or Count
Paskevitch of Poland had given her; he created her Countess of Landsfeld
and, to the amazement of his political advisers, effectively handed over to
Lola the State government. This was an extraordinary passage in Lola's

career: from 1846 to 1848 she was the uncrowned queen of Bavaria, and its conscientious and innovative political leader. Not surprisingly, however, it was arranged by Ludwig's adversaries (principally Prince Metternich) that Lola be expelled, and she fled via Switzerland to England in 1848 leaving insurrection and abdication in her wake.

She married again in 1849, and because remarriage was against the terms of her divorce from James, was promptly arrested for bigamy. To avoid legal proceedings she travelled with her husband to Spain and then, without him, to New York in 1851. Here Lola's new career as a lecturer began, and she toured the country talking on such suitable subjects as 'Beautiful Women' and 'Anecdotes of Love'. She settled for a while in Grass Valley, a mining town in the Sierra Nevada (where she briefly married again) and then took herself over the Pacific to Australia to visit the 'gold diggings' of Melbourne, Ballarat, and Bendigo.

After two more trips to Europe, Lola returned to America and it was there that she died. The last few years of her life took a dramatic turn: she underwent a sort of spiritual revolution, and became a tireless worker for the Church and for an Asylum for Fallen Women in New York, full of desperate repentance for what she considered a gaudy, sybaritic life.

❖

RICHARDSON, Gwen

On the Diamond Trail in British Guiana. London: Methuen, 1925; pp. 243, with 23 halftones and an endpaper map

If anyone deserves to be called an independent woman, it must be Miss Richardson. She was born in Ballarat, Australia, of good Scottish stock, and took herself to England on the outbreak of the First World War to work as a land-girl. After the war she travelled on to South America to stay with friends, and soon became caught up in the romance of the 'diamond rush' in British Guiana, determining to mount her own expedition to the mines of the Illoma, up the Essiquebo river. But even though she could muster her own boat and workmen and was a mean hand with a Colt revolver, she was not allowed by the authorities to travel without a white male escort. Disgusted but acquiescent, she accepted the rather sullen services of one Major Blake, and set off on her personal treasure-hunt. Gwen found two diamonds within the first week, and by the time she was recalled (by a mysterious telegram from New York) six months later she had a little cannister of gems to bring home with her. Major Blake was busy with his own mining concerns, so Gwen had been able to prospect a little on her own; she learned to cradle scorpions in her hand (although she could never quite stomach a centipede) and to rely on her wits and her confidence: it was a thoroughly exciting, thoroughly fulfilling, and all too brief experience. She recommends it.

❦

STANHOPE, Lady Hester Lucy (1776–1839)

[Meryon, Charles Lewis]: *Memoirs of the Lady Hester Stanhope, As Related by Herself in Conversations with her Physician . . . Comprising her Opinions and Anecdotes of Some of the Most Remarkable Persons of her Time* [etc.]. London: Henry Colburn, 1845; 3 vols., 12mo, pp. 394/384/361, with 3 lithographed plates (1 hand-coloured) and a plan

[Meryon, Charles Lewis]: *Travels of Lady Hester Stanhope; forming the Completion of Her Memoirs.* Narrated by Her Physician. London: Henry Colburn, 1846; 3 vols., 12mo, pp. 372/400/423, with 3 lithographs and 20 wood-engraved vignettes

Opinion has long been divided over Lady Hester Stanhope. Some say she was a great explorer, a traveller *extraordinaire* whose life was spent in a heroic search for new lands and philosophies; others argue that she was a privileged, passionate, and frustrated woman whose journeys were not so much crusades as retreats from an increasingly uncomfortable reality of depression and mental illness. On one thing all are agreed, however: Lady Hester Stanhope was a splendid, full-blown English Eccentric.

Her story is well known. She was the eldest daughter of the Earl of Stanhope, left motherless at four and passed around a series of relations before settling with her favourite uncle William Pitt (the younger) in 1803. Until his death three years later Hester acted as his brilliant and witty hostess; it was in many ways the happiest period of her life.

Although by now at the height of her social powers, possessed of a pension of £1,200 per annum, and pursued by the most decorous and eligible of admirers, Lady Hester was restless and soon resolved to look for excitement beyond Britain's parochial shores. In 1810 she evolved a daring plan: to sail to Constantinople by way of Gibraltar and Corinth, to procure a French passport (unavailable in hostile Britain) from the ambassador there, then to go back to France, conquer Emperor Napoleon with her charms, and betray his babbled secrets to her adoring homeland. For several reasons the plan failed as soon as Lady Hester reached Turkey, and so she turned her attentions to Egypt and the Holy Land instead. In 1812 Lady Hester lost nearly everything she owned; it was then that she started (of necessity, at first) to wear male Turkish attire, and a new, exotic phase of her life began.

There followed several years of uninterrupted travel—ostensibly in search of health—for the English aristocrat and a retinue including the doctor Charles Meryon, a diminishing handful of rather jaded suitors, and the faithful lady's maid Mrs Fry. The desert tribes of Syria and the Lebanon were, to put it mildly, rather unused to seeing a 'wondrous white woman' riding like a dervish over the horizon and demanding *as an equal* the safe passage of her infidel self and attendants; most were first shocked and then charmed into compliance with her every whim. Perhaps the highlight of these years was her entry into Palmyra as a modern-day Zenobia in 1813, where she was welcomed by the natives as the first white woman to ride through the forbidden city's gates and crowned Queen (so she said) of the Desert.

Lady Hester eventually came to rest in an old monastery, Dar Djoun, in the foothills of Mount Lebanon. There she lived for some twenty-five years in a hallucinatory haze of *chibouk* fumes, first as a local champion and seeress, but then, as her behaviour grew more and more bizarre, as a hermit. Her native servants and European attendants left one by one (taking treasures with them as they went); her government pension was withdrawn in protest at the debts she had left unpaid; at last, when there was no one left to clean up her squalor and care for her, she walled herself up in Dar Djoun and died, alone.

❦

SWALE, Rosie (b. 1947)

Rosie Darling. London: Pelham, 1973; pp. 200, with 14 halftones

Children of Cape Horn. London: Paul Elek, 1974; pp. 242, with 22 halftones, 3 charts, and an endpaper map

Libras Don't Say No. London: Paul Elek, 1980; pp. 208, with 11 halftones

Back to Cape Horn. London: Collins, 1986; pp. 223, with 15 halftones and 6 maps

Sometimes it is hard to believe that Rosie Swale's early life is not just a figment of some over-excited newspaperman's imagination. The heroine, he might say, is a British Bardot, a pouting blonde who once earned her living posing for dubious magazines, and his headlines run something like this: Orphan Given to Local Postmistress—No School for Gypsy Rose—Child's Affair With Local Romeo—Girl Hitch-Hikes Sixteen Thousand Miles—Our Thirty-Foot Home On the Sea—Baby Ahoy!—Family Outing Round the Horn—and so on. Rosie *was* entrusted to Davos post office soon after she was born in Switzerland: her mother had died, and there was no one else to care for her. But then a grandmother arrived and whisked her off on a tour of Europe before taking her home to Ireland. Grandmother didn't believe in schools and managed to keep Rosie away from them until she was thirteen. Four years later she took herself off to London, where she modelled and typed before leaving with a friend for India, aged twenty, on the first of her remarkable travels.

Rosie hitch-hiked through Europe, Turkey, and Iran to Delhi and back, and then to Scandinavia and Russia, covering over 16,000 miles before her marriage and twice as many again before her divorce. She and her husband acquired in fairly quick succession two cats, a daughter, and a 30-foot catamaran (their home); together they all sailed to Italy where Rosie's second child was born (on board ship in the mouth of the Tiber) and then across the Atlantic and Pacific to Australia, returning round the Horn (babies and all): a crazy, glorious voyage. By now the Swales had run out of money, and spent the next few years in a desperate search for sponsorship, naïvely imagining a lifetime spent sailing *en famille*. The pipe-dream evaporated when their yacht was impounded for non-payment of bills and soon the marriage was over too.

But since then Rosie has emerged as an original and talented traveller in her own right. In 1983 she sailed the Atlantic solo in a tiny 17-foot yacht and a year later began a pioneering journey on horseback through Chile from the Atacama Desert to Cape Horn—an extraordinary task she accomplished with courage and an infectious sense of wonder. She called it an odyssey, during which she learned as much about herself (for the first time) as anything else—which some say is the true traveller's goal.

SYKES, Ella C. (d. 1939)

Through Persia on a Side-Saddle. London: A. D. Innes, 1898; pp. 362, with 32 halftones and a map

Persia and its People. London: Methuen, 1910; pp. 356, with 20 halftones

Through Deserts and Oases of Central Asia [with Brigadier-General Sir Percy Sykes]. London: Macmillan, 1920; pp. 340, with 46 halftones and 2 maps

Ella Sykes, who was to become a distinguished member of the Central Asian Society and the Secretary of the Royal Asiatic Society, was proud of her instinctive love of 'real' travel. She had never been beyond Europe when, in 1894, her soldier-diplomat brother Percy asked her to accompany him to Kerman (now in Iran) where he was to found a British consulate. She surprised herself by assenting straight away and relished the harsh journey across Russia and the Caucasus, the voyage across the Caspian Sea to Tehran, and the subsequent trek into the heart of Persia: its keen hardships cauterized her Western indolence into an acute sense of awareness: 'I had been civilized all my life,' she said, 'and now I had a sense of freedom and expansion which quickened the blood and made the pulse beat high.' She

harnessed this new sense of freedom later on by travelling and working in
Canada as a keen advocate of what was called 'Empire Settlement'; for now,
the experience was one of sheer exhilaration.

Ella spent an aggregate of nearly three years in and around Kerman, and
travelling with the British Frontier Commission (and Percy) along the
Baluchistani border. She wrote two refreshing books on the country, the
first a travel narrative and the second, with the acknowledged help of
Isabella BIRD, a sympathetic study of the people and landscape she
discovered. There was no chance to travel again until 1915, when Percy was
sent to Kashgar (now Kashi) in Chinese Turkestan—one of the Colonial
Office's most inaccessible outposts—to relieve Sir George Macartney as
British representative there. Ella's was the first woman's account of life at
the consulate and beyond to be published: although her story carries on
where Catherine MACARTNEY's left off, Catherine's book was not published
until 1931 and Diana SHIPTON did not even arrive in Kashgar until 1946.
Unencumbered by children and driven by an aggressively questing spirit,
Ella travelled within Chinese Turkestan more than either of the other
women, intoxicated by 'the glamour of the Back of Beyond' and after a while
resenting even the miserly fleshpots of Kashgar: she was a purist for whom
real travel had nothing to do with the familiar.

❖

TULLIS, Julie (1939–1986)

Clouds From Both Sides. London: Grafton, 1986; pp. 306, with 10 colour photographs, 12 halftones, and maps throughout

The first thing one noticed about Julie Tullis whenever she talked of her
brilliant career as a mountaineer was her utter enthusiasm. An excited sense
of awe mixed with a need for challenge drove her at the age of thirty-eight to
begin tackling her favourite Karakoram giants (she travelled as the sound-
recordist for Austrian film-maker and mountaineer Kurt Diemberger), and
by the time of her death a few years later she had become the first woman in
the world to top 26,250 ft. (8,000 m.) with an ascent of Broad Peak
(26,400 ft.) to her name, as well as a staggering array of other high-altitude
climbs, all achieved with the calm and harmonizing influence, she said, of
Japanese martial arts.

Before her central-Asian conquests Julie mountaineered in the Alps and
the Andes, and when not actually aboard a mountain she and her husband
sold climbing equipment near Harrison's Rocks at Tunbridge Wells (Nea
MORIN's training-ground), and taught their sport to others. Her dream was
to climb Mount Everest's deputy, K2, at 28,244 ft. the second-highest peak
in the world. It was her 'mountain of mountains', her special love, finally

claimed when she reached the summit on 5 August 1986. The relationship was consummated a few days later when K2 in a fit of foul weather claimed Julie in turn. 'If I could choose a place to die,' she wrote in *Clouds From Both Sides*, 'it would be in the mountains': the choice is not granted to many.

3

Unfeminine Exploits

Most unfeminine. What sort of well-bred woman spends the Season in khaki up an Indian tree waiting for savage creatures to pass underneath and be shot? Would any *real* lady choose to spend winter in a tent in Lahoul surrounded by a bloody and burgeoning selection of bears' heads? Where is Lahoul, anyway? And what could be less becoming than climbing some cold, wet mountain—taking days over it, sometimes—with only a native guide or two for company? It is almost immoral.

Victorian drawing-rooms must have been stiff with such censure, we imagine, as news of the latest exploits of Mrs BAILLIE, Mrs TYACKE, or even gentle Mrs FRESHFIELD filtered home in gossipy, cross-written letters from fellow-travellers abroad. It is certainly true that such women were thought eccentric in the extreme. But not quite beyond the pale, for one jolly good reason: most of them were rich.

Game-hunting expeditions and safaris were hugely expensive affairs, involving swarms of aristocrats, trackers, porters, gun-bearers, and assorted camp-followers; they occupied months at a time; and, of course, one could only mount the trophies of stricken-looking antelopes, spread the prostrate tigers, and stand the stuffed bears in suites of suitable proportion back at home.

Tackling mountains required more than courage, too. By the end of the nineteenth century, the Alpine regions—already dubbed the 'playground of Europe' thanks to the gentlemen climbers of the 1860s and 1870s—had become fashionably expensive. Guides had to be hired, equipment bought (Meta Brevoort, aunt of the pioneering mountaineer W. A. B. Coolidge, even fitted her constant canine companion Tschingel with its own leather climbing boots), and there was always the odd little extra to be accounted for, like the champagne and sponge cake diet on which Lucy Walker conquered the Matterhorn in 1871 (sadly, neither Miss Brevoort nor Lucy wrote about their travels).

It was obvious from their books, however, with their startling illustrations, carefully researched maps (of what was often a relatively 'blank' area), and exhaustive notes and appendices, that these faddish and moneyed eccentrics took their sport seriously. And so did most of their successors.

The Ladies' Alpine Club was founded in 1907 to foster and encourage the art and science of mountaineering, and the Pinnacle Club, its rock-climbing equivalent, fourteen years later. Of all the sports included in this chapter of unfeminine exploits, mountaineering has perhaps developed the furthest, with women like Nea MORIN and Gwen MOFFAT (the latter appointed Britain's first qualified female climbing guide in 1953) as its vanguards.

Sailing has become a skilled and competitive pastime too, pioneered by the likes of Ann DAVISON (see Chapter 1 above) and Clare FRANCIS, and bears little relation to the utterly leisurely affair 'yachting' had been in the past. And even Miss ERSKINE's humble Lady's Safety Cycle, that ungainly hybrid of suffragette statement and fairground novelty, soon blossomed into a most acceptable and enlightening form of exercise.

That just leaves the big-game hunters. They certainly had their day: between the 1890s and 1920s books were spurting off the presses and on to the bookshelves of like-minded ladies and gentlemen of the shires and the colonies, wooed by the thrill of the chase. But that day was soon over. It belonged to an age of excess, and the richness of such titles as *Days and Nights of Shikar, Rifle and Spear with the Rajpoots*, and, more succinctly, *How I Shot My Bears* soon curdled in the bracing air of the 'health and fitness' Thirties. Now, of course, the works of Agnes HERBERT, Diana STRICKLAND, and Mrs Tyacke would be lucky to get beyond the typescript stage. That sort of thing just is not done any more: *most* unfeminine.

ADAM SMITH, Janet (b. 1905)

Mountaineering Holidays. London: J. M. Dent, 1946; pp. 194, with 32 halftones and 2 maps

Although Janet Adam Smith is acknowledged to have been one of Britain's best women mountaineers, and although she has translated a number of French mountaineering classics into English and has written and edited books and anthologies herself, this is her only personal travel account. And it was written well before her climbing days were over. Summarily, and modestly, it describes repeated seasons' climbs in the Scottish hills and her beloved Alps (particularly the Graian ranges of the Franco-Italian border) first with her family, then with her first husband, Michael Roberts, and finally with her own children, always climbing not for glory but for *fun*. Roberts sadly died two years after the book was published; through her work as a translator and literary editor Janet managed to support her four children and treat them to the 'mountaineering holidays' of her youth (often in company with Nea MORIN's young family), and no doubt the Adam Smith tradition of happy climbing is safe for another generation or two.

❖

BAILLIE, Mrs W. W.

Days and Nights of Shikar. London: John Lane [1921]; pp. 241, with a colour frontispiece by the author

Mrs Baillie seems to have been an incurable big-game hunter. 'Shikar' is the Indian name for such sport, and it was in India—all over it—that the lady did her shooting. She was married to a clergyman stationed in the Deccan; from there (rarely accompanied by the gentle padre) she made expeditions as far afield as Chamba in the Himalaya and Dharwar towards the south, in search of trophies to bring home to England.

This book is less bloodthirsty than many (not a patch on Agnes HERBERT or Diana STRICKLAND, for example); the worst moments in it are not concerned with the poor animals' appointment with death, but with the author's own, which she managed to postpone twice at the very last minute. The first time was on a night shoot from Bilaspur, near Simla. She had placed a goat's carcass as a decoy under her hammock; unfortunately, just as an obliging tiger was approaching, the hammock string snapped and bowled Mrs Baillie into the mess below. Meanwhile the tiger fled away. A closer shave involved a black bear, who caught our heroine as she fell from a rock, first chewing her leg and then making off with her head—for which it luckily mistook her sola topi. The leg healed after five weeks in hospital, and, nothing daunted, Mrs Baillie lived to shoot another day.

DEACOCK, Antonia

No Purdah in Padam: A Story of the Women Overland Himalayan Expedition 1958. London: Harrap, 1960; pp. 207, with 11 halftones and 2 maps

This author, born and educated in South Africa but for a long time resident in Malvern, Hereford and Worcester, was one of three women (with Anne Davies and Evelyn Sims) who left their comfortable English homes behind in 1958 to drive to Ladakh and back. It was not a mountaineering expedition, like Monica JACKSON's; it was rather an epic 16,000-mile trek by Landrover and on foot: what they called an Officers' Wives' Bid for Independence. It nearly ended when two of the said Officers turned up in Padam on their own Expedition in Search of the Brave Little Women, but the men were soon dispatched and Davies, Deacock, and Sims went on to become three of the first Western women to explore eastern Zanskar and, on the way home, the first to cross Afghanistan unescorted. Antonia used her new-found skills later in running an outward bound school in Australia, where she eventually settled.

DIXIE, Lady Florence (1857–1905)

Across Patagonia. London: R. Bentley, 1880; pp. 251, with 12 wood engravings and 2 vignettes

In the Land of Misfortune. London: R. Bentley, 1882; pp. 434, with 15 wood engravings and 2 vignettes

Lady Florence, the Marquis of Queensberry's daughter, was one of the big-game brigade. Her journeys to Patagonia and to South Africa (the Land of Misfortune) were ostensibly for other reasons: to South America because she was suffering from social ennui and there she would be 'free from fevers, friends, savage tribes, obnoxious animals, telegrams, letters and every other nuisance for 100,000 square miles'; to Natal and the Transvaal, surprisingly enough, as the first official woman war correspondent, covering the Boer War for the *Morning Post*. But most travelling time was spent astride a horse, hunting.

One needs a strong stomach for these two books, as the author reels off a single dinner menu of guanaco-head soup (a guanaco is a sort of llama); ostrich slices with rice; guanaco ribs; roast goose and duck; ostrich wings, liver, and fat, and blood pudding. Or when she meticulously describes a hartebeest chase on the veld. The books even proved too much for Lady Florence in the end: after visits to the Rockies to shoot bear and Arabia for gazelle, she apostasized and published two further works, *The Horrors of Sport* (1891) and *The Mercilessness of Sport* (1901). She gave up travel with sport, and seems to have beguiled her later years at home with a variety of increasingly bizarre crusades (including a campaign for total parity of dress for men and women) so that *The Times* in its obituary rather sadly concluded that Lady Florence Dixie would best be remembered as 'a somewhat peculiar woman'.

❖

ERSKINE, F. J.

Tricycling for Ladies: Or, Hints on the Choice and Management of Tricycles with Suggestions on Dress, Riding and Touring. London: Illiffe, 1885; pp. 32

Bicycling for Ladies. London: Illiffe [1896]; pp. 80, with 5 wood engravings

Lady Cycling. London: W. Scott [1897]; pp. 138

Miss Erskine was at the vanguard of a whole new race, the woman awheel, that swerved on to the lanes of England and near-Europe towards the end of the nineteenth century. Each of her books is immensely practical, advising the would-be lady cyclist on every conceivable aspect of the new sport. The first one, on tricycles, is positively majestic: Miss Erskine quotes from Shakespeare's *Troilus and Cressida* at its beginning ('Mark what I say, attend me where I wheel') and goes on to suggest that perhaps a 'sociable' might be best for a real beginner—a sort of double-breasted tricycle for two (do not forget to load down your dress hem with lead shot, by the way, for modesty's sake). By the time her last book was written, things had changed. There were plenty of women's cycling clubs by now; a doughty lady had recently ridden 200 miles in twenty-four hours, and there were even bicycle gymkhanas springing up around the country: nothing better for an afternoon's healthy amusement.

❖

FRANCIS, Clare (b. 1946)

Come Hell or High Water. London: Pelham Books, 1977; pp. 198, with 33 photographs and 2 maps

Come Wind or Weather. London: Pelham Books, 1978; pp. 224, illustrated throughout with photographs, diagrams, and maps

Clare Francis seems to have had a love-hate relationship with challenge all her life. Her passion was sailing; until a legacy gave her enough money to buy a boat of her own, she trotted into her London office every day, wishing she could be on the sea instead. But once *Gulliver G.* arrived on the scene, she left security behind and dedicated herself to the boat. Someone bet that she dared not cross the Atlantic alone in this 32-footer—and so she did. Never again, she vowed, as she cruised off to the West Indies to recover. However, in 1976 she let herself be persuaded to enter the *Observer* Royal Western Single-Handed Transatlantic Race—the subject of her first book: 125 started the race, only 73 finished. Clare came in thirteenth, and was the first woman across the line, breaking the record by three days: never, *never* again. But the next year she was off once more: this time as the only woman skipper in the Whitbread Round the World Race, leading a crew of eleven. After seven months at sea, they finished well, but exhausted: NEVER again. This time Clare kept her promise. Her skills as a writer show how big-race sailing combines the most exhilarating and keen sensations with utter discomfort, and sublime fulfilment with gruelling hard work. For a while, at least, she has left such races behind and conquered two new challenges: that of best-selling novelist, and champion for sufferers—like her—of the debilitating disease myalgic encephalomyelitis (ME).

❖

FRESHFIELD, Mrs Henry (Jane Quintin) (d. 1901)

Alpine Byways: Or, Light Leaves gathered in 1859 and 1860. By a Lady. London: Longman, Green, Longman, and Roberts, 1861; pp. 232, with 8 lithotints and 4 maps

A Summer Tour in the Grisons and Italian Valleys of the Bernina. London: Longman, Green, Longman, and Roberts, 1862; pp. 292, with 4 lithotints and 2 maps

'Without aspiring to exploits which may be deemed unfeminine . . . ladies may now enjoy the wildest scenes of mountain grandeur with comparative ease'. Mrs Freshfield dedicated her two books to showing where dignified and hard-wearing women such as herself might go to find this grandeur. The Freshfield family (her husband Henry was an early member of the Alpine Club and their son Douglas went on to become one of history's great mountaineers) spent every summer in the Alps, priding themselves in keeping off the tourist track. All the routes designed in the books are 'new', in formerly hidden corners of what was rapidly becoming 'the playground of Europe'. Although Jane and her lady companion (who drew the illustrations for the books) could not manage all the menfolk did, they did make several of the smaller Alpine ascents and became particularly fond of those passes marked ominously by Murray 'not for ladies'. Unlike Mrs COLE, Mrs

Freshfield was not concerned with the mechanics of climbing for women; her audience would know the basics, she assumed. The books are more topographical guides, acknowledging (though rarely specifying) difficulties and occasional dangers but finding them far outweighed by the aesthetic and physical pleasures of Alpine travel.

⊰⊱

GARDNER, Mrs Alan (Nora Beatrice)

Rifle and Spear with the Rajpoots: Being the Narrative of a Winter's Travel and Sport in Northern India. London: Chatto and Windus, 1895; pp. 336, with 58 halftones

Somewhere in England at the turn of the century there stood a house with a very special 'feature' inside. The house belonged to Alan and Nora Gardner, and the feature, the first thing visitors would see when the maid admitted them to the hall, was a huge stuffed bear with a silver salver for visiting cards clamped into its rigid paws. Bagging that bear had been the highlight of the Gardners' season in Chamba in the winter of 1892; although there had been stags to stalk, panthers to spear, and even pigs to stick, the bear had been the thing. They had found him after a long trek through the withering cold or syrupy heat of the Himalayan foothills, involving nauseating dandy-rides through the jungles lower down, terrifying rope bridges beset with snarling dogs and hair-raising rescues from chasms and precipices (by way of an unrolled turban): it was quite a journey. As a reward for their sporting struggles the Gardners travelled down to Lahore to attend the races and to Jeypore for an elephant ride, before coming home to arrange the house for its stately new inhabitant.

⊰⊱

H(AMILTON), H(elen)

Mountain Madness. London: W. Collins, 1922; pp. 274, with 13 halftones

Helen Hamilton wrote this book quite late in life, which invests it with a rather wistful sense of nostalgia for 'those happy, carefree, bygone days when we had money to spend and could roam afar'. It concerns four successive Alpine seasons: the whole of Helen's career as a mountaineer.

Her love affair with mountains began one summer in the Dauphiny Alps, whilst she was taking the hotel's dog for a walk. He disappeared up a glacier, and Helen followed him, climbing higher and higher until she found herself teetering on a shelf of ice overlooking the valley below, exhausted and jubilant. Two Alpine guides had been watching her, astonished that she appeared to be not only a woman, but still alive; she obviously had the spirit of a true mountaineer, but lacked the technique. They offered to teach her how to climb properly. She spent that winter at home in England, reading every mountaineering book she could find, and by the end of the next season in the Alps felt ready to tackle Mont Blanc itself. Actually Mont Blanc did not succumb to her ice-axe and hobnailed boots until the next season, after ascents of the Aiguilles du Midi (12,609 ft.), de Grépon (11,447 ft., one of the most perilous of all Alpine climbs), and Verte (13,541 ft.).

The last visit Helen speaks of to the Alps was her fourth, and this time she went there in search of health rather than adventure. Some unnamed illness made her too weak to climb, and left her with none of the high-spirited exuberance that drew her to the mountains before. The book closes rather sadly with a rainy walk in the Lake District at home. Helen Hamilton went on to become something of a writer; had she the choice, I think she would rather have been just a mountaineer.

❖

HERBERT, Agnes (d. 1960)

Two Dianas in Somaliland: The Record of a Shooting Trip. London: John Lane, 1908; pp. 306, with 25 halftones

Two Dianas in Alaska. London: John Lane, 1909; pp. 316, with 29 halftones

Casuals in the Caucasus: The Diary of a Sporting Holiday. London: John Lane, 1912; pp. 331, with 22 halftones

These three books reek gloriously of blood. The 'two Dianas' and the 'Casuals' of their titles were Miss Herbert and her cousin Cicely, both handsome, elegant women in their thirties, who decided to defy popular (i.e. male) opinion and mount their own game-hunting expeditions. They were brought up amongst a sporting family, and had already been to the Rockies, but the Somaliland trip was their first 'solo' show (solo in as much as they were the only Europeans; the native contingent was huge, and they hired 49 camels). It was highly successful—despite the fact that Agnes's leg was mauled by a lion, her arm torn by an oryx horn, a guide was killed by a rhino, a camel-man died from swallowing a newt, and their butler(!) ran away— the gory final photograph showing the whole expedition's bag proves that. Agnes was obviously a born predator.

The ladies were accompanied on their Alaskan trip by a brace of British *shikari*, or hunters, one of whom Agnes calls a 'distant kinsman'. They had already encountered each other in Africa, and now the four of them cruised amongst the islands of Prince William Sound and the Bering Sea felling everything from ptarmigan to walrus, including moose, caribou, sheep, and bear.

The two Dianas came home with more than an overwhelming collection of trophies—Cicely bagged a husband (one of the *shikari*) and Miss Herbert scored at least a hit with the other, who contributed the odd chapter to her book on the expedition. This comes as rather a shock, since Agnes constantly takes the opportunity in her books to air her views on men. She is witheringly cynical, scoffing at those who seek 'the haven of a good man's love', and proud of her single state: 'A man sees in every unmarried woman a walking statistic against his irresistibility'. The only time she loses her robust sense of humour is when she speaks of feminism.

The last book is less a narrative and more a description of the Caucasus. She and Cecily were this time accompanied by a cousin (Cecily's husband had been left behind and there is no mention of Agnes's swain). Miss Herbert went on to write several guidebooks, mostly on Britain, and a series of 'animal autobiographies' on the moose, the lion, and the elephant, but it is the high-spirited 'Diana' books that were always the most popular—one cannot help but admire her panache.

❧

JENKINS, Lady (Catherine Minna)

Sport and Travel in Both Tibets. London: Blades, East, and Blades [1909]; pp. 87, with 25 colour sketches and a map

Lady Jenkins was already an experienced big-game hunter, having shot in Somaliland and in India (where she was 'lucky enough' to bag five tigers), when she started from Bombay on this expedition to seek out the gazelles, antelopes, and wild sheep of Baltistan, or Little Tibet, and the north-west corner of Tibet proper. She organized her trip from Srinagar, where she stayed as a guest of the Maharajah of Kashmir, and from there travelled by houseboat and horseback up to Leh. She employed a caravan of local *shikaris* (or hunters) and coolies, and began her long trek through the Himalayan foothills where even the valley floors are higher than the summit of Mont Blanc. She covered over a thousand miles (not exactly the easiest country for stalking) and, for her pains, collected seventeen matchless trophies. She writes smoothly and coyly, noting how astonished the natives were at her prowess as they wondered how the *Ladysahib*, 'looking like a town lady and not a shikari, and being so horribly weak and thin, could be so strong and walk so well'.

In fact Lady Jenkins needed all the strength she could muster. What may have begun as a rather daring jaunt changed, once beyond the Lanak and Lungnak passes, into a fight for survival. The weather worsened and food supplies began to run out. Blizzards made movement outside the camp impossible, and collapsed the tents inside it. Shooting was out of the question—besides, Lady Jenkins had frostbite—and the baggage yaks, as well as the coolies, were growing weak. At this stage she had no map, only a guide who claimed to know the country; luckily she gave the order to turn back (before fainting with altitude sickness) just in time to save them all from starvation and exposure.

The organization for this little trip seems to have been frighteningly cavalier—like Lady Jenkins's description of it—but then women like her were brought up to expect success, and rarely settled for anything less.

◆

MOFFAT, Gwen (b. 1924)

Space Below My Feet. London: Hodder and Stoughton, 1961; pp. 286, with 11 halftones

Hard Road West: Alone on the California Trail. London: Gollancz, 1982; pp. 198, with 29 halftones and 3 maps

Gwen Moffat has led an uncompromisingly independent life. She joined the Land Army straight from school and graduated to the ATS as a driver and dispatch-rider; at the age of twenty-one she deserted to live rough in Wales (where she first learned to climb) and in Cornwall for six months; three years later she married and lived on a boat with her husband and baby daughter, and in 1953 became Britain's first professional female climbing guide. She has been travelling (vertically, if not horizontally) ever since, sharing her need to climb with an equally urgent need to write, but for all her many books (mountaineering memoirs, including *On My Home Ground* (1968) and *Survival Count* (1972), novels, detective tales with a rock-climbing heroine, and so on) only these two concern her efforts abroad. The first nonchalantly chronicles some staggering ascents in the Alps and Dolomites, while *Hard Road West* is more conventional, describing a long hot drive in the tracks of the old fortyniners from Omaha to southern California—alone, of course.

◆

MORIN, Nea (1905–1986)

A Woman's Reach: Mountaineering Memoirs. London: Eyre and Spottiswoode, 1968; pp. 288 (including an appendix on notable female ascents), with 34 halftones and 3 maps

Nea Morin was one of Britain's foremost twentieth-century mountaineers. With her friends Jo Marples and Miriam Underhill, her sister-in-law Micheline Morin, and her daughter Denise Evans, she pioneered the modern *cordée féminine*, and was a key member of the exclusive *Groupe des Hautes Montagnes* in France, the Ladies' Alpine Club, and the Pinnacle

Club. Her father, an antiquarian bookseller named Barnard, was an enthusiastic climber and trained Nea in his art on the rocky outcrops around their home in Tunbridge Wells, Kent. After her first Alpine season in 1919 ('heaven!'), she rarely missed a summer on the mountains of Europe, climbing often with her French husband Jean Morin and later with both her children. Her first trip to the Himalayas was in 1959, when she joined an ill-fated expedition to 'the Matterhorn of Nepal', Ama Dablam, as the only woman member. Two of the climbers were killed on the mountain and Nea, injured a week before setting out, became too ill and thin to do anything but watch. Her career as a serious mountaineer was ended soon afterwards by arthritis exacerbated by the hardships of the Himalayan trip, but she continued an inspiration to her successors and will always be remembered as a worthy successor herself to the likes of Mrs FRESHFIELD and Elizabeth LE BLOND.

❖

PIGEON, Anna (1833–1917) and ABBOT (née PIGEON), Ellen

Peaks and Passes. London [for private circulation only]: Griffith Farran, Okeden, and Welsh, 1885; pp. 31

Anna and Ellen Pigeon were sisters from south London who started climbing together in the Alps in 1869 and carried on, season after season, until 1876. Their name sounds agreeably fluttery: we imagine two fur-belowed spinsters rambling happily beside the snow-line, collecting flowers and coy little reminiscences along the way. But nothing could be further from the truth. The two Miss Pigeons were fearless and accomplished mountaineers, elders of the climbing fraternity at home, authors of 63 major ascents—including the Matterhorn (14,782 ft.), which they climbed twice, Dom (14,942 ft.) and Mont Blanc (15,782 ft.)—and some seventy Alpine passes. Their book is blunt and business-like, listing all peaks and passes attempted, with their heights (the average is well above 11,000 ft.) and climaxing in the first female traverse of the Matterhorn from Breuil to Zermatt in 1873 and the second ever traverse of the notorious Sesia Joch from Zermatt to Alagna, the triumph of their very first season. A few terse comments are added on the quality of their guides and the variety of Alpine accommodation (ranging from hotel or mountain hut to nothing but snow and stars); and with a summary index of their eight years' achievements, that is that. The two women were still climbing in the 1880s, and in 1910 Anna was elected Vice-President of the Ladies' Alpine Club: a fitting close to her career.

❖

PILLEY, Dorothy (RICHARDS, Mrs I. A.) (1894–1986)

Climbing Days. London: G. Bell, 1935 [second edition with new 'Retrospection': Secker & Warburg, 1965]; pp. 352, with 69 halftones and maps on endpapers

In her book *Women on the Rope*, the climber and historian Cicely Williams described Dorothy Pilley as 'the outstanding woman mountaineer of the early 1920's—and for many years after'. No mean claim, this, of a period burgeoning with the talents of such as Nea MORIN and Joyce DUNSHEATH.

Miss Pilley started climbing during a childhood holiday in Wales, and soon became what she calls 'infected with mountain madness', visiting the Alps most seasons (where she met her husband the scholar-mountaineer I. A. Richards in 1921) and when time and money allowed venturing farther afield to Canada, Ceylon, China, Japan, and the Himalayas. She lived for some years with Richards in the United States—a good chance to comb the Selkirk, Rocky, and New Hampshire peaks—but always considered the Alps her spiritual home. Switzerland was the scene of her first triumphs (traverses of the Charmoz, the Grépon, the Drus, Dent du Géant, and the Petit Clocher de Planereuse in 1920) and of her greatest: the first ascent, with Richards in 1928, of the north ridge of the Dent Blanche (14,318 ft.).

Although she and Richards often climbed in partnership, Dorothy Pilley was a pioneer of the *cordée féminine* and a soloist of virtuosity too. She was elected a member of the Ladies' Alpine Club in 1920 (one of its youngest), and a year later helped found the Pinnacle Club, the first women-only rock-climbing association. Although a car accident in 1958 prevented her from ever seriously mountaineering again, she continued an inveterate hill-walker until her death, and saw in her last new year (over ninety by now) characteristically holed-up in a climbing hut on Skye.

***

PLUNKET, The Honourable Frederica (Louisa Edith) (d. 1886)

Here and There Among the Alps. London: Longmans, Green, 1875; pp. 195, with title-page vignette

Miss Plunket's jolly little book was designed 'to persuade other ladies to depart from the ordinary routine of a Swiss summer tour' and lead them into that 'borderland between the forbidden ground of danger and the beaten paths of safety'. She and her sister spent a doughty summer not exactly mountaineering, but certainly clambering on the slopes of the Upper Engadine and the Austrian Alps in 1874, keeping well away from 'that heterogeneous mass named the British Tourist' (not so easy in late nineteenth-century Switzerland).

Frederica's suggestions to what she modestly calls 'lady pedestrians' include the choice of suitable headwear (a linen mask with holes cut for eyes and mouth and a blue gossamer veil over the top is best), and the points to look for in a travelling companion: strength, health, and a perfectly steady head.

SAVORY, Isabel

A Sportswoman in India: Personal Adventures and Experiences of Travel in Known and Unknown India. London: Hutchinson, 1900; pp. 408, with a photogravure plate and 48 halftones

In the Tail of the Peacock [Tangier]. London: Hutchinson, 1903; pp. 352, with a photogravure plate and 48 halftones

The Romantic Rousillon: In the French Pyrenees. London: T. Fisher Unwin, 1919; pp. 214, with 26 halftones and a map

Miss Savory's travels certainly began with a bang. Her first book is all about a game-hunting expedition she made with friends from Bombay up to Peshawar, to the Khyber Pass, into Chamba, Kashmir, and then down to the Nilgiri Hills in pursuit of almost anything that moved. She spent months revelling in a polite orgy of fox-hunting, pig-sticking, and tiger-shooting, with a touch of Himalayan couloir-climbing and conventional sightseeing thrown in: it must have been somewhat of a come-down when, at the end of 1901, she found herself pony-trekking towards the Atlas mountains of Morocco. And the momentum of *A Sportswoman in India* was lost altogether by the time Miss Savory reached the Pyrenees: unfortunately, one glorious journey (and two tamer ones) do not necessarily add up to a successful career as a travel writer.

SPEED, Maud

A Yachtswoman's Cruises and some other Voyages. London: Longmans, Green, 1911; pp. 266, with 2 coloured plates and 25 halftones

Through Central France to the Pyrenees. London: Longmans Green, 1924; pp. 245, with 4 coloured plates and 4 halftones

Cruises in Small Yachts by H. Fiennes Speed: *And a continuation, entitled More Cruises* by Maud Speed. London: Imray, Laine, Norie, and Wilson, 1926; pp. 355, with 5 coloured plates, 11 halftones or sketches, vignettes throughout, and 4 maps

A Scamper Tour to Rhodesia and South Africa. London: Longmans, Green, 1933; pp. 148, with 6 coloured plates and 5 halftones

Maud Speed had much in common with her contemporary Dorothy Una RATCLIFFE. Both women were married to sailors, and keen amateurs themselves; their love of travel by sea spread to a love of touring on land, and they both wrote breezy, lightsome travel books. Mrs Speed began her sailing career aboard her husband Harry's 'sailing canoe' off the south coast; soon they were yachting to Normandy in a 28-footer and then to Holland in a 6-ton cutter called *Beaver*. The next treat was a steam yacht, in which they started winter cruising across the Channel and then, by way of relaxation, the Speeds took to the big luxurious liners of the early twenties, holidaying afloat as far afield as Morocco and the Turkish coast. Once in a while Maud turned landlubber and took off on her own, and she stopped sailing altogether after Harry's death on board ship in 1926.

❖

SPEEDY, Mrs Charles (Cornelia Mary)

My Wanderings in the Soudan. London: R. Bentley, 1884; 2 vols., pp. 239/264, with 20 wood-engraved vignettes and a map

Cornelia Speedy had always had a 'wild scheme' to go shooting in the Sudan, and when her husband suggested they spend part of their furlough there in 1878 (they had been stationed at Penang for several years before then) she was more than happy to concur. An eminently practical lady, she ran up a tent on her sewing machine before setting out, and then spent a 'delightful' few weeks stalking lions, elephants, hyena, wild asses, and the odd recalcitrant porter through the jungles and deserts between Kassala and the River Setit. This 'holiday' involved day after day of hard, hot marching, nights spent listening to the white ants munching through the leg of the bed (or worse, the tent-poles) and repeated bouts of fever ('most provoking'). But to Cornelia, who proudly claimed herself the first European woman to safari in Sudan since Florence BAKER, it was all quite wonderful: life as it should be lived.

❖

STRICKLAND, Diana

Through the Belgian Congo. London: Hurst and Blackett [1925]; pp. 288, with 16 halftones and an endpaper map

Of all the lady game-hunters, I think Mrs Strickland must be the most unemotional. Her travel book, describing the route she and her party took from Matadi up the Congo and then across to Lakes Edward and Victoria to Mombasa—right through the heart of equatorial Africa—is business-like in the extreme (rather like her only other book, *Love through the Ages* (1933)); it reads more like an official report than anything else. One might think this rather refreshing after the gory passions of Agnes HERBERT or the strenuous intemperance of Mrs TYACKE—but really Mrs Strickland is just as excessive. Her clinical attitude both to travel and to the business of game-hunting spares us nothing. Every mile is cumbersomely accounted for, without involving too much spurious (or interesting) description of the surroundings, and we are given the immediate history of every pelt in the camp in the same deadpan manner. Nothing escapes her unblinking attention, from the foolhardy maggot daring to violate her latest trophy to the exact location of the bullet with which she felled her beautiful, ivoried elephant (a picture of whose carcass, incidentally, is adorned in the book with the caption 'The End of a Perfect Day'). She makes everything sound utterly unromantic, and quite exhausting. But then it probably was.

❖

THOMPSON, Dorothy E. (1888–1961)

Climbing with Joseph Georges. Kendal: Titus Wilson, 1962; pp. 159, with 13 halftones

Dorothy Thompson was another of those prodigious women responsible for putting British mountaineers on the map in the twentieth century. It was Dorothy PILLEY who first introduced her namesake to the Alps—and to the legendary Swiss guide who features so prominently in this book. 'Tommy' Thompson spent successive seasons in the mountains of Europe, culminating in a spectacular traverse of Mont Blanc on a hideously treacherous route till then untried by a woman (via the Aiguille Bionnassay and the Peuteret Ridge); when she could no longer climb high she ski'd, and when a heart condition slowed her down completely in the 1950s, she was happy just to gaze. *Climbing with Joseph Georges* was in fact written whilst Miss Thompson was still at her prime as a mountaineer; sadly, it was published (after thirty years lying fallow) just too late for the author to see it for herself.

❖

TYACKE, Mrs R. H.

How I Shot My Bears: Or, Two Years' Tent Life in Kulu and Lahoul. London: Sampson Low, Marston, 1893; pp. 318, with 13 halftones and a map

Anyone who can give their book a title like this starts at a disadvantage as far as I am concerned. Mrs Tyacke was a dapper little woman of five-foot-one whose favourite pastime was stalking game in the uplands and valleys of Kashmir, Albania, or, in this case, the foothills of the western Himalayas where she eventually settled with her husband Richard Humphrey. He was the author of a best-selling sportsman's handbook, and together they delighted in a bag during their first season in Kulu of 137 pheasants, 321 *chikor* (partridge), 49 cock, 9 snipe, 3 duck, 7 barking deer, 3 musk deer, 3 *goral* (Indian antelope), 6 black bears, and 8 red ones . . . 'and this might have been considerably increased,' she says, 'had we cared to go in for slaughter.' The next season was almost as glorious; when not involved in the chase or the kill, Mrs Tyacke would arrange skins and skulls in a photogenic group for R.H. to record with his camera before sending them off to Cheltenham (of all places) to be cured.

To give the woman her due, she did survive two years largely above the snow-line in a canvas tent, living for the most part off the land and never complaining. Perhaps what makes books like this and Agnes HERBERT's so hard for us to digest now is the authors' uncomfortable knack of anthropomorphizing their objects. One moment Mrs Tyacke is cooing over an 'old lady bear' trying valiantly to shield her family from the gun, and the next she is exultantly picking over the carcass. This strangely Victorian mixture of sentiment and bloodthirstiness is a little too rich for modern taste.

When Penelope CHETWODE visited Kulu in the 1960s, the memory of Mr and Mrs Tyacke was still very much alive. Apparently no one was quite sure if the pair were ever actually married (there was some scandal attached to their sudden arrival together in Kulu); that they were devoted to each other was quite obvious, however, and they hunted the Himalayan foothills together with legendary gusto until the end.

4

An Up-to-Anything Free-Legged Air

'It is so like living in a new world,' wrote Isabella BIRD of travel, 'so free, so fresh, so vital, so careless, so unfettered . . . that one grudges being asleep.' She was perhaps the best-qualified authoress of all to say so: from the age of forty until her death at seventy-three she only touched base long enough to pay duty calls to relations and write her books. The rest of the time she spent the world over, from Hawaii to Tibet, 'up-to-anything' and 'free-legged'. She, like all the women in this chapter, was swept along from boundary to boundary by the momentum of sheer curiosity.

What set them travelling in the first place was a different matter. Some of them were escapees. Isabella herself was: at home she lived a drab and feeble-bodied life, but abroad she blossomed. On a ship at sea, on horseback in the Rocky Mountains or marooned in a Malayan stable there could be 'no bills, no demands of any kind, no vain attempts to overtake all one knows one should do. Above all, no nervousness, no convention'. Travel meant leaving precedent behind and, quite simply, starting again.

It was low spirits (following her father's death) that first sent Emily BEAUFORT away to Egypt in the late 1850s, while the elegant Lady CRAVEN, travelling nearly a century earlier, could not possibly have stayed at home a moment longer: her husband was too boring. Even now, women like Dervla MURPHY and Christina DODWELL are travelling still because they once tried to escape—Dervla from the sort of family situation that was driving her towards mental breakdown and Christina from the enervating rut she had settled into as a working girl in London.

Barbara TOY, another modern travel writer, first set off on a series of travels that have taken her in the driving seat of her Landrover to most of the dangerous corners of the world in quite another spirit.

The whole business has always been a sort of quest for her, the chance to unearth the sort of experiences likely to enrich her life at home rather than replace it. This sense of discovery, stronger than mere curiosity, is shared by several of the travellers listed here— Beryl SMEETON and Rosita FORBES amongst them—but nowhere is it more prolific than in the ample breast of Constance Fredereka GORDON CUMMING. During a travelling life spanning almost half a century 'Eka' visited countless countries writing, painting, and relishing everything she saw. She was insatiable.

She was not alone. Most of these escapees and discoverers, in fact, seem to have been unable to stop travelling once they had started. It was only those sterling souls who set out to *prove* something (but who probably forgot quite what it was in the excitement of being under way) who appear to have been satisfied with just one or two recorded journeys: Marion DOUGHTY, for example, who 'tumbled' her way through the glaciers and valleys of the Himalayan foothills just to show that 'ordinary' women like her could; or the eloquent Miss CRAWFORD, passionately concerned to justify the very act of travelling itself in her 'Plea for Lady Tourists'. Travelling, for them, was a means of practising what they preached—and a surprisingly enjoyable one.

I will leave the last word with the ubiquitous Isabella. She describes in a letter from the Rocky Mountains the moment when she and her companion found themselves (at the end of a hard day's ride) high up above one of the most magnificent vistas in the world. In common with most of the world-wanderers in this chapter, she was able to extract the pure essence of the present moment: 'we stood . . . in the splendour of the sinking sun, all colour deepening, all peaks glorifying, all shadows purpling, all peril past.' Never mind about the future.

AYNSLEY, Harriet Georgina Maria (*c.* 1827–1898)

Our Visit to Hindostan, Kashmir, and Ladakh. London: Wm. Allen, 1879; pp. 326

An Account of a Three Months' Tour from Simla through Bussahir, Kunowar and Spiti, to Lahoul. Calcutta: Thacker Spink, 1882; pp. 83

Our Tour in Southern India. London: F. V. White, 1883; pp. 358

One Season, Mr and Mrs Aynsley decided to go to Hindustan. They had 'for some years had a great desire to visit our Indian possessions', and thus began in 1875 a love affair that lasted twenty-one years. Nearly half of that time was spent actually travelling in India, and although Mrs Aynsley had spent the years between 1850 and 1860 living in Rome, and a fair time in Algeria, most of her published work concerns the subcontinent.

An Account of a Three Months' Tour advises 'sister tourists' on the practicalities of Himalayan travel, recommending the 'Ashantee hammock' and goats' hair ropes for when the going gets really rough. The other books are not so pragmatic. Harriet Aynsley's particular interest was in the local legends and symbolism of India (earning her the singular feminine distinction of being elected an Associate of the Order of Freemasons) and all her travel accounts tend towards the scholarship of her articles for the *Indian Antiquary* and her own posthumously published work *Symbolism of the East and West* (1900): she was a very objective lady traveller.

◆

BALFOUR, Alice Blanche

Twelve Hundred Miles in a Waggon. London: Edward Arnold, 1895; pp. 265, with 38 sketches

The weightiness of the title is misleading—this was essentially a pleasure trip from London to the Cape and on to Mashonaland [southern Zimbabwe], Matabele, Zanzibar, and the Victoria Falls: a jolly expedition of two couples who travelled 'in happy uncertainty' of their destination. The eponymous wagon was a sort of chariot led by fourteen oxen; there was a mule-driven 'spider' buggy included in the sizeable caravan too, just for short sorties. Any journey taken through southern Africa between the Zulu and Boer wars was bound to be an exciting affair, and Alice Balfour made plenty of political references in the original manuscript of her travel account. The publisher, however, thought it wiser not to print them, and the book remains a light-hearted narrative of Victorians touring self-consciously 'off the beaten track'.

◆

BEAUFORT, Emily A. (Viscountess STRANGFORD) (1826–1887)

Egyptian Sepulchres and Syrian Shrines including Some Stay in the Lebanon, at Palmyra, and in Western Turkey. London: Longman, 1861; 2 vols., pp. 411/484, with 6 chromolithographs and a map

The Eastern Shores of the Adriatic in 1863: With a Visit to Montenegro. London: Richard Bentley, 1864; pp. 386, with 4 chromolithographs and a photograph

The Viscountess Strangford—née Emily Beaufort—is perhaps best known for her work organizing nurses during the Crimean War, and her subsequent interest in the nursing profession at home in England. A book she published in 1874 (*Hospital Training for Ladies*) was full of good sense, claiming that to be a well-bred dilletante (as she had been) was not enough for today's 'modern' lady: 'We lonely women want more and more means of helping ourselves'—and nursing, she said, was one way of usefully occupying 'idle hands and unsatisfied hearts'.

Another way, which Emily discovered along with hundreds of moneyed Victorian women in need of distraction, was travel. She opens the account of her first tour with an explanation: 'Circumstances [i.e. the death of their father Sir Francis 'Beaufort Scale' Beaufort] having rendered change of scene and climate imperatively necessary for my sister and myself, we hastily determined, at the close of the year 1858, to leave England for Egypt . . . wanting no society.' The two sisters' tour—beautifully illustrated by Emily—is straightforwardly recounted so that young ladies reading it at home by the fire might realize 'in what ease they may travel—even alone'. An appendix lists the *sine qua non* of feminine tourism, including smelling-salts and aromatic vinegars for 'bad odours', high-laced boots against mosquito 'stings', and of course a portable tin bath (one's dragoman supplies the iron bedstead). The second book was written after the author's marriage and is altogether more serious. The route the Strangfords took, visiting Albania, Montenegro, Dalmatia, and Corfu, was not yet hackneyed; Emily describes it conscientiously. And the turbulently exciting politics of the eastern Adriatic merit their own chapter, contributed by the Viscount and simply headed 'Chaos'.

❖

BIRD, Isabella Lucy (Mrs BISHOP) (1831–1904)

[Anon.]: *An Englishwoman in America.* London: John Murray, 1856; pp. 464

The Hawaiian Archipelago: Six Months Among the Palm Groves, Coral Reefs and Volcanoes of the Sandwich Islands. London: John Murray, 1875; pp. 473, with 10 wood engravings and a map

A Lady's Life in the Rocky Mountains. London: John Murray, 1879; pp. 296, with 8 wood engravings

Unbeaten Tracks in Japan: An Account of Travels on Horseback in the Interior Including Visits to the Aborigines of Yezo and the Shrines of Nikkó and Isé. London: John Murray, 1880; 2 vols., pp. 348/383, with 42 wood engravings and a map

The Golden Chersonese and the Way Thither. London: John Murray, 1883; pp. 384, with 10 halftones and a map

Journeys in Persia and Kurdistan, including a Summer in the Upper Karun Region and a Visit to the Nestorian Rajahs. London: John Murray, 1891; 2 vols., pp. 381/409, with 13 halftones, vignettes throughout, and 2 maps

Among the Tibetans. London: Religious Tract Society, 1894; pp. 159, with 21 vignettes

The Yangtze Valley and Beyond: An Account of Journeys in China, Chiefly in the Province of Sze Chuan and Among the Man-tze of the Somo Territory. London: John Murray, 1899; pp. 557, with 116 halftones and a map

Of the three women explorers known to everyone—Lady Hester STAN-HOPE, Lady Mary WORTLEY MONTAGU, and Isabella Bird—this last is the best loved, most prolific, and most ubiquitous of all. She is also one of the best documented of the species, with several biographies to her name and a proud position at the prow of all the anthologies: the archetypal Victorian Lady Traveller.

Isabella Bird did not begin her travelling career until quite late in life (apart from a trip to Canada and North America in her twenties), for until she was forty she was occupied as the family spinster in caring for her parents, and by the time they had died Isabella herself had succumbed to a debilitating spinal complaint, depression, and acute insomnia. Travel was prescribed—and she became addicted. 'I have only one rival in Isabella's heart, and that is the high table-land of Central Asia', sighed the local doctor John Bishop, who spent the years between 1877 and 1881 regularly asking Isabella to marry him. She refused as long as her beloved and ailing sister was still alive, but when Henrietta died she consented at last. Grief for Henrietta and the obedience of a fifty-year-old bride to her groom kept Isabella at home for a few years (she had thought of visiting New Guinea, but that was hardly the sort of place to which one took one's husband). But then, just five years after the marriage, Bishop himself expired and there remained nothing whatever to keep her from travelling—even her own ill health blew away with a good sea-breeze.

During her itinerant career Isabella visited and wrote about countless countries all over the world. It was in recognition of the vast range of her journeys and observations that she was awarded the singular honour of being the first woman to address a meeting of the élite Royal Geographical

Society in 1892. An even greater distinction followed in the same year when she became the first of fifteen pioneering women elected to fellowship of that august institution. She continued travelling until the year of her death, just as thrilled by the excitement and discovery of it all at seventy-three as she had ever been.

'Travellers are privileged to do the most improper things', she once said, and in her 'up-to-anything free-legged air' she certainly had her moments: falling in love with Rocky-Mountain Jim, a rough Canadian desperado, for example; casually taking up with a passing British Intelligence Officer in Luristan (south-west Iran) and playing her own small part in the Great Game, or just pottering about the banks of the Yangtze developing her photographs in its yellow waters. She was a born traveller, and a lasting inspiration.

❧

BLESSINGTON, Marguerite, Countess of (1789–1849)

[Anon.]: *Journal of a Tour through the Netherlands in 1821*. London: Longman, 1822; pp. 171

The Idler in Italy. London: Colburn, 1839–40; 3 vols., pp. 400/565/372, with frontispiece portrait

The Idler in France. London: Colburn, 1841; 2 vols., 12mo, pp. 356/270

The glamorous Irish Countess of Blessington, author of a clutch of novels and the popular *Journal of Conversations with Lord Byron* (1832), is perhaps best known for her *grande affaire* with the impossibly handsome Count d'Orsay, her stepdaughter's husband, with whom she lived from 1829 until her death.

That was the last of three unions: she first married (at fourteen) an Irish soldier whom she left after three months; after his death in 1817 she became the wealthy wife of a doting husband, the first Earl of Blessington. It was at this stage that the 'idling' began: tiring of the London Season and with her appetite whetted by a short tour to the Netherlands, she removed with Blessington to the Continent, living and travelling in France and Italy for the next ten years or so. She went again to Paris in 1849, this time with her lover to escape his debts, and there she died, just a few months afterwards, leaving d'Orsay desolate.

Her three travel accounts, all *langeur* and sardonicism and with nothing of the distasteful practicalities of getting from A to B, did much to popularize the Grand Tour amongst 'women of quality'. Continental travel became a coveted accomplishment, suave, a little *risqué*, and utterly fashionable.

❧

BOSANQUET, Mary (1918–1969)

Canada Ride: Across Canada on Horseback. London: Hodder and Stoughton, 1944; pp. 222, with 18 halftones and endpaper map

Journey into a Picture. London: Hodder and Stoughton, 1947; pp. 196, with colour frontispiece and 14 halftones

Mary Bosanquet's epic journey from Vancouver to New York was an impulsive affair. She was only twenty when she started it in June 1938, fresh from her degree in modern languages at Bedford College, London, and had never been further than Germany in her life. The idea had come to her one rainy evening as she was 'bucketing down the Bayswater Road in a number seventeen bus': she would sail for Halifax the next spring, take the train to Vancouver, buy a horse, and ride across Canada. It was as easy as that, as long as Mummy and Daddy said yes.

They did. When she reached Vancouver, Mary bought Timothy (almost the first horse she saw) and together they crossed the continent in eighteen months. They stayed at convenient farms, or camped amongst the grizzlies, and happily travelled with anyone who happened to be going their way. Their 3,000-mile route took them over the Selkirk mountains to Calgary, on to Winnipeg and the Great Lake 'Wilderness' beyond. Mary was forced to winter at Dayton, a small town between Sault-Ste-Marie and Montreal; the weather was closing in and Timothy was growing too lame to continue. She earned her keep on a farm, and resumed the journey through Vermont and Connecticut eight months later.

The war, which had broken out during her journey, was in full swing by the time Mary returned to England. In 1944 she was sent out to Italy, to work with the Young Men's Christian Association there. She was appointed 'Head of Club Activities' and asked to tour the country lecturing on Italian art and history. This artistic pilgrimage, as she called it, is described in her second book. Mary settled at home in England after the war, raising a family and collecting material for her celebrated biography of Dietrich Bonhoeffer, which was published just four months before her untimely death at the age of fifty-one.

❖

BRIDGES, F. D.

Journal of a Lady's Travels Round the World. London: John Murray, 1883; pp. 413, with 10 wood engravings and 8 vignettes from sketches by the author, and a map

To Mr and Mrs Bridges, travel was simply a matter of education. Both felt that anyone with the time and the money also had the obligation to travel, to benefit themselves and others from the Wonders of the World. So, finding themselves with both means and opportunity in the summer of 1878, they set off.

The first leg of their world tour took them through Greece and Egypt to India. After a little sightseeing, they trekked up to Kashmir and over the Bhutkhul Pass to Leh; Mrs Bridges rested here whilst 'H' (obviously a respected military man) made a dangerous and rather hush-hush journey to Yarkand (Shache) in Chinese Turkestan. After this little digression the pair moved on to Burma and through the straits of Malacca to Singapore and Java. By the time they reached Japan, 'H' was ready for another expedition of his own; this time he went to Peking, leaving his wife alone to explore Kobe. On his return, they both left for North America, crossing the continent in a leisurely fashion (by train) before sailing home to England in 1881.

Mrs Bridges must have been more than a match for her adventurous husband: unlike most female world-tourists of her day, she hated to be hauled about by natives ('*anything* is better than being carried!'), and avoided 'English' food when abroad; the ways of the simpering sightseer were anathema to her. 'The sooner one falls into the ways of a country the better', she advised: that was the only way really to *learn* from travel.

❖

CHAPMAN, Olive Murray (d. 1977)

Across Iceland, the Land of Frost and Fire. London: Bodley Head, 1930; pp. 193, with 8 colour illustrations, 46 halftones, and a map

Across Lapland with Sledge and Reindeer. London, John Lane, 1932; pp. 212, with 8 colour illustrations, 62 halftones, and a map

Across Cyprus. London: John Lane, 1937; pp. 256, with 2 colour illustrations, 34 halftones, and a map

Across Madagascar. London: Burrow [1943]; pp. 144, with 17 halftones and a map

Mrs Chapman went abroad to sketch, to photograph, and to study the people she found there. She illustrated her own books and on all but the first 'Across . . .' expeditions, carried a 'cinematographic camera'—newfangled machine—to film her subjects too. She was a Fellow of the Royal Geographical Society, regularly reporting on her enterprising tours and showing her films there. But she was no travel snob; although she only rode a pony in Iceland, she was not above taking a car in Cyprus and Madagascar, or taking the bus. And when she got home, she wrote popular descriptions of the places she visited that made them sound so easily accessible (which of course they were not): they are very encouraging travel accounts indeed.

⬦

CHETWODE, Penelope (1910–1986)

Two Middle-Aged Ladies in Andalucia. London: John Murray, 1963; pp. 153, with 24 halftones and a map

Kulu: The End of the Habitable World. London: John Murray, 1972; pp. 233, with 39 halftones and a map

The two middle-aged ladies of Penelope Chetwode's first book are herself, she says, and her discontented donkey La Marquesa. Together they spent a month at the end of 1961 ambling through the villages of southern Spain on a gentle journey (despite La Marquesa's sullen demeanour) inspired by the works of George Borrow and Richard Ford. Miss Chetwode—actually Lady John Betjeman—knew little Spanish, and even less donkeymanship, but managed to get comfortably by in the time-honoured fashion of smiling at everyone and surreptitiously pouring dubious liquors and soups into geranium pots.

The journey to Kulu which followed in 1964 was a harsher affair. Lady Penelope grew up in northern India as the daughter of the British Commander-in-Chief, and now she planned to revisit its high slopes and rich history on a trek from Simla to the head of the Rohtang Pass. She returned to the Himalayas in 1965 and increasingly often until her death there in 1986, leading tourist treks or just wandering on horseback amongst the familiar foothills.

⬦

CLOSE, Etta (d. 1945)

A Woman Alone in Kenya, Uganda and the Belgian Congo. London: Constable, 1924; pp. 288, with a map

Etta Close is rather refreshingly undoughty. She was a well travelled tourist turned 'explorer' who, she tells us, was the first woman to safari 'alone' in north-west Kenya. In fact she had a companion-guide on her travels amongst the Masai in 1922, a certain Dutch gentleman rejoicing in the name of Trout, and although she went alone on a subsequent journey to the Kilo Gold Mines of the Belgian Congo, she met many a solicitous European escort on her way. The route up to the mines along the Victoria Nile was a comparatively well provided one by now, involving steamers, trains, and even a Ford motor-car ride, but it was still an unhealthy country. The final march beyond Masindi during the rainy season proved too much for Miss Close who soon found herself 'clean off my head' with malaria. She recovered only after several weeks at a mission hospital, and then scuttled straight and safely home, to write up her detailed diary with pride.

◆

CRAVEN, Lady Elizabeth (later Margravine of ANSPACH) (1750–1828)

A Journey through the Crimea to Constantinople. In a Series of Letters . . . to His Serene Highness the Margrave of Brandebourg, Anspach and Bareith [etc.]. London: G. G. J. and J. Robinson, 1789; quarto, pp. 328, with 6 engraved plates and a map

Memoirs of the Margravine of Anspach, written by herself. London: H. Colburn, 1826; 2 vols., pp. 430/406, with frontispieces to each volume

Elizabeth Berkeley, later Lady Craven and later still Margravine of Brandenburg, Anspach, and Bayreuth, Princess Berkeley of the Holy Roman Empire, was a society beauty, famous for her published plays, her excruciating verse—and notorious for what Walpole called in her the 'infinamente indiscreet'. Her extensive travels began in 1783 (bored with her promiscuous husband, she went in search of adventure) and never really stopped until 1817, when she bought the Neapolitan villa in which she died.

In her long-winded and thickly anecdotal accounts and memoirs Lady Elizabeth hardly ever talks about the art of travel: it seems unquestionably natural that she should have visited almost every country in Europe at a time when it was still unusual for a woman to visit *any* as a private individual. Confident and well-connected, she became the guest of the most fashionable courts—Paris, Vienna, Cracow, St Petersburg—and found herself 'protected by sovereigns and ministers, and treated with respect, and care, and generosity' wherever she went, whether travelling *en suite* in her liveried coach, or side-saddle and alone. This Progress (much more than just a 'tour') took her as far afield as Greece and Turkey before she came to rest in Bavaria at Anspach, home of the rich and admired Margrave of Brandenburg. She used his castle as a base for further peregrinations until kind fortune disposed of both his wife and Lord Craven, and they were able to marry in 1791.

They returned to England soon afterwards, by way of Spain and Portugal, but because of her long absence and widely advertised adultery, none of Lady Elizabeth's six children would acknowledge her and, even worse, nor would Queen Charlotte receive her at court. None daunted, the new Margravine bought properties in Hammersmith and in Berkshire, funded by the sale of her husband's principality to the King of Prussia and a lottery win, and gave herself a rest. But as soon as the Margrave died in 1806, she was off again, this time through revolutionary France (travelling under Napoleon's protection) to her last home in Italy.

Travel kept Lady Elizabeth young; once settled, the glittering quality of her life tarnished, and there is a sad account of her in 1828 recorded by fellow English exile Countess BLESSINGTON 'spade in hand, in very coarse and singular attire, a dessicated, antiquated piece of mortality'.

❖

CRAWFORD, Mabel Sharman

Life in Tuscany. London: Smith, Elder, 1859; pp. 337, with 2 lithotints

Through Algeria. London: R. Bentley, 1863; pp. 362, with a lithotint frontispiece

Mabel Sharman Crawford's first book, on Italy, is rather formal and impersonal. It is fully descriptive and very informative, but it could have been written by anyone. We learn the reason for this anonymity in the introduction to her second book: poor Miss Crawford had been desperately trying to avoid the common fate of women travellers and writers: 'the butt of wit and witling, the satirist's staple theme ... that ideal Gorgon, the "strong-minded woman", from whose wooden face, hard features, harsh voice, blunt manners and fiercely-independent bearing, society shrinks in horror'. She can contain herself no longer.

Through Algeria is prefaced by a 7-page exordium, a 'Plea for Lady Tourists' in which she passionately argues against the injustices of a world that deems women only fit to travel with 'a gentleman'. As a result, the book itself is far more relaxed and entertaining. Miss Crawford was no pioneer, nor even very adventurous, but she was a thorough and appreciative tourist of independent means and without domestic ties (of necessity, she says, not choice), who found an honourable occupation in the act of travel. 'And if the exploring of foreign lands is not the highest end or the most useful occupation of feminine existence, it is at least more improving, as well as more amusing, than the crochet-work or embroidery with which, at home, so many ladies seek to beguile the tedium of their unoccupied days.' So there.

❖

DIBBLE, L(ucy) Grace (b. 1902)

Return Tickets. Ilfracombe: Stockwell, 1968; 2 vols., pp. 445/453, with 50 halftones

More Return Tickets. Ilfracombe: Stockwell, 1979; pp. 373, with 16 colour photographs

Return Tickets to Southern Europe. Ilfracombe: Stockwell, 1980; pp. 400, with 10 colour photographs

Return Tickets to Scandinavia. Ilfracombe: Stockwell, 1982; pp. 160, with 16 colour photographs

Return Tickets to Yugoslavia. Ilfracombe: Stockwell, 1984; pp. 249, with 16 colour photographs

Return Tickets Here and There. Ilfracombe: Stockwell, 1988; pp. 271, with 8 colour photographs

Miss Dibble is a traveller of the old traditional school: the cheerful maiden lady whose inveterate world-wending habits are harnessed to a clear sense of Christian duty, and whose prodigious books both edify and entertain. She would be quite at home with the likes of Constance GORDON CUMMING or Charlotte CAMERON, even though there is a good generation between them. Miss Dibble's career as a teacher opened gently in rural Kent in 1931; over twenty years later she found herself Principal of the Ilesha Training College in Nigeria, having already spent a year working in Canada, eight in India and another eight in West Africa. Holidays were spent either visiting home (by way of the world), catching a bus across the Sahara, or perhaps just trekking to the Cape of Good Hope and back. Her retirement in 1965 heralded a series of cruises, safaris, and general moseyings around the globe that continued well into her eighties. And still the books keep coming . . .

⟨⟩

DODWELL, Christina (b. 1951)

Travels with Fortune: An African Adventure. London: W. H. Allen, 1979; pp. 316, with 13 halftones and 3 maps

In Papua New Guinea. Yeovil: Oxford Illustrated Press, 1983; pp. 256, with 26 colour photographs and a map

The Explorer's Handbook. London: Hodder and Stoughton, 1984; pp. 191, with diagrams throughout

A Traveller in China. London: Hodder and Stoughton, 1985; pp. 160, with 29 photographs and a map

A Traveller on Horseback in Eastern Turkey and Iran. London: Hodder and Stoughton, 1987; pp. 191, with 41 colour photographs and 5 maps

Travels with Pegasus: A Microlight Journey across West Africa. London: Hodder and Stoughton, 1989; pp. 208, with 36 colour photographs, a map, and a plan

Christina Dodwell's travels began in 1975 when, at the age of twenty-four, she dropped out of a boring round of unfulfilling jobs (and relationships) by answering an advertisement in a travel magazine. With a New Zealand nurse and two unknown young men, she set off in a Landrover for Nigeria to Have Fun and Find Adventure. Half-way across the Sahara Desert, however, the enterprising young men managed to sell the Landrover—without telling the girls—and soon afterwards they disappeared, leaving Christina and her companion to travel on alone. Christina had never even carried a rucksack before—but she soon learned. *Travels with Fortune*, her account of the subsequent 3-year journey first with Lesley the nurse and then alone, reads like the best of those intrepid lady travellers' accounts of two generations back. She well-nigh covered the length and breadth of Africa by any means available. She walked; she rode horses, camels, and zebroids; she cadged lifts on anything with a wheel or two and even spent several weeks paddling along the Congo in a dug-out canoe (only to be arrested as a spy at her eventual destination). She faced 'tick-bite fever', rabid jackals, and a diet at one stage of fresh blood-and-urine yoghurt with equanimity, all in the cause of Discovery.

Christina's later journeys, usually with a horse (but once with a microlight aircraft) in Papua New Guinea, Greece, Turkey, Iran, and West Africa, with unchronicled visits to Greece, the USA, and Mexico in between, all follow the same pattern: the books she wrote about them are discreet and sensitive, rather exhausting for the armchair traveller, but obviously challenging and enriching for the real one and for those she meets. And she has condensed all the practical techniques she has learnt along the way into a hair-raising manual of solo travel. *The Explorer's Handbook* includes everything, from what to pack in your rucksack and how to carry it, through 'tested exits from tight corners', to first aid and bush cookery (in Papua she used to start each day with a breakfast of good old bacon and eggs—except that the bacon was a pig's ear and the egg a crocodile's). She is a real explorer—especially of the art of travel itself.

❖

DOUGHTY, Marion

Afoot through the Kashmir Valleys. London: Sands, 1902; pp. 276, with 12 halftones and 28 vignettes

Even now, the Kashmir Valley sounds quite remote and exotic; in 1902, when the perfectly named Miss Doughty visited it, it was still the summer preserve of those stout-hearted men and women whose winters were spent in British India, or the even stouter-hearted few who tramped through the flowery foothills to shoot game. What a botanizing and curio-collecting matron like Marion was doing there just being a *tourist* is hard to tell. She makes the glaciers and peaks of the Himalaya sound as tame as the South

Downs, giggling indulgently at her tendency to 'take a tumble' or two when treading the ice and snow in her sandals, or mildly complaining at a foot-full of blisters at 9,000 ft. And after cataloguing the pleasures and perils of a Kashmiri holiday she has the effrontery to insist that any 'idle person' like her could do the same! These Victorian lady travellers . . .

❖

FORBES, (Joan) Rosita (1893–1967)

Unconducted Wanderers [Pacific Islands and Far East]. London: John Lane, 1919; pp. 198, with 72 halftones

The Secret of the Sahara: Kufara. London: Cassell, 1921; pp. 350, with 76 halftones and a map

Adventure: Being a Gipsy Salad: Some Incidents, Excitements and Impressions of Twelve Highly-Seasoned Years. London: Cassell, 1923; pp. 309, with 4 coloured plates

From Red Sea to Blue Nile: Abyssinian Adventures. London: Cassell, 1925; pp. 368, with 61 halftones and a map

Conflict: Angora to Afghanistan. London: Cassell, 1931; pp. 302, with 48 halftones

Eight Republics in Search of a Future [South America]. London: Cassell, 1933; pp. 340, with a map

Forbidden Road—Kabul to Samarkand. London: Cassell, 1937; pp. 289, with 76 halftones and a map

India of the Princes. London: Gifford, 1939; pp. 319, with 32 halftones

A Unicorn in the Bahamas. London: H. Jenkins, 1939; pp. 244, with 15 halftones and maps on endpapers

Gypsy in the Sun [autobiography]. London: Cassell, 1944; pp. 382, with 55 halftones and 7 maps

Appointment with Destiny [autobiography]. London: Cassell, 1946; pp. 304, with 55 halftones and 2 maps

Rosita Forbes went everywhere, saw everything. She is a little like Mrs Alec TWEEDIE in this respect: it is not the travels but the traveller that counts. Born plain Joan, a Lincolnshire lass, she was to become a glamorous figurehead of daring modern womanhood and one of our most prolific travel writers.

The journey that made her famous was a trek through the Libyan desert to the 'forbidden' city of Kufara in 1920. After six months' careful preparation, she joined a camel caravan dressed as an Arab (but with a camera secreted under her veil) and thus became the first white woman to broach its sacred gates. Not only this: she had also taught herself the rudiments of surveying, and from Kufara travelled on to chart a 'new' route into Egypt. Strong stuff, and despite subsequent dramatic journeys she never quite matched the sensationalism of this Saharan episode.

The journey 'From Red Sea to Blue Nile', across Ethiopia, was a journalistic venture: she took with her a cameraman and together they mapped 1,100 miles on mule-back 'in search of photographic material'. All her other journeys were searches too. Except for India, where she went for the game, and the Caribbean where she went to retire (and, in 1967, to die), she travelled as a sort of self-appointed envoy of Political Truth. In Morocco, western Arabia, Iraq and Persia, Afghanistan, Russian Central Asia, and South America she covered staggering distances to interview as many political leaders as possible, reporting their conversations through her books and offering herself as a mediator between them. Her political altruism never quite rings true, however, and inevitably it is Rosita Forbes the books are really about. To her adoring public she became a female Richard Halliburton, irresistible copy for the gossip columns and society magazines of the 1920s and 1930s as a glamorous, twice-married adventurer whose lipstick never smudged in jungle or desert.

She wrote best when giving a first-hand account of her travels: they were considerable expeditions after all, and earned her Fellowship of the Royal Geographical Society and a handful of other awards besides; unfortunately she was inveigled by her own image into producing such books as *Women of All Lands*, a ghastly anthology of photographs and sugared prose on the secrets of feminine allure, and the uncomfortably egotistical reminiscences of Important People I Have Met in *Women Called Wild* (1935), *These are Real People* (1937), and *These Men I Knew* (1940). Too much glamour and self-aggrandizement have made it easy to forget how serious and worthwhile a *traveller* Rosita Forbes really was.

❧

FRANKLIN, Lady Jane (1792–1875)

The Life, Diaries and Correspondence of Jane, Lady Franklin, 1792–1875. Edited by Willingham Franklin Rawnsley. London: Erskine Macdonald, 1923; pp. 231, with 13 halftones and a map

Korn, Alfons L.: *The Victorian Visitors: An Account of the Hawaiian Kingdom 1861–1866, including the Journal Letters of Sophia Cracroft, Extracts from the Journals of Lady Franklin, and Diaries and Letters of Queen Emma of Hawaii.* Honolulu: University of Hawaii Press, 1958; pp. 351, with 1 halftone and a map

If any Victorian woman deserves to be called indomitable, it is Lady Jane Franklin. She started travelling when a girl, accompanying her father to

Russia, Scandinavia, western Europe, and Spain. With these travels began her writing—long and exhaustive journals chronicle everywhere she ever went and everything she did.

In 1828, aged thirty-seven, she became the second wife of arctic explorer John Franklin, who had just been appointed captain of a frigate in the eastern Mediterranean. Jane spent the next few years in Asia Minor, travelling independently (with her husband's niece Sophy Cracroft) and joining Sir John (as he now was) on board his ship whenever possible. Next followed a stint in the antipodes, when Franklin was sent out to Van Diemen's Land as Governor in 1837. Jane soon swept through Tasmania, taking Sophy with her as companion, and went on to conquer most of New Zealand (becoming the first woman to climb Mount Wellington) and large chunks of south-eastern Australia too.

By 1844 Jane's considerable energies were taken up with preparing for Franklin's greatest task, the command of a long-planned expedition to the arctic to discover the fabled North-West Passage. She saw her husband off in 1845, and to while away the years till his return launched into a series of voyages to America, the West Indies, and back to Europe. Franklin never did return. His ship was lost, and despite five search expeditions commissioned and largely financed by Jane between 1850 and 1857, nothing more than the barest outline of what happened to him was found.

Her indefatigable efforts to trace Franklin, enlisting amongst others the help of Lord Palmerston, the President of the United States, and Napoleon, resulted in Jane being directly responsible for an important increase in our knowledge of the northern polar region. In recognition she was awarded the coveted Royal Geographical Society's Gold Medal in 1860—the first woman to be thus honoured.

Her own travels took a frenetic turn after Franklin's loss. Still with the faithful Sophy in tow she tore around the world, running a commentary in her closely-written diaries, and making only brief visits home to London. She died on one of these far-flung jaunts in 1875, two weeks before the monument she had planned for her husband was unveiled in Westminster Abbey. She was eighty-three.

❖

GORDON CUMMING, Constance Fredereka (1837–1924)

From the Hebrides to the Himalaya: A Sketch of Eighteen Months' Wanderings in Western Isles and Eastern Islands. London: Sampson Low, Marston, *et al.*, 1876; 2 vols., pp. 376/364, with 27 wood-engraved plates and 19 vignettes

At Home in Fiji. Edinburgh: Blackwood, 1881; 2 vols., pp. 293/324, with 7 halftones and a map

A Lady's Cruise in a French Man-of-War [Samoa, Tonga, Tahiti]. Edinburgh: Blackwood, 1882; 2 vols., pp. 304/309, with 8 halftones and a map

Fire Fountains: The Kingdom of Hawaii, its Volcanoes, and the History of its Missions. Edinburgh: Blackwood, 1883; 2 vols., pp. 297/279, with 8 halftones and 2 maps

Granite Crags [California and the Sierra Nevada]. Edinburgh: Blackwood, 1884; pp. 384, with 8 photo-engravings and a map

Via Cornwall to Egypt. London: Chatto and Windus, 1885; pp. 361, with frontispiece

Wanderings in China. Edinburgh: Blackwood, 1886; 2 vols., pp. 382/370, with 10 halftones and a map

Two Happy Years in Ceylon. Edinburgh: Blackwood, 1892; 2 vols., pp. 438/442, with 19 wood-engraved plates and a map

Memories [autobiography]. Edinburgh: Blackwood, 1904; pp. 487, with 7 photogravures and 17 halftones

'Eka' Gordon Cumming was the perfect candidate for a life of leisurely travel. She was of a noble and wealthy Scottish family, so she had the means. As a spinster with 14 brothers and sisters and 50 cousins scattered around the world she had the opportunity, and with sound health and few responsibilities (her parents died in her youth), she had the freedom. Once she began travelling (and writing), it became difficult to stop.

Contemporary critics, probably flabbergasted by the sheer bulk of her work, likened Miss Gordon Cumming to Isabella BIRD and dubbed her one of the greatest of what *Vanity Fair* called 'our wonderful lady travellers'. In fact she was one of that tribe Isabella Bird once professed to despise: the 'globe-trotteress'. Her first visit abroad was in 1868, via Egypt to India, where her sister and brother-in-law were stationed. This was in fact one of the few times she took advantage of family connections; for the rest of her travelling life she relied on her social position and growing reputation to win her invitations. Consequently her globe-trotting took on the air of a series of rather far-flung social calls. After a year's tour in India she progressed to Ceylon: the parish priest at home in Scotland had recently been appointed Bishop of Colombo. The next expedition was to Fiji. Fiji had just (in 1874) been presented to Queen Victoria by its chiefs, and 'Eka' was invited by its first Governor, Lord Stanmore, and his lady, to accompany them on their inaugural posting. This trip also involved a few months' sojourn at Sydney. Whilst in Fiji she met the Bishop of Samoa, who was engaged in a tour of his diocese on board a French man-of-war, and she gladly accepted the invitation to accompany him to the South Sea Islands. Next stop was San Francisco and the Yosemite Valley (her 'Granite Crags'), then to China, where she became involved with Hill Murray's mission to the blind, and then Japan (which is described only in her autobiography). She sailed back to Britain in 1879, staying only long enough to accept General and Mrs Ulysses Grant's invitation back to California; thence to Hawaii, squeezing in visits to Henry Longfellow and Oliver Wendell Holmes on the way home to Scotland, where she eventually settled.

It all sounds rather breathless, but Miss Gordon Cumming was never less than decorous, dignified, and polite—which does not always make for exciting reading. She was (of course) a neat and accomplished artist, and over the years amassed a collection of a thousand paintings of foreign parts, which she likened to her friend Marianne NORTH's. Her declining years in Scotland were spent arranging them for an exhibition, and supervising the publishing and reprinting of her books and countless articles. She died at home at the admirable age of eighty-seven—one more example of the strange law amongst women travellers that the further you go, the longer you live.

HARRIS, Audrey

Eastern Visas. London: Collins, 1939; pp. 392, with 69 halftones, a map, and further maps on endpapers

'I could not begin to understand life . . . with only the knowledge of the adolescent and materialised West . . . I was only learning how to travel . . . and did nothing that anyone else could not have done with the time, a good digestion, and freedom from the slavery to familiar comforts.' Parts of this wordy and intense book might have been written by any young student today, retreading the old 'hippy trail' through India and central Asia in search of 'the spirit of the surroundings' and a philosophy for life. In fact Audrey Harris was travelling over fifty years ago—but for just the same reasons. In much the same manner, too. Her journey—financed by tolerant parents—was part of her self-imposed education. Its itinerary was punishing: by the Trans-Siberian Railway to Korea, to Japan, to China and Jehol in central Asia, to the islands of Indo-China and on to Calcutta, to Nepal, Sikkim, Tibet, and finally Afghanistan. She travelled alone, or with companions acquired along the way, and took whatever means of transport presented itself. She loved the thought of danger and deliberately sought out the unconventional and unfamiliar. This modern Grand Tour lasted for sixteen months, and when Miss Harris eventually arrived home she was quite happy to admit that she had travelled enough to last her a lifetime: she had Seen The World: no need to see it again.

HICKMAN, Katie (Catherine Lucy) (b. 1960)

Dreams of the Peaceful Dragon: A Journey through Bhutan. London: Gollancz, 1987; pp. 192, with 23 colour photographs and a map

There is still romance in travel. There must be, if a young English woman can be invited to a tiny Himalayan kingdom by one of its princesses, given guides, horses, Jeeps, and whatever else she needs to explore the misty and long-forbidden valleys of the secret country, if she can meet startled hill people who think she is a fairy, fall into a frightening fever, start writing her first book, and then come home and marry her travelling companion. This all happened to Katie Hickman on her journey through Bhutan with the photographer Tom Owen Edmunds in 1985. Their trek from the capital, Thimphu, across to the eastern mountain strongholds of the Bragpa people (whom Katie was one of the first Western women to visit) took five years to arrange. But Katie's travelling childhood as the daughter of a diplomat and her recent tendency to take holidays where the rest of us would only consider expeditions—Ecuador or Antarctica, for instance—must have helped in the preparations. Her practice as a free-lance travel writer stood her in good stead, too, when it came to writing the book she felt she must as soon as she set foot in Bhutan. Tom Owen Edmunds took the photographs (and has since produced his own illustrated book on the journey): this partnership of travel and romance looks set to prosper.

❧

HOLMES, Mrs Dalkeith (E. Augusta) (d. 1857)

A Ride on Horseback to Florence Through France and Switzerland. Described in a Series of Letters. By A Lady. London: John Murray, 1842; 2 vols., pp. 346/436

A curious procession set out from Liverpool one spring morning in 1838. The man and woman on horseback were obviously wealthy: their horses were gleaming, their dress of the highest quality, and all four had a confident air. Yet they had no retinue, save for a rather surly Irish groom who carried some luggage. And, strangest of all in this modern age of perfectly good diligences, it was rumoured that they were actually going to ride *all the way to Italy*! It was quite true: the Holmeses rode to Dover, and from Calais to Paris; after staying there for a while to rest Fanny and Grizzel (the horses) they set out to trot over the Alps to Florence.

Each stage of the journey is carefully described (out via Geneva, the Simplon, and Milan and back via the Apennines and Mont Cenis), with copious historical notes and lots of local colour. The Rhône valley had just been flooded when they rode through it, and the storms they met on the

Simplon Pass made the going very rough, what with landslides, impromptu waterfalls, and fallen bridges. Stopping at an inn did not provide much comfort either. Mrs Holmes complained that most were full of 'bad women and bad beds'—bad manners too: she once heard a landlord asking Mr Holmes in hushed tones whether or not his wife was an Amazon. In fact she often slept more comfortably on the hoof.

The whole journey took eight months (almost as long as it takes to read the book) and seems to have daunted neither Holmeses nor horses. Fanny and Grizzel were awarded a long rest (and an epic poem composed in their honour by our authoress) whilst their owners settled down to enjoy the sparkling pleasures of a Paris winter season.

❖

LLOYD, Sarah

An Indian Attachment. London: Harvill Press, 1984; pp. 244, with 9 halftones and 6 plans

Chinese Characters: A Journey through China. London: Collins, 1987; pp. 270, with 20 halftones

Sarah Lloyd is a well-travelled writer who writes not about places, but about people. In India, where she first went in 1979, she formed an 'attachment' to a Sikh living with his family twenty miles from the Pakistani border, and her first book describes the unquiet months they spent living together—with time out to travel and to recover from amoebic dysentery—both then and during a return visit in 1980.

Her second book is a little less personal. Tired (again) of her English life as a free-lance landscape architect she spent eight months wandering around the people of China, recording their reactionary conversations and her own intense sympathy in a series of sketches. And meanwhile, she writes, the Sikh is still vainly wondering if she will come back to the Punjab as his wife.

❖

MURPHY, Dervla (b. 1931)

Full Tilt: Ireland to India with a Bicycle. London: John Murray, 1965; pp. 235, with 31 halftones and 2 maps

Tibetan Foothold. London: John Murray, 1966; pp. 206, with 26 halftones and a map

The Waiting Land: A Spell in Nepal. London: John Murray, 1967; pp. 216, with 16 halftones and a map

In Ethiopia with a Mule. London: John Murray, 1968; pp. 281, with 24 halftones and a map

On a Shoestring to Coorg: An Experience of South India. London: John Murray, 1976; pp. 261, with a map

Where the Indus is Young: A Winter in Baltistan. London: John Murray, 1977; pp. 266, with 9 halftones and a map

Wheels Within Wheels [autobiography]. London: John Murray, 1979; pp. 236, with 15 halftones

Eight Feet in the Andes. London: John Murray, 1983; pp. 274, with 15 halftones and a map

Muddling Through in Madagascar. London: John Murray, 1985; pp. 274, with 21 halftones and a map

Cameroon with Egbert. London: John Murray, 1989; pp. 282, with 14 halftones and a map

It is tempting to think of the traditional Victorian lady traveller as a spirited spinster, whose youth is spent at home caring for her elderly, eccentric, or invalid family with a mixture of devotion and frustration, who nourishes her instinctive desire for self-expression by gazing at the atlas and planning imaginary journeys for herself, and who is rewarded for her pains when at the age of thirty or so, her dependants die and suddenly she is free. There may be a little tension at first between feelings of guilt or depression and hedonistic self-indulgence, but usually, after the first impossibly wild journey or two, she blossoms.

All this is relevant to this thoroughly modern and highly original woman Dervla Murphy, one of the most joyful travellers and successful authors around. She has written books on her native land of Ireland, on what the English politely term 'inner-city problems', and on the nuclear arms race: she is not just a travel writer. But *primarily* she is, and as such her life has followed the maidenly pattern of her Victorian predecessors remarkably closely—or at least it did, up to her second or third book. After cycling to the Himalayan foothills in a sort of daze following her parents' death, and discharging her sense of duty at a Tibetan refugee camp in Dharmsala for six months, she really began to explore the art of travel (becoming a cheerfully rabid travel snob in the process—woe betide the ordinary tourist encountered by Dervla abroad) and she looks likely to carry on charming herself and her readers as long as her stout Irish legs will carry her.

Dervla Murphy always travelled alone (except for Roz the bicycle or Jock the donkey) and rather resented 'civilized' company on her treks into India, Pakistan, Nepal, and Ethiopia—until Rachel came along. Rachel is her daughter, who has been Dervla's companion and touchstone on almost all of her travels since her birth in 1968. When she was five, Dervla took her to southern India for Christmas, a fairly tame experience which only involved Rachel's contracting brucellosis and then being rushed to hospital with an infected foot; a year later mother led daughter along the teetering paths of the soaring Indus valleys on a pony; they tackled the Andes when Rachel was nine and five years later they backpacked and caught hepatitis together in Madagascar before plunging into the dank and tropic depths of west Africa in 1987. Rachel has become as much a feature of her mother's sensitive, evocative, and riotous books as Dervla is herself.

MYERS, Wendy (b. 1941)

Seven League Boots: The Story of my Seven-Year Hitch-Hike Round the World. London: Hodder and Stoughton, 1969; pp. 223, with 43 halftones and endpaper map

Bumming around. That is what they called the summer students often spent between school and college or work, usually in Europe or perhaps a kibbutz. The hope was that the student would return suitably enlightened or shriven to face life at home as an adult. Wendy Myers followed the pattern, hitch-hiking alone to Turkey at eighteen, soon after leaving school—only she did not come home for seven years. She had not been tempted down the endless 'hippy trail', or captured for the white-slave trade: she'd simply fallen in love with travel. Setting out from Yorkshire with her parents' blessing and a hundred pounds sewn inside her bra, she proceeded to work her passage through Asia Minor to India, Africa, the Far East, Australia, and North and South America, living for a year here with some hospitable family or a month there in a convent . . .

Having served a hardy apprenticeship during her seven years away by working as a broadcaster for Sri Lankan Radio or a Saharan bandit's accomplice, crouching under fire in Vietnam or recovering from hepatitis in Peru, Wendy went on to train as a nurse as soon as she got home and has since worked with various charities in the Pacific islands and in Africa, combining travel with a hard-earned and well-won livelihood.

RATCLIFFE, Dorothy Una (1891–1967)

To The Blue Canadian Hills: A Week's Log in a Northern Quebec Camp. Leeds: North Country Press [1928]; pp. 90, with 6 halftones, line-drawn vignettes, and a map

The Babes of the Sea: Being an Account of their First Voyage in the Sea Swallow. Leeds: North Country Press [limited edition of 150 copies, 1929]; pp. 140, with 14 halftones and a map

South African Summer: 5,000 Miles with a Car and a Caravan-trailer. London: Country Life [1933]; pp. 190, with 15 halftones and a map

Equatorial Dawn: Travel Letters from North, East and Central Africa. London: Eyre and Spottiswoode, 1936; pp. 305, with a map

Swallow of the Sea: Pages from a Yacht's Log. London: Country Life, 1937; pp. 159, with coloured frontispiece and 22 line drawings

News of Persephone: Impressions in Northern and Southern Greece with a Car, a Kettle and Cameras. London: Eyre and Spottiswoode, 1939; pp. 224, with 46 halftones and map on endpapers

Icelandic Spring. London: Bodley Head, 1950; pp. 128, with sketches throughout and map on endpapers

Although not born there—nor even living there until she had been 'finished' at Weimar and Paris and married a Leeds business man—Dorothy Una Ratcliffe was of hearty Yorkshire stock, and spent most of her writing life recording the county's history, folklore, and dialects. She was an inveterate holiday-maker too, never taking herself seriously enough to be a real traveller and yet becoming in the 1920s and 1930s quite a well-known adventurer abroad.

Apart from a spell in Canada (where she had accompanied her first husband Charles Ratcliffe on a business trip), a honeymoon caravan tour in Cape Province and Natal (husband number two), and an extended safari in Kenya, Greece and Iceland were as far afield as Dorothy went. She rarely travelled alone, and never minded revisiting old haunts. Sailing leisurely in her yawl *Sea Swallow* between Brittany and the Baltic usually satisfied her modest wanderlust. Perhaps her light and approachable style made her books popular—certainly, nothing sensational or exotic ever happens. Or perhaps she just represented a welcome new breed of lady travellers: those for whom a journey was nothing to do with achievement—simply with pleasure.

❧

SELBY, Bettina (b. 1934)

Riding the Mountains Down [Himalayas]. London: Victor Gollancz, 1984; pp. 232, with 22 colour photographs and 3 maps (1 on endpapers)

Riding to Jerusalem. London: Sidgwick and Jackson, 1985; pp. 216, with 17 colour photographs and 4 maps

Riding the Desert Trail [Africa]. London: Chatto and Windus, 1988; pp. 240, with 19 colour photographs and 4 maps

Many Victorian lady travellers were spinsters who did not start travelling until well into middle age. Only then did they have the money and the time (due usually to the demise of their parents) to escape the confines of home.

Bettina Selby rather echoes the pattern, except that she waited until she had saved up enough time and money of her own (she was a primary-school teacher and a free-lance photographer, and her children had all grown up) before setting off to the Himalayas at the age of forty-seven. She had always been a keen cyclist, and had a special steed expressly designed for the four-month spin from Karachi to Kathmandu. She travelled in search of freedom (difficult for a woman in Muslim areas of Pakistan and India, she found—even a white one in a shapeless cycling-suit) and in search of the fabled Himalayas, whose peaks were in sight for most of the journey. The journey was a success, encouraging Mrs Selby to plan another one a year later to the Holy Land, following the path of the medieval pilgrims and Crusaders across Europe to Turkey and through Syria (Lebanon being out of bounds due to fighting) to Jerusalem. One of her latest adventures was inspired by Florence BAKER and Amelia EDWARDS, leading her along the Nile (on her bike) from its delta to the Mountains of the Moon in Uganda. She is getting a taste for travel, it seems.

◆

SIMPSON, Myrtle Lillias (b. 1931)

Home is a Tent [Spitzbergen, Iceland, Surinam]. London: Gollancz, 1964; pp. 192, with 4 halftones, 6 diagrams, and 4 maps

White Horizons [Greenland]. London: Gollancz, 1967; pp. 191, with 28 halftones and 2 maps

Due North [North Pole and Greenland]. London: Gollancz, 1970; pp. 171, with 27 halftones and a map

Greenland Summer [for children]. London: Gollancz, 1973; pp. 96, with 30 halftones and a map

Armadillo Stew [Surinam, for children]. Glasgow: Blackie, 1975; pp. 96, halftones throughout, and a map

Vikings, Scots and Scraelings [Greenland]. London: Gollancz, 1977; pp. 189, with 22 halftones and a map

'I do not run away from responsibilities and commitments, but perhaps from tin gods, like the television and artificial worries about what the neighbours think. I do not retreat, but advance into a world where to survive one must come to terms with the natural elements.' So says Myrtle Simpson, explorer, mountaineer, skier, biographer, and travel writer. All her travel books relate the challenge—whether it be skiing across the Greenland ice-cap (she was the first woman to do it), towards the North Pole (she reached the furthest north for a woman in 1968), or climbing in the Himalayan mountains (she was a member of the British Ladies' Himalayan Expedition in 1974)—the challenge of meeting those 'natural elements' on their own terms. She started travelling young, being educated in India and then spending her early twenties climbing in New Zealand and Peru, and working her way around Australia. In 1959 she married Dr Hugh Simpson, whose medical researches in Greenland, Surinam, and on the polar ice-cap gave Myrtle the chance to indulge her passion for aggressive travel even further. Four children were born in fairly quick succession who, according to age and size, were as often as possible packed into rucksacks or slotted into tiny kayaks and taken along too. Mrs Simpson lives now in her native Scotland, a string of awards and honours to her name, passing on through her work as a ski instructor and writer her challenging philosophy to a new generation of travellers.

❖

SMEETON, Beryl (1905–1979)

Winter Shoes in Springtime. London: Rupert Hart-Davis, 1961; pp. 264, with 1 halftone and 2 maps

Miles Smeeton's books are famous. He wrote a whole collection of them describing his travelling life in the mountains and on the sea, and in them all his wife Beryl is the heroine. She was 'always looking for some wilder country, some more difficult task to set herself', he says: a born traveller, who relished their vagrant lifestyle (home, for twenty years, was a 46-foot ketch in which they explored the coasts and islands of the world) and she thrived on nothing more corporeal than a spirit of adventure.

Beryl's only published book describes the years immediately before she and Miles married: she was a well-tried traveller even then. She met Smeeton in India during the breakup of her first marriage, and after he had taught her how to climb by shinning up a radio mast and she had outridden him spectacularly on various jaunts with the Poona and Kirkee Hounds, they decided (at Beryl's suggestion) to travel overland together from Quetta to London. This was in 1933. The next year she made the same journey alone, and then accompanied her brother—again overland—to Hong Kong before returning to the West via China and the forbidden Kunming Highway to Burma. Any spare time Miles might cadge from the army was spent climbing with Beryl, who was usually *en route* somewhere, in the Himalaya or the Alps and Tyrol.

Beryl wrote a second book, as yet unpublished, describing the astonishing solo ride she completed just before the outbreak of World War II through the southern Cordillera of Patagonia. Peter Fleming of *The Times* had suggested she explore the mountains dividing Chile from Argentina in search of a 'prehistoric' tribe said to haunt a lost Andean city there. Actually the tribe and its archaeology were of secondary interest to Beryl: it was the challenge of travelling alone through uncharted country that attracted her to South America. She married Miles on her return to England in 1938, and was making her way out to join him in India, where he had been immediately posted, when war was declared.

When it was over, and after Beryl had squeezed in a quick trip to Tibet with her friend Ella MAILLART, the Smeetons left for British Columbia where they farmed for ten years and planned the voyages to come. It was in Canada—Alberta—that they eventually settled in the 1970s, choosing a perfect resting-place under the irresistible shadow of the Rocky Mountains.

❧

TOY, Barbara

A Fool on Wheels: Tangier to Baghdad by Land-Rover. London: John Murray, 1955; pp. 256, with 15 halftones and a map

A Fool in the Desert: Journeys in Libya. London: John Murray, 1956; pp. 180, with 22 halftones and a map

A Fool Strikes Oil: Across Saudi Arabia. London: John Murray, 1957; pp. 207, with 16 halftones and a map

Columbus was Right! Rover Around the World. London: John Murray, 1958; pp. 242, with 30 halftones and a map

In Search of Sheba: Across the Sahara to Ethiopia. London: John Murray, 1961; pp. 242, with 33 halftones and a map

The Way of the Chariots: Niger River–Sahara–Libya. London: John Murray, 1964; pp. 161, with 32 halftones and a map

The Highway of the Three Kings: Arabia—from South to North. London: John Murray, 1968; pp. 188, with 24 halftones and a map

Rendezvous in Cyprus. London: John Murray, 1970; pp. 141, with 17 halftones and a map

Even though she has travelled so far and so often, Barbara Toy has never been a restless type. Just inexhaustible. With her beloved Landrover as her constant companion and a sense of humour wiry enough to cope with most eventualities, she has usually travelled with a very definite purpose in mind. First of all it was to rouse in herself the sense of discovery she found lacking in her life as a playwright and theatre director in England. And then, having successfully completed her inaugural drive as a fully fledged Lady Traveller from Tangier to Baghdad, she motored to Saudi Arabia to get to know the desert better and to study the influence of 'black gold', or oil, on its sparse inhabitants. By now she had become thoroughly intoxicated by travel: when the time came to visit relations in her native Australia she decided to drive there, going one way round the world and coming back the other. There followed a series of journeys in search of semi-mythical archaeological remains, first to Ethiopia, during which she became the first Westerner to reach the summit of Wahnu Mountain (she was winched up there by helicopter and left alone for the night), then to the Sahara again, and back to Arabia. Her last book describes an innocuous visit to Cyprus, timed unfortunately to coincide with the unrest and massacres of November 1967; since then she has driven to Mauretania ('rather like one's first visit to Cornwall!'), followed the Trans-Siberian trail from Moscow to Japan and South Korea—and feels now that it may be time to revisit Saudi again. As I said: inexhaustible.

5

In Search of the Picturesque

THE women in this chapter had nothing to prove, no point to make, no mission to fulfil: they went abroad merely to look at what was most worthy to be seen. They were just tourists.

It is not the fashion for *real* travellers to be 'just tourists' these days: sightseeing is too easy. But it was not ever thus: Lady Grisell BAILLIE's epic 'holiday' involved no end of effort. Moving an aristocratic family and its retinue any distance at all through familiar country in 1731 was a major and complicated undertaking; in the strange expanses of France, Italy, and the Netherlands the problems were exacerbated by the responsibility of having to uphold one's standards abroad in the face of inedible food, vulgar accommodation, and a disconcerting preponderance of Foreigners.

Things were a little easier by the time Marianne COLSTON made her 'wedding journey', or honeymoon, on the traditional European Grand Tour circuit in 1819. The tour took three years, and was no doubt cushioned at every turn by the confident attentions of Marianne's compatriots making similar journeys, or else temporarily resident in the civilized cities and watering-places of the Continent.

Forty years later, when Switzerland had become fashionable and the lower slopes of the Alps rang with the well-bred trillings of ladies abroad, touring had almost become one of those accomplishments, like singing at the pianoforte and fine stitchery, to which every 'nice' young woman aspired. And as sketching and the keeping of an edifying journal were similarly appreciated, a glut of thickly feminine travel books began to emerge, all illustrated with what look now like stock-footage views of the same peasants, villages, and mountains.

These lightsome books were largely responsible for giving the woman traveller a bad name. They begged not to be taken seriously, with their sighing eulogies, florid effusions, and illustrations of trim little figures picking their way gingerly over the *Mer de Glace* in the utmost height of whaleboned and high-heeled fashion, shepherded by indulgent-looking husbands and brothers. 'Petticoated pilgrims' and 'pic-nic pioneers', the critics called them.

It was an image which infuriated certain stouter-hearted women like Emily LOWE or Ellen BROWNING. Emily had no time for chaperones and escorts: she described herself proudly as an 'unprotected female', and maintained that men were only good for carrying luggage (and as she travelled light on her jaunts through Norway and Sicily, there was obviously no need for a man at all). Miss Browning, meanwhile, chose to visit Hungary, alone, wearing 'cloth knickers under my gown', because all those things were precisely what the usual 'hysterical female' avoided.

Some tourists went even further afield than Hungary: Helen PEEL scandalized her sister débutantes by sailing to the Polar seas of Russia and Adelaide GERARD went to Spitzbergen, while the STUART WORTLEYS, mother and daughter, between them went almost everywhere else. The United States of America tended to attract those blessed with enough money and stamina to afford long-haul sightseeing; the main talking point there seems to have been the unwonted splendour of the transcontinental railways. But yachting was the ultimate touring experience. In a yacht, especially a 217-ton leviathan like the MARCHIONESS OF WESTMINSTER's, one could drift wherever one wished, see the most outlandish sights, and still come 'home' every evening to civilization on board ship. It was like travelling with all the advantages of never having left home.

BAILLIE, Lady Grisell (1665–1746)

The Household Book of Lady Grisell Baillie, 1692–1733. Edited with notes by Robert Scott Moncrieff. Edinburgh: University Press, 1911; pp. 443, with 11 halftones

Lady Grisell Baillie had an illustrious and well documented childhood. As twelve-year-old Grisell Hume she was the brave go-between for Robert

Baillie, imprisoned in Scotland in 1683 for high treason, and her sympathetic father the Earl of Marchmont. When Baillie was executed the Earl fled to Holland taking his wife and daughter with him. Almost immediately afterwards Grisell was sent back to England alone to collect her younger sister and bring her to Utrecht. Later she married Baillie's son and began to keep the detailed household account book which, amongst other domestic minutiae, chronicles an epic family visit to the Continent between 1731 and 1733. The Baillies were well-heeled Whigs now, and expected to spare no expense in 'doing' France, Italy, Austria, and the Low Countries properly. There were music lessons to be paid for, portraits to be commissioned, language tutors to be engaged, dinners to be given, excursions to be planned ... To be a tourist at this time was more or less synonymous with being an aristocrat; it was a wonderfully rich life—and wonderfully insular.

❖

BODDINGTON, Mrs M.

[Anon.]: *Slight Reminiscences of the Rhine, Switzerland, and A Corner of Italy.* London: Longman, Rees, Orme, *et al.*, 1834 [according to the author's preface an 'incorrect edition' was published in Paris by private subscription, 1830; copy not seen]; 2 vols., 12mo, pp. 339/332, with 2 engraved plates

[Anon.]: *Sketches in the Pyrenees, with some remarks on Languedoc, Provence, and the Cornice* [sic]. London: Longman, Rees, Orme *et al.*, 1837; 2 vols., 12mo, pp. 440/486

'I hate church and palace-hunting,' complains Mrs Boddington, 'but what can one do?' Little else, it seems, until one arrives at the grandeur of the Rhine, the Alps, the Italian Lakes or the Pyrenees. There the spirit trembles (in true Udolphian fashion), the heart swells, and the body faints, overcome with romance and noble melancholy. Never mind miserable France, with its 'lower classes of females' and 'perfection of bad taste'; never mind drab Germany where 'the dames looked coarse and the damsels heavy'; even Lucerne has 'nothing alive in it but the river': it is Nature Mrs Boddington has come to see. She was one of a flock of slightly published literary ladies of the 1820s and 1830s—votaries of the Goddess Nature, all of them—who enjoyed ephemeral success with such books as these (and in her case a work called *The Gossip's Week* published in 1836 and some *Poems*) but who are now no more than names in a library catalogue. She went to the Continent because one did in her position, and wrote about it as one should: a traveller *à la mode*.

❖

BROMLEY, Mrs (Clara Fitzroy)

A Woman's Wanderings in the Western World: A Series of Letters Addressed to Sir Fitzroy Kelly, M.P., by his Daughter. London: Saunders, Otley, 1861; pp. 299, with 4 lithotint plates

Clara Bromley travelled to North and South America with a young lady friend in 1853, to recover her health and spirits after—as she somewhat stiffly puts it—being 'severely shaken by domestic losses during the preceding year'. The schedule she designed was punishing—20,000 miles in ten months—but it achieved its object.

The two women based their wanderings on the Caribbean islands; from there they went as far afield as New York and Montreal in the north, and Panama and Lima to the south. The holiday nearly came to a tragic end during one visit to Barbados, when Clara the tourist turned nurse. A young victim of the recent yellow-fever epidemic was billeted (for want of a hospital) in her hotel, and since no one else would touch him, the tender-hearted young widow cared for him herself. He died within a week, leaving Clara and her companion depressed and weak. But not for long: their itinerary would not allow it, and as soon as it was certain they were clear of the fever themselves, they rattled off again to another port of call. The busier she was the calmer Clara felt, no doubt proving some cliché or other, and by the close of her book she was able to declare that 'the year I spent on the other side of the Atlantic was the happiest and most peaceful period of my life.' The reader, meanwhile, is utterly exhausted.

❖

BROWNING, Ellen

A Girl's Wanderings in Hungary. London: Longmans, Green, 1896; pp. 332, with 10 halftones, vignettes throughout, and a map

Miss Browning is definitely a 'modern' girl. She decides to visit Hungary alone, to 'bring her round' after her father's death, knowing no one nor any of the language there. She duly has a mildly adventurous and thoroughly enjoyable time, growing more and more independent as the journey progresses through the picturesque Magyar countryside until she is able exultantly to describe herself as 'an eccentric young person . . . I wear cloth knickers under my gown and feel equally contemptuous towards an "hysterical female" and a dowdy bas bleu. Their day is over!'

❖

CARTER, Anne

[Anon.]: *Letters from A Lady to her Sister During a Tour of Paris, in the months of April and May, 1814.* London: Longman, Hurst, Rees, *et al.*, 1814; pp. 170

'It is with the greatest diffidence and anxiety that the Writer of these Letters submits them to the public eye.' So Miss Carter prefaced her book, in true feminine fashion—but she need hardly have worried that it was not going to be popular. The literature might not be great, but the timing was perfect. Miss Carter and her party arrived at Boulogne at the beginning of a whirlwind tour just two weeks after the abdication of Napoleon, and reached Paris in time to see the greatest spectacle for decades, the formal entrance of King Louis XVIII. They had tickets for the ceremonial service at Notre Dame, seats at the Opéra the night the Duke of Wellington was there, and altogether spent a brilliant fortnight (that is all it was!) basking in reflected British glory.

CATLOW, Agnes (*c.* 1807–1889) and Maria

Sketching Rambles: Or, Nature in the Alps and Apennines. London: Hogg, 1861; 2 vols., pp. 374/368, with 20 lithotint views (including 2 hand-coloured)

Armed with their sketch-books and hot on the heels of Mrs COLE, the Catlow sisters—natural historians both—set out for the Alps in pursuit of 'sublime scenery, grandeur and the picturesque'. They were not strong enough for the mountains, they thought, and contented themselves with exploring the gentler slopes of Switzerland and Italy. Their illustrations are gentle, delicate landscapes; it is a very seemly book, this, written to encourage others in similar maidenly circumstances to travel too. There is some unhealthy excitement when, during a stay near Vesuvius, the Catlows' pension sways with the distant rumbles of the Great Earthquake of 1857, but happily the sisters are spared to continue their amiable journey.

CLERK, Mrs Godfrey (Alice M.)

The Antipodes and Round the World [etc.]. London: Hatchard, 1870; pp. 633, with 7 lithographs (1 hand-coloured), and wood-engraved vignettes throughout

Alice Frere was a one-off 'globe-trotteress', a round-the-world tourist who really did Tour. With her father and a servant she left Bombay (where they had been living for some years) and took the long way home to England via

The Rest of the World. The little woodcut sketches throughout the book are like snapshots—Alice down an Australian mine, or tottering up a tree-trunk staircase in Ceylon; firmly palanquinned in China or eating elderly beef with a pencil (*sic*!) in Japan. The well-connected trio took in New Zealand, too, staying with Bishop Selwyn in Auckland, and visited Tasmania by courtesy of the Governor. On arrival in Washington, via California and pre-canal Panama, Alice was presented to President Johnson: it was all quite an excursion. She arrived home in England a year after setting off, determined to face the 'troublesome and ungrateful task' of writing up her travelogue in as much detail as possible, to serve those other tourists lucky enough to follow in her wake.

❦

COLSTON, Marianne

Journal of a Tour in France, Switzerland, and Italy, during the Years 1819, 20 and 21. Illustrated by Fifty Lithographic Prints from Original Drawings taken in Italy, the Alps, and the Pyrenees [etc.]. Paris: Galignani, 1822 [Galignani was a 'pirate' publisher; the book was not published in London until the following year, by G. & B. Whittaker]; 2 vols., pp. 404/332, with a separate atlas for the plates

The tour so prettily (and impersonally) chronicled here was Mrs Colston's wedding journey. It was an extravagantly lengthy affair involving all the old Continental highlights, diligently and dutifully enthused over and sketched by the talented author. The chapter headings read like an itinerary for the grandest of Grand Tours, panoramic and in excellent taste, but (for young men, at least) quite unoriginal: it was the usual society circuit of Paris, Geneva, the Italian Lakes, Florence, Rome, and Naples, with a quick visit to the exotic mountains bordering France and Spain. The lithographs, of course, are in the best contemporary tradition of the romantic and picturesque.

❦

D'A(lmeida), Anna

A Lady's Visit to Manilla and Japan. London: Hurst and Blackett, 1863; pp. 297, with chromolithographed frontispiece

Mrs D'Almeida does not claim to be a serious writer: her book is just a light account, she says, of 'some months in Japanese waters'. She and her husband William took their holiday cruise at the end of a diplomatic residence in Java; they based it on Hong Kong and from there visited the Philippines, Macau, east-coast China, and Japan. Although most of the travelling time was spent aboard some vessel or other—be it government steamer, cargo tramp, or native canoe—there were some excursions inland: through the spidersome jungles of Luzon to the lakes, for example, to an opium den in Macau, or a Chinese prison (it was the time of the Taiping rebellion). Anna D'Almeida seems to have welcomed everything with relish—even the two thrilling typhoons she witnessed in Victoria—and closed the tour (and her book) only with considerable regret.

❧

DAMER, The Honourable Georgiana L. (d. 1848)

Diary of a Tour in Greece, Turkey, Egypt, and the Holy Land. London: H. Colburn, 1841; 2 vols., pp. 324/304, with 13 lithographs

Most nineteenth-century authoresses justified the publication of their travel journals by claiming that their friends had persuaded them to it; Mrs Damer's justification is different. The book will make money, she says, for 'an object of great public charity and utility in this country'—although I cannot help feeling it might have helped sales if she had mentioned what that object was. Her *Diary* describes a typically exhaustive (and exhausting) Grand Tour to the Holy Land. She and her husband saw *all* there was to see *en route*, from Lazarus' tomb to the harem of Nourri Effendi: the Honourable Georgiana must have been quite addled. Especially when one of her guides in Jerusalem turned out to be both deaf and dumb (but a magnificent pointer), and the whole party constantly seemed to be shunted in and out of quarantine because of a particularly sordid variety of Turkish fever prevalent at the time. Still, a little discreet bribery and corruption did not go amiss on their last stop, where Mr Damer managed to pay off a Maltese official and so skip the customary weeks' incarceration in the island's lazaretto, and home the travellers went, tired but triumphant.

❧

DUFFUS HARDY, Lady (Mary) (c. 1825–1891)

Through Cities and Prairie Lands: Sketches of an American Tour. London: Chapman and Hall, 1881; pp. 320

Down South. London: Chapman and Hall, 1883; pp. 276

The chief interest of this first account lies in the exuberant description the author gives of the great transcontinental railways of North America. Going 'Down South' was fun, but the indolent charms of Florida and Georgia never matched the thrill of Lady Duffus Hardy's epic ride through the cities and prairie lands further north. She travelled with a companion from Quebec to Toronto and the Niagara Falls, down to New York, and from there to Chicago, Omaha, Sherman in the Rocky Mountains, Salt Lake City, and over the Sierra Nevada to San Francisco. After the usual spot of gold-prospecting and Christmas on a Californian ranch, the two women returned by a more southerly route through Colorado and Kansas to Baltimore and Philadelphia. The scenery was magnificent, and the people fascinating, but nothing was as impressive as the trains. They were 'travelling hotels'—great palaces of luxury lumbering through the American countryside. A first-class passenger like Lady Duffus Hardy could expect to be ensconced in a spacious private suite, and dined in a gleaming restaurant carriage from an elaborate menu and highly creditable cellar with the finest china, silver, and crystal. Periodic stops would be made at railside hotels for the stretching of legs (or, in one case, for an *en route* wedding to take place), and nothing would be spared to afford the utmost comfort. Most decidedly, it was the *only* way to travel.

❧

EDGCUMBE, Lady Ernestine and **WOOD,** Lady Mary

Four Months' Cruise in a Sailing Yacht. London: Hurst and Blackett, 1888; pp. 307, with 7 lithotints by Lady Mary and a map

This book could not have been written but by a true Victorian aristocrat: it is redolent of all the assured and stately complacency of its age. The Ladies Edgcumbe and Wood and their party were following in the honoured tradition of Lady BRASSEY and the MARCHIONESS OF WESTMINSTER when they decided to take their holiday afloat. They travelled first through France to Marseilles and then by steamship to Algiers, where the sleek 380-ton schooner *Ariadne* was awaiting them. Then followed a smooth and thoroughly luxurious saunter through the sun-dappled seas of north Africa, Malta, the Ionian isles and the Adriatic coasts. At each stop the party managed both to sightsee and to socialize (it being the Season): a delightful time was had by all. The book and its illustrations are beautifully evocative of the utterly comfortable, Elysian pleasures of a *real* holiday: the stuff nostalgia is made of.

❧

EGERTON, Lady Francis (Harriet Catherine, Countess of Ellesmere)

A Journal of a Tour in the Holy Land, in May and June 1840. For Private Circulation Only: for the Benefit of the Ladies' Hibernian Female School Society. London: Harrison, 1841; pp. 141, with 4 lithographed plates and vignettes

This is a prime example of the fashion amongst titled philanthropic ladies of producing extremely boring travel accounts under the aegis of some charitable institution. The book would as often as not be published by subscription, and its well-meaning readers limited to the same circle of worthies who had probably done the very trip themselves. As a contemporary reviewer said, 'She travelled with all the comfort and protection which station and wealth could secure her, and the smooth ways of pilgrimage now permit': it is interesting to note that even in 1840 de-luxe travel could be derided.

❧

FORDE, Gertrude

A Lady's Tour in Corsica. London: Richard Bentley, 1880; 2 vols., pp. 263/272

Gertrude Forde's is one of those accounts written by an 'unprotected lady' who has surprised herself by not only surviving but enjoying a holiday in an unfrequented corner of the globe. And Corsica was quite unfrequented enough for the writer and her two companions: 'it cannot be denied that the Corsicans, though polite and even friendly to strangers, have a weakness for shooting one-another.' The fabled bandits of the hills do exist, she finds, but there are no brigands to be found (a nice difference) and the snakes are of the non-frightening variety. So come and tour Corsica, she urges: for bold women like her 'it is the gipsy's life, robbed of its discomforts, but not of its primitive ease and romance.'

❧

GERARD, Adelaide M.

Et Nos in Arctis. London: Ballantyne, 1913; pp. 150, with 6 original photographs and a map

A rather pretentious book, this is, but interesting anyway as an account of an early tourist voyage to Spitzbergen. Adelaide Gerard was one of six ladies on a Bergenske and Nordenfjeldske Company cruise to the Arctic Circle, all of whom travelled in the breathless spirit of pioneers. The book suffers somewhat from the prevailing weather condition on the cruise, an almost unremitting blanket of freezing fog, but whenever a glimpse of land or sea *is* spotted, it is voraciously photographed and described in careful detail. To compensate for the necessary lack of topographical interest, we are provided with a history of travel to Spitzbergen, which concentrates on those few women who preceded Adelaide there.

❧

GROVE, Lady (Agnes) (1863–1926)

Seventy-One Days' Camping in Morocco. London: Longmans, Green, 1902; pp. 175, with photogravure portrait and 32 halftones

Agnes Grove achieved a certain celebrity around the turn of the century by virtue of her popular books on social fads and fashions. She was admired as a witty and well-read beauty, friend to George Bernard Shaw (what witty and well-read beauty was not?) and inspiration to Thomas Hardy (his elegy 'Concerning Agnes' was written for her). This sole travel book shows her good breeding at its best. It is delicately blasé, mentioning the mundane pursuits of a tour abroad *en passant*—the usual things like pig-sticking, polo, paper-chases, champagne picnics—and concentrating instead on an alto-gether more stylish holiday jaunt. She and her husband Sir Walter, together with his sister and brother-in-law, hit upon the idea of a camping trek to pass the long lazy days in Morocco. It was the most 'proper' tour I have come across. Lady Agnes wore an embroidered white riding-habit—'a most suitable but becoming outfit'—and rode a magnificent grey, and, remem-bering the salutary tale of one mysterious 'Miss A.', who 'went out there young and pretty and returned in two months, burnt up and damaged beyond repair', always kept a pith hat, an umbrella, and a battery of face-creams at the ready. At Casablanca they stayed at the best hotel and at Marrakesh met the Sultan in his palace; the whole undertaking was a thoroughly seemly affair. Lady Agnes, incidentally, is at the vanguard of lady travellers whose books are illustrated chiefly with pictures of themselves.

GUSHINGTON, The Honourable Impulsia (DUFFERIN, Lady Helen Selina Sheridan) (1807–1867)

Lispings from Low Latitudes: Or, Extracts from the Journal of the Hon. Impulsia Gushington. London: John Murray, 1863; pp. 98, including 23 wood-engraved plates

This is not really a travel account, but an uproarious and affectionate parody of one, by the illustrious granddaughter of Richard Brinsley Sheridan. Lady Selina's son published a book in 1856 about his travels in Iceland called *Letters from High Latitudes*, and this is her answer to it.

The Hon. Impulsia Gushington (authoress of several humorous ballads and ditties) is the apothegm of Silly Women Travellers, with too much time and money on her aristocratic little hands than she knows what to do with: 'I feel an inclination to go somewhere immediately. It must be so beneficial to the mind.' She chooses Egypt (as so many did) after reading Alexander Kinglake's book *Eothen*, and once there falls victim not only to the d——

natives but to that even more dastardly bunch, the British (and Irish!) Abroad. She is successively done out of her hotel, her luggage, and all but the ears and tail of her lap-dog Bijou by the likes of the MacFishys, the O'Whackers, and the Fitzdoldrums, before being rescued in true romantic fashion by an unctuous and nattily dressed gentleman of colour named de Rataplan; and all ends happily, just as it should.

❧

HOWARD, Lady Winefred (Fitzalan) of Glossop (d. 1909)

Journal of a Tour in the United States, Canada and Mexico. London: Sampson Low, Marston, 1897; pp. 355, with 37 halftones

Like Lady Mary DUFFUS HARDY, Lady Winefred was well pleased with what she found 'across the pond'. She set out with her brother in the autumn of 1894 (having nothing much on that season, she said) aboard the Cunarder *Lucania* from Liverpool to New York, and proceeded to 'do' all the usual places in Canada and the United States before moving down to the more exotic climes of Mexico. They concentrated on the Aztec antiquities of Jalapa and Vera Cruz, and exhausted the museums of Mexico City before travelling up to Florida and through the Blue Ridge Mountains of Virginia. Here the brother was summoned home, and Lady Winefred continued alone to Washington, DC and the coast beyond. She thoroughly enjoyed herself—whether languishing in a luxurious Pullman car or striding out with alpenstock and sketch-book in hand—and her text is full of the 'glorious', 'wonderful', 'beautiful', 'romantic', 'fascinating', and 'enchanting'. For all that, however, she is happily not subject to the usual affliction of the Victorian aristocratic lady traveller: she does not gush.

❧

LADY, a

A Sketch of Modern France. In a Series of Letters to A Lady of Fashion. Written in the years 1796 and 1797, During a Tour through France. Edited by C. L. Moody. London: for T. Cadell and W. Davies, 1798; pp. 518

We know little about the author of these letters, except that she travelled with her husband ('a military gentleman'), and that she was herself decidedly a Lady of Fashion, or, as the French officials suspiciously remarked, 'une aristocrate'. She was brave to go to France at all: Louis XVI was just three years dead at the guillotine, and it would be another six years before the Treaty of Amiens was signed, bringing temporary peace between

France and England. This Lady was unafraid, however, remarking happily as she trundled from Calais to Paris (a three-week carriage journey) that 'levity and inconsistency still mark the French character' and that the inns are definitely not what they used to be.

There was much sightseeing to be done in the capital; not just the depleted statuary and sad Notre-Dame ('now as poor as it once was rich') but, by virtue of her husband's privileged position, the École Militaire and the sessions of the two new councils legislating for the Republic. After a fortnight in Paris the travellers waded on through Burgundy—a wilderness with neither roads nor horses to ease the journey—and through Dijon to the mountains of Jura. They reached Lausanne in Switzerland for Christmas, and then doubled back over the border to Grenoble, where the letters close with their author indefatigably contemplating another trip across the Alps to Geneva. Several times the travellers were stopped on the track by diligent Republicans (spy-hunting being a current national sport), but the reams of documentation they were obliged to collect at Calais and Paris saw them through.

Appended to the main body of letters is a highly informative series of essays by the same Lady on the general state of France, covering every aspect from the military to morality.

❧

LAYARD, Mrs Granville (Gertrude)
Through the West Indies. London: Sampson Low *et al.*, 1887; pp. 168

This is a no-nonsense account of a six-month tour to the Caribbean, written to prove that it is possible—even when travelling so far afield—completely to avoid the bane of a lady tourist's life, the 'difficult journey'. Mrs Layard sailed to Tobago on a comfortable Royal Mail steamship, changing there for a colonial steamer which took her in leisurely succession to Trinidad, Georgetown (on the mainland of British Guiana), St Vincent, Barbados, and Jamaica. In each place she stayed long enough to collect local goodies to be sent home—dead birds, for example, and tamarinds and sugar cane— and to meet the local colonials. At this time, any English traveller in the West Indies (and not just there) automatically belonged to an exclusive sort of club: its members were the British Abroad and its rules included unlimited dispensation of hospitality and the provision, as far as possible, of a home from home. Consequently Mrs Layard rarely had to experience the notorious squalor of Caribbean hotels and was never lost for a suitable travelling companion: it was all too, too easy.

❧

LONDONDERRY, (Frances Anne Emily), Marchioness of (1800–1865)

Narrative of a Visit to The Courts of Vienna, Constantinople, Athens, Naples, &c. London: Henry Colburn, 1844; pp. 342, with lithographed portrait

There are two tours chronicled in this book: the first by steamer in 1839 to Portugal and Spain, and the second overland in 1840 to Austria, Turkey, and Greece. Both were family affairs. Frances Vane-Tempest was the second wife of Charles Stewart, third Marquis of Londonderry; when they married in 1819 he was acting British Ambassador in Vienna. He had also been Wellington's adjutant-general on the Iberian peninsula: he knew the ground they were covering well. There are not many surprises in the book; the marquis and marchioness visited all the usual places—except that they stayed with kings and courtiers rather than hoteliers. They may have chartered their own diligences from time to time, or taken a yacht when tired of the steamer, but like any other tourists they still gaped at the Alhambra in Granada, or tut-tutted over Countess BLESSINGTON's decaying villa in Naples. They were not even above the temptation of carving their names on the pillars of the Parthenon, 'to be perhaps read by our children, and children's children', as a souvenir.

❖

LOWE, Emily

[Anon.]: *Unprotected Females in Norway: Or, The Pleasantest Way of Travelling There, Passing through Denmark and Sweden: With Scandinavian Sketches from Nature.* London: G. Routledge, 1857; pp. 295, with 3 hand-coloured wood engravings and vignettes throughout

Unprotected Females in Sicily, Calabria, and on the Top of Mount Ætna. London: Routledge, Warnes and Routledge, 1859; pp. 265, with 4 chromolithographs

'The only use of a gentleman in travelling is to look after the luggage', claimed Emily Lowe, and as she and her mother always limited themselves (most unusually) to one carpet-bag each, they were able to leave the gentlemen, along with 'crochet and scandal, to the watering-places', and set off to the wilds of northern and southern Europe by themselves. They got on perfectly well; whenever the pair appeared over the horizon, the daughter astride her mule or pony, and the mother more decorously riding side-saddle, they were treated with unfailing courtesy, and all the respect due to English ladies abroad. The peasants queued to be their guides, and the farmers' wives to put them up for the night. There were none of the high passions and dudgeons of travelling with gentlemen, and one could flirt with whom one pleased (Emily even tried a monk in Sicily): nothing could be more gratifying.

❖

(MORRELL), Miss Jemima (1832–1909)

Miss Jemima's Swiss Journal: The First Conducted Tour of Switzerland. London: Putnam, 1963; pp. 114, with 32 halftones

Although it chronicles a very important event in the history of tourism, Miss Jemima's journal is very light-hearted, and full of fun. The original, in two beautifully hand-written and illustrated volumes, was found in a tin box in the ruins of a blitzed Thomas Cook office just after the Second World War. It bore a wryly impressive title-page: 'The Proceedings of the Junior United Alpine Club 1863 . . . by the artist to the expedition' and was identified as being the work of a cheery spinster from Selby, Yorkshire, travelling on the world's first ever conducted tour (conducted by none other than Thomas Cook himself) with her brother, cousin, and two friends. It was a three-week trip, during which the party of one hundred and twenty went by rail to Geneva, by diligence to Chamonix, on foot or by mule to the fabulous *Mer de Glace*, then on to Interlaken, Lucerne, and Neuchâtel. Their first morning of leisure, gasps Jemima, was not until 10 July (two weeks into the tour)— that earnest little man Mr Cook was quite indefatigable! At the end of her account, she lists the essentials of a package holiday abroad: a Gentleman's outfit for fourteen days, a Lady's for seven, and the total cost of the whole excursion: £19.7s.6d.

❖

MOSS, Lady (F. N.)

A Scamper Round the World. London: John Ousley [1910]; pp. 127, with 37 halftones

Lady Moss and her husband had always wanted to visit the Far East. After many years of travel they decided, in 1909, to trot the globe and linger in the hotels east of Suez. And a catalogue of hotels is just about all this book contains. They graced the rooms and restaurants of the Eastern and Oriental in Penang, of the Astor House in Tientsin, of Tokyo's splendid Imperial—and many other such shrines to the British abroad. In between stays, Lady Moss managed to notch up 27,280.09 miles (all carefully accounted for in an appendix to the book), and returned home aboard the *Mauretania* more convinced than ever that 'there is no country like my own, nor one in which I would sooner live.'

❖

NIGHTINGALE, Florence (1820–1910)

[Anon.]: *Letters from Egypt.* London: For Private Circulation only, by A. & G. A. Spottiswoode, 1854; pp. 334

Florence Nightingale in Rome: Letters Written . . . in Rome in the Winter of 1847–1848. Edited by Mary Keele. Philadelphia: American Philosophical Society, 1981; pp. 322, with 17 halftones and 2 sketch-maps

One does not usually think of Florence Nightingale as a travel writer. She never published a narrative of the 'lady of the lamp' years spent at Scutari and Balaclava between 1854 and 1856, and although she wrote copiously on India, Algeria, and Syria, we know she never visited them. But languishing amongst her phenomenal output—over 200 books, pamphlets, and articles, and 12,000 letters—are two of the most refreshing travel accounts of the period, the first published for private circulation only (and not reprinted until 1987) and the second gleaned from family letters and published over seventy years after their writer's death. What makes them so charming, I think, is their unfashionable lack of romanticism: 'I am ill at description', Florence demurred, and she never had much time for Art. They were genuinely written for family and friends' eyes only: cheerful, spontaneous, and familiar.

Travel was nothing new to Florence. She was actually born *en route*, as it were, in the city that gave her her name. The appearance of their eldest daughter interrupted one of the Nightingales' frequent and leisurely Grand Tours around Europe (their younger daughter arrived at Naples and was promptly christened Parthenope). Between 1837 and 1838 the family toured Europe again, and ten years later Florence wintered in Rome (all the rage) with her lifelong friends the Bracebridges. In 1849, again following fashion with the same companions, she sailed to Alexandria and up the Nile as far as Thebes.

That was the last holiday abroad she ever took: on the way home, during a stay with the nurses of a progressive Lutheran hospital at Kaiserwerth in Germany, she dedicated herself to the vocation that made her a legend.

❧

PEEL, (Agnes) Helen (b. 1870)

Polar Gleams. An Account of a Voyage on the Yacht Blencathra [etc.]. With a preface by the Marquess of Dufferin and Ava and contributions by Capt. Joseph Wiggins and Frederick G. Jackson. London: Edward Arnold, 1894; pp. 211, with 15 halftones and 2 maps

Helen Peel was a well-brought-up young lady, who obediently did the rounds of the 1892 Season in London but having failed (presumably) to secure a husband thereby, decided to Do Something Different next year. *Very* different: she and her companion became the first two women to sail across the Siberian Kara Sea (navigated for the first time in 1874) to the mouth of the Yenisei river and back.

Quite how the voyage came about we are not told: all we know is that when the convoy schooner *Blencathra* set sail from Appledore in July 1893 to accompany a cargo of rails bound for the building of the Trans-Siberian Railway, Miss Peel was aboard. Miserably seasick, she saw little of the voyage up past the Shetlands and Norway, preferring to sit below making 'comforters and petticoats for the poor in England', and she only started taking notice of her surroundings on the 'placid and icy' Kara sea itself. The convoy put in at various Samoyede settlements (once to drop off Frederick Jackson, the romantic British explorer preparing for an assault on the North Pole, who had hitched a lift) and these visits, along with the odd reindeer ride and walrus hunt, varied the 'perfectly delightful' days of the arctic voyage. The whole trip, including three weeks' anchorage off Golchikha, took three months, during which time (we are proudly informed) Miss Peel never changed her costume of 'a blue serge skirt, jacket to match . . . a red flannel shirt, and a straw sailor hat'.

There was a fine line to be drawn between originality and impropriety at this time—I suspect that if Helen had voiced any wish to repeat her escapade, then things would have been different. But as it was, her grit and style were greeted on her arrival home with indulgent admiration: 'That a last year's débutante', chuckled the marquess in his preface, 'should thus exchange the shining floors, wax lights, and valses of a London ball-room for the distant shores of Novaia Zemlia and the Taimyr Peninsula . . . exhibits the untamable audacity of our modern maidens.'

❖

ROBERTS, Violet, Mildred Compton, and another
A Jubilee Jaunt to Norway. By Three Girls. London: Griffith, Farran, Okeden and Welsh, 1888; pp. 204

This is a book of the Jolly Japes Abroad variety. The 'three girls' are named as Goggles (aged seventeen), Scrappit (her sister, whose first trip abroad this is) and The Counsellor (an older and wiser German Fräulein). They are, they say, renowned for their 'mishaps', and perhaps to save Queen Victoria's national pride, decide to spend the Golden Jubilee summer abroad, fashionably cruising the Norwegian coast. It is a twenty-day cruise to the North Cape and back to Trondheim, for which each has paid £30 (and the price of some good Jaeger underwear: 'it has the advantage of requiring to be so seldom washed'). From Trondheim the girls make a few trips into what they dramatically refer to as The Interior in a neat little carriole, regally dispensing silver pins from Gorringes to the clamouring natives. Then follows a slow train journey to Gothenburg, passed in giggly phrase-book blundering and gossip, and, at last, the voyage home.

❖

STALEY, Mrs

Autumn Rambles: Or, Fireside Recollections of Belgium, The Rhine, The Moselle, German Spas, Switzerland, The Italian Lakes, Mont Blanc, and Paris. By a Lady. Rochdale: E. Wrigley, 1863; pp. 217

This is a pleasant and homely account of a six-week jaunt (it's Thursday—must be Baden Baden) around Europe. It was published for private circulation only, in aid of the Rochdale Relief Fund, to be bought by Mr and Mrs Staley's Lancashire friends and relations. Their itinerary was highly organized, and seems to have unfolded without a hitch—and this a whole year before Thomas Cook dared take Jemima MORRELL and her party to Switzerland on the world's first ever 'package tour'. Perhaps Mrs Staley missed her vocation: her confidence and jollity would have made her a wonderful holiday guide.

❖

STUART WORTLEY, Lady Emmeline (1806–1855)

Travels in the United States, &c. during 1849 and 1850. London: Richard Bentley, 1851; 3 vols., 12mo, pp. 307/351/316

&c. [North America, Caribbean and Peru]. London: Bosworth, 1853; pp. 450, with a lithographed frontispiece

A Visit to Portugal and Madeira. London: Chapman and Hall, 1854; pp. 483, with wood-engraved frontispiece

The Sweet South [Spain]. London: G. Barclay [for private circulation], 1856; 2 vols., quarto, pp. 380/504, with a frontispiece plate [copy not seen]

STUART WORTLEY, Victoria (1837–1912)

[Anon.]: *A Young Traveller's Journal of a Tour in North and South America During the year 1850.* London: T. Bosworth, 1852; pp. 260, with 16 tinted wood engravings

Lady Emmeline Stuart Wortley, the second daughter of the Duke of Rutland, was an obsessive traveller. Most of her considerable literary output related to her travelling habit, although these three are the only narratives she produced: the rest was poetry. Right from her wedding journey through Europe until her death at Aleppo, Lady Emmeline was quite captivated with the business of sightseeing abroad. According to her letters (and to the book *Wanderers* written by her granddaughter Mrs Henry Cust in 1928), she travelled first of all for sheer enjoyment, but later, after her husband's and her youngest son's death in 1844, something of desperation entered into the increasingly punishing itineraries she set for herself and her daughter (later Lady Gregory, whose only travel book was written at the age of twelve). When she and Victoria visited the United States they did not confine themselves to the east coast, like good English aristocrats should, but insisted on exploring Mexico, crossing the isthmus of Panama, and then sailing down to Lima in Peru. And when they left for Portugal and Madeira soon after returning from the Americas, they travelled far further than the regulation destinations of Lisbon and Funchal. Their last journey together took them from the Crimea, where they dropped in on Victoria's soldier brother, to Egypt and the Holy Land. There their enthusiasm turned sinister: they refused guides and porters and set off for Aleppo in the blazing desert heat, quite alone. Their maid caught sunstroke and died, and both mother and daughter suffered dysentery. Lady Emmeline was killed by it, and Victoria was left to crawl away towards help as best she might. That was an episode Victoria could never bring herself to write about afterwards, and, perhaps not surprisingly, her second-hand passion for travel was laid to rest with Lady Emmeline.

❖

THOROLD, Mrs Arthur

Letters from Brussels, in the summer of 1835. London: Longman, Rees, Orme *et al.*, 1835; pp. 287

This little confection, written in the florid tradition of the Ladies' Annuals and Keepsake books of the early nineteenth century, is dedicated to the author's widowed mother who is apparently in financial difficulties: the readers' 'assistance in forwarding the sale of this work', urges Mrs Thorold, is most earnestly exhorted. Half the book is a guide to Brussels *à la* STARKE, while the rest, for some obscure reason, is all about the language of flowers. Thus we learn one moment that the fare from London to Ostend is £2, and the next that the potato is a sign of benevolence; a Belgian passport need only cost 7s. and nettles, by the way, mean naughtiness. All this makes for bizarre and somewhat unsettling reading.

❖

TUCKETT, Elizabeth (Lizzie)

[Anon.]: *How We Spent the Summer: Or, A 'Voyage en Zigzag' in Switzerland and Tyrol, with some Members of the Alpine Club: From the Sketch Book of One of the Party*. London: Longman, Green, Longman, *et al.*, 1864; oblong quarto, ff. 43, lithographed throughout (including a map)

[Anon.]: *Beaten Tracks: Or, Pen and Pencil Sketches in Italy*. London: Longmans, Green, 1866; pp. 278, with 42 line-drawn plates

[Anon.]: *Pictures in Tyrol and Elsewhere: From a Family Sketchbook*. London: Longmans, Green, 1867; pp. 313, with 28 line-drawn plates

[Anon.]: *Zigzagging amongst Dolomites*. London: Longman, Green, Reader and Dyer, 1871; oblong quarto, ff. 41, lithographed throughout (including a map)

Lizzie Tuckett's special talent was for humorous sketching: her brother Francis's was for climbing. So when he and a group of his Alpine Club friends invited Lizzie and other members of the Tuckett family to join them in Switzerland and the Tyrol in 1864, it was naturally assumed that Lizzie would be the expedition artist. What resulted from that first trip was the light and popular *Voyage en Zigzag*; later seasons amongst the mountains inspired three more books in much the same vein. The ladies of the cheerful Tuckett party kept strictly to the gentler hills and valleys of the region, stopping frequently to picnic or sketch, whilst the gentlemen went about their pioneering business in the higher Alps. The 1871 Dolomite season seems to have been Lizzie's last: after that, she married a certain Mr William Fowler only to die, sadly, at the birth of their first child.

<hr />

WESTMINSTER, Elizabeth GROSVENOR, Marchioness of (1797–1891)

[Anon.]: *Narrative of a Yacht Voyage in the Mediterranean during the years 1840–41*. London: John Murray, 1842; 2 vols., pp. 363/378, with 26 lithographs

Diary of a Tour in Sweden, Norway, and Russia, in 1827, with Letters [etc.]. London: Hurst and Blackett, 1879; pp. 297

When the Grosvenors did something, they did it in style. So when the marchioness says a 'Yacht', she really means a 217-ton vessel replete with sixteen crew, a maid, and a manservant (and a bath), and when she speaks of a 'Tour' in Sweden, Norway, and Russia she is really talking about a sort of royal progress. The voyage lasted for over a year; it was supposed to be a rest for Lady Elizabeth who by then had borne thirteen children (amongst them the first Duke of Westminster). She took her four eldest daughters with her, and with a sketch pad on her knee and Rap the spaniel at her feet, sat back on deck and enjoyed herself. It was only in the introduction to her second book, published thirty-seven years later, that she casually mentioned she had never been a very good sailor—no wonder the narrative was a little strained. One gets the impression that the Tour was a more relaxed affair: the marchioness was actually able to see the countryside of northern Europe (which thrilled her in true gothic style) and to meet several of her friends *en route* (not least Anne DISBROWE with whom she stayed in St Petersburg). Perhaps the marchioness might have been a *real* lady traveller, had she been allowed.

❧

WHITWELL, Mrs E. R.

Spain: As We Found It in 1891. London: Eden, Remington, 1892; pp. 160, with 9 sketches

Through Corsica with a Paint Brush. Darlington: Wm. Dresser [1908]; pp. 67, with 16 woodcut plates and a map

Through Bosnia and Herzegovina with a Paint Brush. Darlington: Wm. Dresser [1909]; pp. 76, with 16 woodcuts

It became fashionable during the late Victorian and Edwardian ages for travel writers to give their books titles based on a simple formula 'To *x* with a *y*'. Annie HORE went to Lake Tanganyika with a Bath Chair; with Pole and Paddle Mrs CURRIE oozed up the Shiré and Zambesi; Nora GARDNER rampaged through northern India with a Rifle and Spear, while Mrs Whitwell from Yarm-on-Tees travelled with a Paint Brush. Her books are suitably unadventurous, involving nothing more threatening than a skirmish with the lower classes in Seville ('repellent in the extreme'), a coach-ride amongst the unseen bandits of Corsica, and a yacht voyage down the coast of what is now Yugoslavia ('La la! The cold!'); as long as the view was pretty enough to paint, then she was satisfied.

6

The Means to an End

So far I have spoken only of those women who chose to travel. Although the impulses that sent them away may have been different, none pretended to be anything other than a voluntary traveller of some sort, be it pioneer, explorer, sportswoman, wanderer, or just plain tourist. But for a great many women writing books about their experiences abroad, the fact that those books might be regarded as *travel* books was incidental. This chapter concerns them—not the missionaries and professional authors making up the greatest part of 'vocational' travellers (they have chapters of their own)—but the rest: the scientists, governesses, artists, almost anything from opera-singers to spies, for whom travel was a means to an end.

Some of the genre's most colourful characters find their way in here. I suppose one has to be pretty peculiar in the first place even to contemplate a career in 'fish and fetish' in West Africa, when one has scarcely stepped beyond the threshold of one's childhood home alone before. Mary KINGSLEY managed it, though, and produced two books (*Travels in West Africa* and *West African Studies*) acclaimed for their glistening wit and solid scientific content alike. She is joined in the ranks of surprising scientists by the lepidopterist Margaret FOUNTAINE, who sped the world in vigorous pursuit of butterflies and men, and the quaint, bespectacled figure of Marianne NORTH, the botanist whose mission in life was to paint the tropical flowers and exotics of the world as and where they grew, which meant lugging her easel and oils from Chile to Borneo, and far beyond.

Such women must have been regarded by their Victorian audience as twofold eccentrics: for seeking some other fulfilment in life than the traditional one involving the family and the fireside, and for scratching around in the dank corners of the world in the search. A globe-trotting lady scientist could not possibly hope to be taken seriously either as traveller, lady, *or* scientist. Perhaps in rueful recognition of this, many such writers spiced their books with a distracting mixture of humour and self-mockery.

No such technique was needed by those women travelling as governesses. There was something rather noble about a poor young gentlewoman—which is what most governesses were; one could only admire the efforts of Emmeline LOTT, for example, to earn a living under the various (and rather delicious) disadvantages to be met with in the schoolroom of an Egyptian harem. And what Mrs LEONOWENS went through to keep her fatherless children from starvation in Siam has become a celluloid legend. Both women knew how to capitalize on the entertainment value of their stories, and their books were bestsellers.

Just as exciting is the work of such diverse journeywomen as Kate MARSDEN, who rode 'On Sledge and Horseback to Outcast Siberian Lepers' to feed them a magic curative herb and Mary LESTER, the English schoolmistress who travelled from Australia to Central America on the dubious promise of land and employment given by an alcoholic entrepreneur in a dog-collar.

Not all these 'vocational' travellers were strangely occupied and entertaining Victorian spinsters. Amongst their descendants are Isobel HUTCHISON, Jenny VISSER-HOOFT, and Jane GOODALL— career-women all, and travellers as well as authors. And for a common ancestor they have no less brilliant and mysterious a figure than Aphra BEHN, the seventeenth-century lady of letters whose visit to Flanders (where *The Fair Jilt* (1688) is set) had little to do with literature or wanderlust. It was something altogether more intriguing . . .

ANLEY, Charlotte (*c.* 1796–1893)

[Anon.]: *The Prisoners of Australia.* London: J. Hatchard, 1841; pp. 191

This book is dedicated to the London Committee of the British Ladies' Prison Visiting Association, and is principally a treatise on the position of female prisoners in New South Wales. Charlotte Anley, an author of 'moral tales' for children and adults, was commissioned to undertake what would these days be called a 'fact-finding tour'—but what then amounted to a pioneering journey—by the great reformer Elizabeth Fry, and she carried out her task with brave aplomb, inspecting womens' prisons and penal 'factories' around Sydney and Paramatta. She is typical of a certain breed of self-effacing lady traveller: unacknowledged, serious, and dismissive of the hardships of travel as a means to an end.

❖

BATES, Daisy Mary (1861–1951)
The Passing of the Aborigines. London: John Murray, 1938; pp. 258, with 17 halftones and a map

Daisy Bates's travel book, subtitled *A Lifetime spent among the Natives of Australia*, is a combination of serious anthropological study and autobiography. She was a diminutive and determined Englishwoman who came to devote herself entirely to the natives of a country half a world away from home.

Her relationship with the aborigines of the southern and western parts of Australia began soon after her marriage to an Australian cattle-farmer; when he died she returned to England and took up journalism before setting out again in 1899 as *The Times*'s correspondent investigating allegations of cruelty to the aborigines by white settlers. She stayed for nearly forty years (spending half of them in a small white tent with the birds and 'little burrowing creatures of the earth' for company) learning the language, habits, and needs of her adopted people. *Kabbarli*, they called her— 'grandmother'—and she soon became well known and loved as their nurse, healer, and their gentle champion.

Daisy's nomadic life involved several remarkable journeys. Once, she drove nearly eight hundred cattle 3,000 miles in six months, riding side-saddle all the way; another time she rode across the Great Australian Bight on a camel-buggy to attend a scientific congress in Adelaide; whenever anthropological curiosity or plain compassion called, she would answer by jamming on her hat, seizing up her brolly, and trekking wherever she was needed. As her influence began to grow—both with the dwindling aborigines and the burgeoning white Government of Australia—Daisy Bates's position as go-between became increasingly prominent, and her career reached its zenith when she was honoured (on behalf of her people, she said) by being made a Commander of the British Empire in 1933.

❖

BEHN, Aphra (1640–1689)

Oronooko, or the Royal Slave. London: Will. Canning, 1688; pp. 239

The Fair Jilt; or, the History of Prince Tarquin and Miranda. London: R. Holt, 1688; pp. 120

The Histories and Novels of the late Ingenious Mrs Behn [edited by C. Gildon]. London: S. Briscoe, 1696; pp. 416 (with a biographical notice of Aphra Behn)

There has been much scholarly discussion on the origins of Aphra Behn; where she was born, and to whom, are still uncertain, for all the biographies and critical studies of her work. This adds to the romance of the woman best known as the premier playwright and novelist of her sex in the seventeenth century. She was not just a writer. She was a traveller, too, and—most mysteriously of all—a spy.

Oronooko, Mrs Behn's most popular novel, tells the exotic story of a 'noble savage' of Surinam. 'To a great part of the Main I myself was an Eye Witness', claims the author—a common enough literary device, to be sure, but in her case it could well have been true. Snippets of external evidence and Aphra's own first biographer suggest that she spent some time in the Dutch East Indies between 1658 and 1659 with her mother and sisters. Her father, recently appointed Lieutenant-Governor of Surinam, had died on the voyage out and his family were allowed to stay in St John's Hill, a magnificent marble house there, until a passage home could be arranged. In the novel Aphra mentions the flowers and birds of the island with detailed familiarity, and is said to have made a collection of its insects which was presented to the King's Antiquary on her return.

The Fair Jilt was published in the same year as *Oronooko*, and was again set in familiar territory, this time closer to home in Antwerp. Behind its location lies an intriguing story, which begins with Aphra's husband, a Dutch merchant who became a favourite courtier of King Charles II. On Behn's death in 1665, Aphra found herself in urgent need of funds, and through her late husband's contacts in Flanders and at the English court a curious commission was arranged for her. She was to travel to Antwerp as a British agent to report on a suspected traitor there—which she did, by all accounts, with discreet aplomb, spending several months 'under cover' and supplying information to the government in a series of letters (written under the code-name 'Astrea') which still exist. For some dark reason, however, all her pleas for payment were ignored and she was forced to return to London at her own expense, even more in need of money than before. So she turned to writing—and the rest we know.

❧

BLENNERHASSETT, Rose and SLEEMAN, Lucy

Adventures in Mashonaland by Two Hospital Nurses. London: Macmillan, 1893; pp. 340, with a map

That part of what is now Zimbabwe lying between Salisbury and the Mozambiquan border used to be called Mashonaland. In 1891, newly acquired by Cecil Rhodes for the British Empire, it was considered a potential treasure-house: rich in gold, diamonds, game, and glory. Rose Blennerhassett (who narrates this book) and her colleague Lucy Sleeman were invited there by the Bishop of Bloemfontein, anxious to set up a hospital for colonists and natives at Umtali, half-way between the coast and the capital. They were obviously suitable candidates: both were maiden ladies from 'good' families who had been working in British hospitals before volunteering to nurse victims of the 1890 typhoid outbreak in Johannesburg; they had already been introduced to the rigours of African travel, and had managed to survive a year of heat and disease. With nothing better to do, the nurses signed their two-year contracts and set off from Beira (now Sofala) at the mouth of the unexplored Pungwe River for the uplands of the interior.

The 300-mile journey to Umtali took far longer than expected: local fighting, leaky boats, belligerent crocodiles, and constant calls on the women's medical services along the route all helped delay progress. And then part of the route was overland—an appalling thought to Rose who insisted that 'women had never walked in Africa' before. At last they reached Umtali, and settled there until the hospital had been built and relief nurses sent out in 1893 to take their places. During their stay they entertained several imperial luminaries on their way up to Salisbury—the celebrated shot Frederick Selous, for example, and even Rhodes himself on one occasion. Explorers and colonists came and went like malaria—only the natives were constant, obligingly filling the hospital to keep the nurses busy.

Rose closes the book with a brief account of the women's journey home, via mission hospitals in Dar-es-Salaam and Zanzibar, and the hope that one day—bitten as they are by the African bug—they will return.

◆

BROWN, Lady Richmond (Lilian Mabel) (d. 1949)

Unknown Tribes and Uncharted Seas. London: Duckworth, 1924; pp. 268, with 52 halftones

The Caribbean journey this book describes was embarked upon solely to satisfy the author's aristocratic wanderlust. She was no born traveller, she said, and was only a woman, after all: how could she hope to be taken—and take herself—seriously? But as the journey progressed, so did Lady Brown's enthusiasm and curiosity, and what might just have been another flimsy feminine travel account became a carefully researched and much-respected ethnographical study. And all because the journey was so much worse than she ever imagined. When asked if she would do it again, she replied: 'Weighing the intense thirst and burning heat, the fever and the mosquitoes, the not being able to take off clothes for days on end, even the shortage of food, I can truthfully say "Yes"—for I was not the same being—sex had disappeared.' Which all boils down to the same thing, expressed by so many women travellers in so many different ways: that to travel is to be free.

⋯⋯

CHEESMAN, (Lucy) Evelyn (1881–1969)

Islands near the Sun: Off the Beaten Track in the Far, Fair Society Islands. London: H. F. & G. Witherby, 1927; pp. 236, with illustrations throughout and a map

Backwater of the Savage South Seas. London: Jarrolds, 1933; pp. 285, with 18 halftones and 2 maps

The Two Roads of Papua. London: Jarrolds, 1935; pp. 286, with 50 halftones and a map

The Land of the Red Bird [Dutch New Guinea]. London: H. Joseph [1938]; pp. 300, with 22 halftones and a map on endpapers

Six-legged Snakes in New Guinea: A Collecting Expedition on two Unexplored Islands. London: Harrap, 1949; pp. 281, with map on endpapers

Things Worth While [autobiography]. London: Hutchinson, 1957; pp. 330, frontispiece portrait

Time Well Spent [autobiography]. London: Hutchinson, 1960; pp. 224, with one halftone

Evelyn Cheesman travelled professionally. She had wanted to stay at home and be a vet, but women were denied that training in her youth. So instead she chose entomology, and became an insect keeper at Regent's Park Zoo. In this capacity she was invited to join a zoological expedition to the Marquesas and Galapagos Islands in 1924—the beginning of a lifetime of travel in the islands of the south-west Pacific.

Miss Cheesman spent a total of about twelve years in the field, mostly on solo collecting expeditions financed by her books. There is a photograph of her in *Time Well Spent*: a frail old woman in multi-pocketed breeches, with walking stick and butterfly net at the ready, and looking perfectly relaxed. She does seem to have been quite unflappable, always sleeping in her own string hammock (often alone, as native porters mistrusted jungle spirits) and rarely taking supplies with her, preferring to rely on local resources instead. Once, during a two-year stint in the high tropical forests of New Guinea, she survived an exclusive diet of yam for five weeks, and she frequently suffered malaria (but never succumbed to it).

The success of her travels depended on Evelyn's considerable popularity amongst local tribesmen—even the most primitive of New Hebrideans. She approached them all with respect and a genuine desire to work with and learn from them. She wrote sixteen books in all: two for children, with terrible tales of cannibals, seven technical works, and the above travel accounts, and was awarded the OBE for her entomological achievements. This last was the greatest ordeal of all, she said, and it was jolly lucky it did not rain. Otherwise the Queen would have been confronted by Miss Cheesman in her only waterproof—a fetching and well-worn cape of Papuan palm leaves.

❦

CRICHTON, Kate

Six Years in Italy. London: Chas. Skeet, 1861; 2 vols., pp. 326/320, with lithotint frontispieces

Miss Crichton and her mother left for Milan in 1852. To Kate it was a thoroughly romantic journey: she had just made her début as an opera-singer in London, and was on her way to study at La Scala. Travelling there was less difficult than finding a suitable apartment once arrived, but eventually they settled down and Kate's singing lessons began. Then everything suddenly changed. Kate fell victim to Migliara (or scarlet) fever and was so seriously ill the doctors feared for her life. She did recover, but it was soon clear that she would never be strong enough to sing again, and as she couldn't yet weather the English climate there was only one thing left to do: travel. The two women covered ground not only in Italy (Genoa, Venice, Pisa, Siena, and so on) but in Germany, on the Riviera, and in Provence too, before returning to the ordinary life they thought they had left behind, at home.

❦

EYRE, Mary

A Lady's Walks in the South of France in 1863. London: Richard Bentley, 1865; pp. 436, with a wood-engraved frontispiece

Over the Pyrenees into Spain. London: Richard Bentley, 1865; pp. 361

Mary Eyre was poor and genteel, a spinster with breeding but no means, who travelled because she had heard it was cheaper to do so than to stay at home. With her bony dog in tow, she set out in the autumn of 1862 to spend the winter traipsing from one miserable boarding-house in the south of France to another. While the weather was warm enough to eat as she walked she survived on bread, peaches, and grapes; when winter came she shared a dish of meat a day with the dog. For a while this whaleboned vagabond life was rather delightful—an opinion shared by Richard Bentley, who not only accepted her description of it for publication but also commissioned a similar book on Spain. But there the novelty wore off: she was regaled not with the admiration and sympathy she might expect but with 'hooting and insult', and the more wretched she looked, as her boots began to rot and her dress to fray, the harder the children pelted her with stones. At last the time came to hobble home, stiff with lack of sleep and insect bites, in the hope that her new book's sales might buy her a modest cottage and she need not travel again: 'Huzza! Thank God!'

❧

FOUNTAINE, Margaret (1862–1940)

Love Among the Butterflies: The Travels and Adventures of a Victorian Lady. Edited by W. F. Cater. London: Collins, 1980; pp. 223, with 20 colour plates, and halftones throughout

Butterflies and Late Loves: The Further Travels and Adventures [etc.]. Edited by W. F. Cater. London: Collins, 1986; pp. 141, with halftone frontispiece

Margaret Fountaine's diaries are irresistible. She was the highly independent, highly original, and highly sexed daughter of a Norfolk parson, left enough money in 1889, when she was twenty-seven, to enable her to indulge her three great passions: travel, butterflies, and Unsuitable Attachments. From her sixteenth birthday she kept a diary; this beautifully written and illustrated record was continued until her death in 1940. Margaret was a globe-trotteress *par excellence*—perhaps the best-travelled and most zestful of them all. The pursuit of butterflies (and romance) took her to Europe, to North and South America, to Africa, the Middle and Far East, to India and Tibet, China and Japan, Australia and New Zealand. She revisited most of these, and travelled within any country as much as possible, nearly always alone but for guides and porters for her equipment. She was a serious entomologist, a great authority on her subject, but it is the romance in her life that makes her diaries so rich.

One of her first journeys was to Ireland, in pursuit of a Norwich Cathedral chorister with whom she was desperately in love. The subsequent journey to Switzerland was largely a matter of getting over the chorister, who had proved sadly unreliable. An opera-teacher in Milan, an ageing professor on the Italian Lakes, a handsome baron in Palermo, a hotel-keeper's son in Messina, a Hungarian doctor, and a Greek guide, all begging her favours, had all to be repulsed—or encouraged—before Miss Fountaine reached her destiny in Damascus in 1901. There she was wooed and won by a deliciously Eastern dragoman, Khalil Neimy: he was 24, she 39. They planned marriage, and continued planning it for the next twenty-five years, travelling the world together with a butterfly net and a bicycle each. Meanwhile it transpired at one stage that Khalil was in fact already married; Margaret began travelling alone then, and left Khalil (who seems to have become a little addled by her devotion) at their part-time home in Australia, or back in Damascus. He died in 1929.

Margaret was heart-broken, and her subsequent travels up the Amazon, on the Orinoco, to Kenya, Uganda, Singapore, and Cambodia (Campuchea) were brightened only by the new butterflies she found (although she was losing heart for that too), and the occasional nostalgically romantic encounter. She took in her stride by now the usual succession of hair-raising adventures that seemed to follow her wherever she went. Her own death came in her seventy-ninth year, in Trinidad. In her will she left her vast butterfly collection to the Castle Museum at Norwich, and a locked trunk containing her diaries, with a note: 'not to be opened until April 15th, 1978'—a hundred years after they were first begun.

GOODALL, Jane (Baroness von Lawick-Goodall) (b. 1934)

My Friends The Wild Chimpanzees. Washington: National Geographical Society, 1967; pp. 204, illustrated in colour and halftone throughout

In the Shadow of Man. London: Collins, 1971; pp. 256, illustrated in colour and halftone throughout

Jane Goodall is primarily a zoologist, and most of her publications are not on travel but on the science of ethology, or animal behaviour. But she must count not only as a traveller but as a pioneer: the work for which she became famous has meant living almost constantly on safari in east Africa for the past thirty years.

Like many others in this book, Jane always knew she would travel. If anyone asked the Bournemouth girl clutching a toy chimpanzee what she was going to do when she grew up, she would answer that she was 'going to Africa to study animals'. Her chance came in 1958, when a school friend living in Kenya invited her to stay. Before leaving Jane organized a temporary secretarial job for herself in Nairobi (she had been working as a secretary in London) so that she need not rely on hospitality to stay in Africa, and in 1960, when she was twenty-six, her real career began. She asked for a job with the curator of the Kenyan National Museum, who just happened to have been looking for an assistant to carry out field work with chimpanzees at Gombe Stream Game Reserve on the shores of Lake Tanganyika.

Previous candidates had balked at the prospect of living alone in the wild; Jane jumped at the chance. For the first three months at Gombe her mother lived in camp with her, but after this trial period (and the first of several bouts of malaria) Jane was allowed to remain alone—and has never really left. She became known as the woman who lived with the chimps—but what is so astonishing about her work is that she managed to live almost *as* a chimpanzee. She observed them, empathized with them, and even behaved like them, eating leaves and 'grooming' in their company, until they accepted her as one of themselves. She was jealous of her solitude, but reluctantly agreed to allow a photographer to record the work she was doing with the chimps in 1962. He was Baron Hugo von Lawick; they married in 1964 and later published a book about their son (*Grub, the Bush Baby*, 1970).

Jane rarely left Gombe, but did spend six terms (spread out over as many years) taking a doctorate in ethology at Cambridge, and travelled to the Ngorongoro Crater Conservation Area farther west in Tanzania whilst working with her husband on a book about the jackals and hyena there (*Innocent Killers*, 1970).

After divorce and remarriage Jane moved to Dar-es-Salaam and published her first major scientific monograph *The Chimpanzees of Gombe* in 1987. But for six months of the year she still swaps four walls and a roof for a tent on the shores of Lake Tanganyika: Gombe is her real home.

HILL, Rosamond (1825–1902) and Florence Davenport
What We Saw in Australia. London: Macmillan, 1875; pp. 438, with wood-engraved frontispiece

The title of this book smacks of the schoolgirl essay: one expects it to be written by two fluffy and twittering sisters, gushing their way through 'the land of the kangaroo'. Not a bit of it: the authors were the middle-aged daughters of a renowned judge, the Recorder of Birmingham, who had instilled in both a keen sense of social justice, and their holiday was very much a working one. Both sisters spent their lives in what would now be called the social services, as committed educational and prison reformers. Rosamund was especially fervent, travelling either with her father or alone on tours of inspection (she published *A Lady's Visit to the Irish Convict Prisons* after one such visit in 1873) and to congresses and conferences throughout Europe. When the opportunity to go further afield was offered by relatives in Australia, both sisters jumped at it. Their time was divided between Adelaide, where their family lived, and trips to Sydney, Melbourne, and

Tasmania. Whilst anyone else would have gone to 'see the sights', the Hill sisters visited every orphanage, prison, and reform school they could find. As they cheerfully admit, nothing alarming happened to them at all, even though they travelled completely unattended most of the time, and after sixteen months abroad (including a donkey ride in the Syrian desert and a fortnight in Bombay) they arrived home hale, hearty, and eager to put their new ideas into philanthropic practice.

❧

HOWARD, Ethel

Potsdam Princes. London: Methuen, 1916; pp. 295, with 12 halftones

Japanese Memories. London: Hutchinson, 1918; pp. 288, with 42 halftones

Miss Howard was a high-class 'Home Instructor', rather in the mould of Miss LOTT or Mrs LEONOWENS: a governess to foreign royalty. She was plucked from her profession as a teacher in 1895 by a letter from the court of Kaiser Wilhelm II, who had somehow heard of her masculine prowess at classics and mathematics; at first it was thought to be a practical joke, but not long afterwards Ethel found herself firmly established at Potsdam with the education of the Kaiser's six sons—Queen Victoria's beloved grand-children—in her care. Soon after leaving Bavaria because of ill health, she received an even more exotic invitation, this time to the household of the orphaned Princes of Satsuma in Tokyo. She spent seven lonely but satisfying years in Japan, taking every opportunity, as she had in Germany, to travel. The hostilities of the Russo-Japanese War somewhat curtailed these travels, but in 1907 she was entrusted with the youngest boys on a lengthy visit to Korea and China, and she frequently made the journey from Tokyo to the Satsuma country residence at Kamakura before her accolated retirement from governessing in 1908.

❧

HUTCHISON, Dr Isobel Wylie (1889–1982)

On Greenland's Closed Shore: The Fairyland of the Arctic [etc.]. With a preface by Dr Knud Rasmussen. Edinburgh: Blackwood, 1930; pp. 395, with 23 halftones and a map

North to the Rime-Ringed Sun: Being the Record of an Alaska-Canadian Journey made in 1933–34. London: Blackie, 1934; pp. 262, with 4 colour plates, 16 halftones, and a map

Stepping Stones from Alaska to Asia. London: Blackie, 1937 [reprinted as *The Aleutian Islands*, 1942]; pp. 246, with 4 colour plates, 16 halftones, and a map

Dr Hutchison was one of a small breed: the lady plant-hunter. She was a gifted Scottish botanist, commissioned by both Kew Gardens and the

British Museum to collect specimens of her favourite Arctic flora. Most of her writing comprised scientific papers and monographs, and earned her high respect amongst polar and botanical experts alike; these three travel books brought her into the public eye too.

Like other great plant-hunters—Hooker, Cox, or Kingdon Ward—it is obvious that the hunting was just as important as the plant. Isobel lived in an Eskimo village in north-west Greenland on her first solo expedition, in an area closed to all visitors except accredited scientists (hence the book's title). Her hunting excursions were made in company with the local doctor on his rounds or the village pastor in his far-flung parish. The Alaska–Canadian journey she made next was the most spectacular: she took a cargo vessel from Manchester to Vancouver, then a mixture of river boat, train, and four-seater aeroplane up to Nome, just below the Arctic Circle; she cadged lifts from a tiny 10-ton trading vessel and a local motor boat to Sandspit Island, north of Point Barrow, and then, because the ice had closed in early that year, joined a husky team to her destination of Herschel Island, camping the while in perhaps 62 degrees of frost. On the way home—just for fun—she spent two weeks at a reindeer camp in the Mackenzie Delta.

The third book is dedicated to the United States Coast Guard, who (most irregularly) allowed her aboard its 2,000-ton cutter *Chela* during a hydrographic reconnaissance of the Aleutian Islands. She 'worked afield' on as many of the islands as she could, and also visited Kodiak and the Pribilofs before moving on to Japan and home.

Not all Isobel Hutchison's time travelling was spent in plant-hunting: she collected what she called 'curios' for ethnological museums; she painted, and published poetry and volumes on arctic folklore. But her life was most simply summed up, she said, in the words of someone she met on the shores of the Bering Sea: she just went about the world pickin' flowers.

KINGSLEY, Mary H(enrietta) (1862–1900)

Travels in West Africa: Congo Français, Corisco and Cameroons. London: Macmillan, 1897; pp. 743, with 2 lithotint plates, 16 halftones, and 29 vignettes

West African Studies. London: Macmillan, 1899; pp. 639, with 28 halftones and 2 maps

The Story of West Africa. London: Horace Marshall [1899], in the Story of the Empire series; pp. 169, with a map

Notes on Sport and Travel by George Henry Kingsley: With a Memoir by His Daugher Mary H. Kingsley [autobiography]. London: Macmillan, 1900; pp. 1–206

Mary Kingsley is often thought of as the comedienne of women travellers, and the doyenne of that archetypal school who bustled briskly about the distant corners of the Empire in drawing-room dress ('you have no right to go about Africa in things you would be ashamed to be seen in at home'), having hair-raising adventures and generally being jolly.

She encouraged the image, making herself the butt of frequent jokes and delighting in the incongruity of a button-booted spinster paddling up a crocodiley river with only a naked cannibal for company, or wading through mangrove swamps to emerge with a ruff of leeches about the neck like an astrakhan collar. Her books, articles, and lecture notes abound in cheerful stories of appalling mishaps—when she fell through a camouflaged animal trap on to the twelve-inch ebony spikes below, for example ('It is at these moments you realise the blessings of a good thick skirt'), or when she innocently emptied the contents of a native purse into her hat to find 'a human hand, three big toes, four eyes, two ears, and other portions of the human frame. The hand was fresh, the others only so so, and shrivelled.' But Mary Kingsley wasn't just an eccentric out to entertain. Far from it. She was a lonely young woman, who travelled because she could not bear to stay at home, and squeezed into those travels all the spirit denied her here in England.

She had been designed by her upbringing solely for a life of domestic duty. Her father George (brother of Charles (*Water Babies*) Kingsley) loved to travel: he was physician to various peregrine noblemen, and often deputed Mary to take his place at home as best she could. She was given no formal education whatsoever during the first eighteen years of her life; there was considered to be no need for it in a life devoted to nursing an invalid, depressive mother and caring for a sickly young brother. Later on Mary was allowed to take German lessons, so that she could translate certain articles that interested George, and to subscribe to the *English Mechanic*, the better to do the plumbing and other odd jobs around the house. Otherwise, she taught herself.

When freedom came, she did not know what to do with it. 'My life has been a comic one', she wrote to a friend in 1899. 'Dead tired and feeling no one had need of me any more when my Mother and Father died within six weeks of each other in '92 and my brother went off to the East, I went down to West Africa to die.' The scant eight years between her parents' death and her own in many ways echoed the thirty years that had gone before: she cooked and kept house for her brother whenever he was in England; if any relations needed nursing then she would tend them; only after all duty had been done did she indulge herself in 'going home' to fever-sodden West Africa. Calabar, Fernando Po, Libreville, and Lambarene were all familiar to her—far more familiar than any society salon in London; she had escaped to them through the explorers' accounts in her father's library when a child. She would escape to them again now, and make herself useful (as she was trained to) by carrying on her father's studies of tropical natural history and tribal religions: 'fish and fetish'.

After a short sojourn in the Canary Islands (half-way there), Mary made her first visit to West Africa, or 'the Coast', in 1893, and spent six months travelling between St Paul de Loanda (now Luanda) and Freetown, Sierra Leone. She arrived back in England in January 1894, surprisingly still alive. Later that same year she went out again, this time for a full year, and lived as a trader swapping cloth and tobacco for ivory and rubber amongst the cannibal Fang tribe (she calls them Fans, to avoid a lurid association of ideas). Her achievements during these two short expeditions were enormous. She discovered amongst the 'oil rivers' of the Bight of Benin a new genus of fish, six new species, another only seen once before, an unknown snake, and West Africa's rarest lizard—all of which were brought home in pickling jars and proudly presented to the British Museum. She made the most extensive ethnological field studies there to date, and became a highly respected spokeswoman on the behalf of tribal Africans, damning missionaries (apart from her friend Mary Slessor) and certain colonialists alike for the 'second-hand rubbishy white culture' they forced down African throats. She surveyed the upper reaches of the Ogowé river by canoeing herself up its white waters, and then made an unprecedented trek through the jungles between that river and the 'next one up', the Rembwé (the reason *Travels in West Africa* has no map is that there *was* no map to cover her domain in enough detail, and she did not have the time to make one herself). And as a sort of holiday jaunt, she made the first ascent of Mount Cameroon (13,350 ft.) via the north-east face. The two travel accounts she produced were immediate best sellers, both for their serious scientific content and their exuberant raciness. They are masterpieces.

Mary did not die, as she expected to, in West Africa, but in the south. She had volunteered to go out there and nurse during the Boer War, and within three months of arriving at Simonstown in 1900, was dead. Her last wish was to be buried at sea, so that her spirit might drift up towards 'the Coast' and rest amongst the rivers of her home.

LEONOWENS, Anna Harriette (1834–1914)

The English Governess at the Siamese Court: Being Recollections of Six Years in the Royal Palace at Bangkok. London: Trübner, 1870; pp. 321, with 16 wood-engraved plates

The Romance of Siamese Harem Life. London: Trübner, Boston: Osgood, 1873 [first published as *The Romance of the Harem*, Philadelphia, 1872]; pp. 277, with 16 wood-engraved plates and a vignette

Life and Travel in India: Being Recollections of a Journey Before the Days of Railroads. Philadelphia: Porter and Coates, London: Trübner, 1884; pp. 325, with 24 wood-engraved plates

In 1945 Margaret Landon wrote a book called *Anna and the King of Siam*. It was based on an English governess's memoirs of life with the royal family of Bangkok, and immediately caught the public imagination. A musical followed, and then a film, *The King and I*, starring Deborah Kerr and Yul Brynner: Anna Leonowens, the governess, became a full-blown romantic heroine. Now it is difficult to tell the truth about Mrs Leonowens from the legend. She was a remarkable woman, but no more remarkable perhaps than many contemporary Englishwomen abroad; she was a fanciful writer and given to pride and prejudice—but for all that, there *is* something quite irresistible about her story.

Anna Crawford was born in Wales in 1834. As soon as she had finished school she travelled out to join her mother and stepfather living in India, and by the time she was eighteen she had already visited the Levant with friends, started learning Sanscrit, Hindustani, Persian, and Arabic, and married a British major, Thomas Leonowens. After a honeymoon tour of the Deccan she and her husband settled in a house on Malabar Hill, behind Bombay. In 1852 their first child died, and Anna began to grow ill. A change of climate (as usual) was prescribed, and the Leonowens chose Australia. It was an unhappy choice: they were shipwrecked off the Cape of Good Hope, and soon after eventually arriving in New South Wales another child was born and died. They tried London next, and then Singapore in 1856. Two years later Anna was widowed, and forced to support her two surviving children by opening a school in Singapore for officers' children. This is what led her to the court of King Somdetch Phra Paramendr Maha Mongkut of Siam (or Rama IV for short) in 1862: being an educated man himself (he had been twenty-five years a monk before becoming King), and fashionable, he demanded a European education for his favourite children. Anna was heard of, investigated, and summoned. She was given space in the harem to set up her school, and presented with sixty-seven Royal children and a floating population of wives and slaves to be taught the wisdom of the West.

The next six years were spent coping with her pupils, with the capricious king, who commanded her services as private secretary, translator, and occasionally (unsuccessfully) as concubine, and trying to exert that most precious of a governess's powers: a Moral Influence. It was an uphill task: at first Anna was frightened of the king, considering him cruel and his advisers corrupt, but she never lost the chance to petition him on behalf of downtrodden wives or slaves, and between them there grew an exasperated fondness and—as the King put it—a 'large respect'. In 1867 Anna's health broke down again, and reluctantly she was allowed to leave. She went to America and settled there for the rest of her days, regularly corresponding with her beloved royal pupils, and watching with satisfaction the abolition of slavery, new religious freedom, and sense of human justice beginning to blossom in the country whose new king owed his education to 'that tiresome, naughty and meddlesome Mem Leonowens . . . a good and true lady'.

———❦———

LESTER, Mary

A Lady's Ride Across Spanish Honduras. By Maria Soltera. Edinburgh: Blackwood [originally published in *Blackwood's Magazine*], 1884; pp. 319, with 6 halftones

I wish Miss Lester, or 'maiden Mary', had written an autobiography. The scant details she gives of her life in *A Lady's Ride* are tantalizing: whether modesty forbade her to write her full story, or something more mysterious, we shall never know.

Mary Lester was born in the Pyrenees, the daughter (and sister) of British soldiers, and grew up fluent in Spanish and English. Whilst she was still a young woman her immediate family died and like mordant Janet ROBERT-SON before her, she found herself suddenly 'heavy of heart and light of purse', and alone. Disapproving of self-pity, she took off to Fiji (of all places) and lived with a planter's family there as a governess. Later she moved to Sydney and established a modest school which somehow came to the notice of an entrepreneur in Spanish Honduras planning to establish a European colony in San Pedro Sula. The area was full of the richest pickings (he said), just ripe for the harvest: he and Mary would set up a school for colonists' children. She accepted the position of schoolmistress, with its attractive salary and promise of 160 acres of plantation land, and set off from Sydney for Central America in May 1881.

The book opens at San Francisco, where Mary is debating whether to take the steamer to Panama and reach San Pedro Sula the 'easy' way from the Atlantic coast, or to sail down the Pacific coast and cross the country on horseback. It is Hobson's choice: she could not afford the Panama route anyway. After lengthy negotiations to secure a servant and a muleteer, she set off from the port of Amapala with little more than a hammock, a mosquito-net, and a precious side-saddle to her name. The 219-mile journey north over the Honduran mountains was beset with problems. On a diet of 'nothing better than drowned hen' and sleeping a miserly few hours a night, she faced crazed mules, swollen rivers, and sullen innkeepers with admirable equanimity. After all, she thought, her school was waiting for her, and her land: all would be well when she arrived.

In fact things got worse and worse. Rumours of the Sula Colony's demise began floating up the valleys towards her as she approached: the land was sour and the pickings anything but rich. The entrepreneur who had engaged her—a Roman Catholic priest somewhat dubiously named Dr Pope—turned out to have been lately defrocked and a con-man, and the school was nothing but a figment of his drunken imagination. All Mary could do was beg the money for a fare home to England, and leave.

Almost as remarkable as the story itself is the way Mary tells it. Only once does she lose her sense of humour ('My strength is gone, there is neither fight, nor struggle, nor travel in me') and she remains impossibly positive, arriving home after her six-month ordeal 'poorer (God help me!) but wiser, and happy'. Perhaps inspired by the hapless Dr Pope, she published a temperance novel in 1888 and a year later collaborated on a Colonial Handbook for South America—and that, sadly, is the last we hear of brave 'maiden Mary'.

❧

LOTT, Emmeline

The English Governess in Egypt: Harem Life in Egypt and Constantinople. London: R. Bentley, 1866; 2 vols., pp. 302/301, with engraved frontispiece

Nights in the Harem: Or, The Mohaddetyn in the Palace of Ghezire. London: Chapman and Hall, 1867; 2 vols., pp. 308/350

The Grand Pacha's Cruise on the Nile in the Viceroy of Egypt's Yacht. London: T. Cautley Newby, 1869; 2 vols., pp. 290/320

These three deathless books are all about the same adventure: Miss Emmeline Lott's two-year appointment as (take a deep breath) 'Governess to His Highness the Grand Pacha, Ibrahim, son of His Highness Ismael Pacha, Viceroy of Egypt'. She took the position up in 1863—by what means we are not told—and after an unedifying month's delay spent 'imprisoned' in the Cairo quarters of an Egyptian banker (one of the viceroy's lackeys) and his mistress, was eventually admitted into the 'Mansion of Bliss' at Ghezire.

Emmeline's charge was the viceroy's five-year-old heir, a spoilt and cruel child who lived in the harem with his mother; she was responsible not only for his education, but—in a palace full of poisoners and jealous step-mothers—for his life. She found the whole episode utterly distasteful, and spends nearly two thousand delicious pages telling us so. The harem was filthy, its inhabitants sluttish and stupid, and its sole amusements (whoredom and hashish) somewhat removed from the usual habits of an English governess. During visits to other harems farther up the Nile in Cairo or Alexandria, Miss Lott had every opportunity to observe and comment on the thoroughly *louche* behaviour of the viceroy and his subjects, whether they be slaves, wives, or even (horror!) the governess herself—she was once enticed to His Highness' bedchamber under false pretences, to find him clad only in 'pajama drawers' and a lewd smile. Even the Royal eunuchs were suspect ('I doubted their infirmity of body and kept a watchful eye over them').

Such vigilance eventually exhausted Miss Lott, along with 'intermittent fever and cholera and a poor diet', and she was granted dispensation from her contract during a visit to Constantinople, on the grounds of ill health. She was unable to resist dedicating the literary fruits of her appointment to the viceroy and his odious son; they were of course immediate best sellers, and the romance of travel was born again.

❧

MARSDEN, Kate (1859–1931)

On Sledge and Horseback to Outcast Siberian Lepers. London: Record Press [1893]; pp. 243, with 2 facsimile letters tipped in, 25 halftones, and a map

My Mission to Siberia: A Vindication. London: Edward Stanford, 1921; pp. 54, with 8 halftones

It was Kate Marsden's misfortune that people always found it hard to take her seriously. *On Sledge and Horseback to Outcast Siberian Lepers*—even the title of her book sounds like some elaborate and gothic spoof. And the story, laced with her irrepressible sense of humour, seems impossibly far-fetched: an English spinster, in her thirties, with a history of consumption and nervous illness, realizes a dream to ride for two thousand miles across the most inhospitable regions on earth to find a magic herb reputed to cure leprosy. The journey itself smacks of Bunyan or Dante, with its images of forests aflame with subterranean fires, nights of unutterable cold lit with wolves' eyes peeping from the woods like stars, and sweltering days dark with mosquitoes. But it was all absolutely true.

Kate Marsden was born into a comfortable north-London household whose fortunes changed as she grew older (six of her seven brothers and sisters died of tuberculosis) and by the time she was eighteen, she was expected to make her own living. She started working at the local hospital and after only eight months' experience was sent out to Bulgaria to nurse casualties of the Russo-Turkish war. It was there that she first came across the sufferings of lepers, and what she called her 'Mission' was born: one day, she would try to change their lives.

Before the Mission could be realized, Kate was called to New Zealand to care for her dying sister, and spent some time nursing first at Wellington and then Nelson. A mysterious few months followed, during which she suffered, according to her biographer Henry Johnson, 'a most trying mental illness'. From her own account it seems to have been a nervous breakdown, a 'black night of the soul', which nearly cost her her faith and consequently the Mission. But she survived, and soon sailed home to make her preparations.

After enlisting support from the Princess of Wales and the Empress of Russia, and help from Florence NIGHTINGALE and Louis Pasteur, she took herself off to leper hospitals at Alexandria, Jerusalem, and Scutari, and from the latter travelled over the Caucasus to Moscow. There she gathered years' worth of stores (principally plum pudding) and supplies, found herself a gullible chaperone (one Miss Field, who spoke fluent Russian) and someone to hoist her into the sledge (too stiff as she was with Jaeger underclothes and countless layers on top to bend), and set off for Yakutsk, two thousand miles away across Siberia. It was February 1891.

Eleven months later she was back, without the chaperone (who had given up at Omsk on the way out), without the magic herb, and, she said, 'black and blue from my waist down. The knee and anklebones had come through from the constant struggling to keep on the horses. I was a mass of filth and vermin, my clothes were half glued to the skin from the constant chafing and bleeding from sores and bites . . . in fact, I was more like one of the poor lepers I had visited than anything else.'

The real purpose of the Mission had been to seek out and comfort the pitiful colonies of outcast lepers around Yakutsk, and in this she succeeded. But nothing would persuade the public of this—not the lecture-tours she undertook at home and in America, not the Fellowship of the Royal Geographical Society she was awarded (a singular honour for a lady in 1892!), nor even the open admiration of Queen Victoria herself. The story was just too fantastic to be believed. She had never gone to Siberia at all, some said; if she had, she certainly had not ridden there; and if she *had*, then she must have been having an affair with one of the horse boys. This public distrust was the cross Kate bore to the end of her life. She never travelled east again; although she had plans to visit Kamchatka, the first Mission journey had left her too weak to make a second, and she died an invalid, in Middlesex, in 1931.

❖

MARTINEAU, Harriet (1802–1876)

Society in America. London: Saunders and Otley, 1837; 3 vols., pp. 364/369/365

Retrospect of Western Travel. London: Saunders and Otley, 1838; 3 vols., pp. 318/292/293

How to Observe Morals and Manners. London: Chas. Knight, 1838; pp. 238

Eastern Life, Present and Past. London: Edward Moxon, 1848; 3 vols., pp. 336/321/344

Harriet Martineau, political economist, philosopher, and journalist, was also a serious traveller. She wrote *How to Observe* as a guide for other such travellers to explain what one should look for in a foreign country and people (its religion, moral nature, domestic state, and so on) and how one should go about one's search (the 'Mechanical Requisites' and 'Mechanical Methods' of travel).

Miss Martineau practised what she preached, and never travelled idly. Her two-year visit to America from 1834 to 1836 was a study tour: she wanted sociological evidence of the gulf between Principle and Practice (she said) in this new Land of Promise, and found it in the slave plantations of Baltimore and Georgia. *Society in America* discusses the politics of the visit, and *Retrospect of Western Travel*—by popular demand—the 'mechanics'. Throughout, travelling the length of America from Niagara to Texas, Miss Martineau hardly had a moment's discomfort (although she never courted luxury) and was equally proud to be fêted as a literary celebrity or damned as an Abolitionist wherever she went.

Two years after her return to England, she embarked on a European tour, escorting an invalid cousin, but soon Miss Martineau herself became ill and was hurried home to spend the next five years on what she confidently expected to be her deathbed.

By 1846 she felt strong and daring enough to accept some friends' invitation to join them in a trip to Egypt and the Holy Land. She was now well into her forties, almost stone deaf (there are several ear-trumpet anecdotes), and well aware of the perils likely to beset an Englishwoman in the East. 'Egypt is not the country to go to for the recreation of travel,' she advised. 'It is too suggestive and too confounding to be met with but in the spirit of study.' So in that devout spirit she defied 'face-ache', fainting fits, vermin, and robbers to venture down the Nile to Thebes and back, to Petra, then in Moses' footsteps to Palestine and Jordan—eight months' travel in all. It was an academic experience she highly appreciated as historian and philosopher, but not—nor had travel ever been for her—a thing of pleasure.

MONTEIRO, Rose

Delagoa Bay: Its Natives and Natural History. London: G. Philip, 1891; pp. 274, with lithographed frontispiece, 5 halftones, and 14 vignettes

This is a skittish tale of butterfly-hunting in the hills of Lourenço Marques (Maputo). Mrs Monteiro was familiar with the country (and rather frustratingly assumes the reader is, too): she had lived there with her husband in the 1870s, and together they built 'Butterfly Cottage', a holiday haunt overlooking Delagoa Bay. But Mr Monteiro died before they had a chance to stay there, and after five years' widowhood in England, Rose bravely returned to spread her wings alone. The book is much more about natural history than natives (although token drunken Kaffirs and sluttish, pipe-smoking laundry-women make their appearance); it is essentially a catalogue of the trials and tribulations of an amateur African entomologist.

◆

MURRAY, The Honourable Miss Amelia Matilda (1795–1884)

Letters from the United States, Cuba and Canada. London: John W. Parker, 1856; 2 vols., pp. 320/317, with a map

Poor Miss Murray was an ill-used lady. She was an enthusiast whose talents and circumstances constantly mutinied against her. She might have been a great artist, for example, and to prove it she published in 1869 the first two parts of an elephant-folio collection of vast, blowsy aquatints describing the Odenwald in Germany. But lack of money and interest strangled the project at birth and it was never finished. There is a hopeful note at the close of her *Letters from the United States* offering a series of illustrations to accompany the text on application—but nobody applied. She might have been a famous traveller, too. It was still a brave and vaguely eccentric thing to do, to sail to America in 1854 expressly 'to study Botany and Social Questions' (as one contemporary reviewer put it).

The botanizing went beautifully, as did the fish- and fossil-collecting, and the actual journey presented no problems. As a Lady of the Bedchamber to Queen Victoria, Miss Murray could be sure of exalted hospitality wherever she went, and she took advantage of her favoured position from Ottawa to Havana. It was when she arrived home and published the book that the trouble began: she should never have tackled those Social Questions. She had decided, you see, that slavery was a good thing. 'Slaves and masters do not quarrel with their circumstances; is it not hard that the stranger should interfere to make both discontented?' Although she wrote much else to temper her view, it was this central statement that made the book a best seller—for all the wrong reasons, and for a very short time—on both sides of the Atlantic. As a member of the Queen's household, Miss Murray had no business to be so outspoken—and an anti-Abolitionist at that: she was dismissed from the court forthwith.

Amelia spent the rest of her life in quiet retirement, a suitably cautious supporter of ragged schools and female education (she was, after all, a bishop's daughter), practising her painting and composing a pathetically slim volume of 'Recollections'. There is no mention of her travels in it at all.

◆

MURRAY, Mrs Elizabeth (*c.*1815–1882)

Sixteen Years of An Artist's Life in Morocco, Spain, and the Canary Islands. London: Hurst and Blackett, 1859; 2 vols., pp. 352/344, with 2 chromolithographs

Elizabeth Murray was the daughter of portraitist and engraver Thomas Heaphy, and achieved moderate success as a water-colourist herself, exhibiting at the Royal Academy every season from 1834 until the year she went abroad in 1843. She seems to have been quite a free-spirited young lady: 'A vagabond from a baby, I left England at eighteen [more like twenty-eight]. I was perfectly independent, having neither master nor money. My pencil was both to me.' Lured by a colourful advertisement, she took passage on the *Royal Tar* for Tangier (how daring!) and within the year, was married to the British consul there. Her book briskly describes the local sights and history of Morocco and Tenerife, where the Murrays moved (taking in Cadiz and Seville on the way) in 1853.

❧

NORTH, Marianne (1830–1890)

Recollections of A Happy Life: Being the Autobiography of Marianne North. Edited by her sister Mrs. John Addington Symonds. London: Macmillan, 1892; 2 vols., pp. 351/337, with 3 halftones

Some Further Recollections of A Happy Life: Selected from the Journals of Marianne North Chiefly between the Years 1859 and 1869. Edited by her sister Mrs John Addington Symonds. London: Macmillan, 1893; pp. 316, with 3 halftones

A Vision of Eden: The Life and Work of Marianne North. Exeter: Webb and Bower, in collaboration with the Royal Botanic Gardens, Kew, 1980; quarto, pp. 240, with colour illustrations from M.N.'s paintings throughout

'Three globe-trotteresses all at once!', exclaimed a guest at a London party where, by some quirk of timing, Isabella BIRD, Constance GORDON CUMMING and Marianne North all happened to be present. They were the heroines of the hour: eccentric, adventurous, and patriotic women. And along with Mary KINGSLEY, who flourished a little later, they have settled into history as the most intrepid of that peculiar species, the Victorian Lady Traveller.

Like the others, Marianne did not start globe-trotting in earnest until her domestic duties were done. After her mother's death in 1855 she became her father's housekeeper and companion, accompanying him on holiday visits to the Continent and the Holy Land and sketching avidly the while. Mr North died in 1869, when Marianne was nearly forty; it took a year, some of which was spent resting in Tenerife, for her to decide what she should do for the rest of her life.

In 1871 she set off on her famous quest: to paint as many of the world's tropical plants and flowers as she could in their own natural habitat, or, as she put it, 'at home'. She was, after all, a talented artist; she had been encouraged by her scientist friends (who included Charles Darwin and Joseph Hooker) to pursue her interest in botany, and by her traveller friends (Francis Galton, for example, and Lucie DUFF GORDON) that nothing was more enlightening than travel. So she spent the next fifteen years going around the world with easel and umbrella, with the odd brief period in London between expeditions. She circled the globe twice, in different directions, and made single or repeated visits to every plant-bearing continent. Her favourite places for painting were Borneo (where she was the unexpected but welcome guest of Ranee Margaret BROOKE), Ceylon, Jamaica, and the Seychelles—but everywhere she went, her ingenuous enthusiasm and slightly dotty appearance seem to have brought out the best in people and she unfailingly 'fell into kind helpful hands'.

In 1882 the North Gallery was opened in Kew Gardens: a building commissioned and designed by Marianne to house her growing collection of paintings. An exhaustive catalogue was compiled, prefaced by the Royal Botanical Gardens' Director, Joseph Hooker; and whenever a gap in the canon was pointed out to her, she immediately dispatched herself to the appropriate corner of the world to paint its picture. The Gallery is still a highlight at Kew: above a dado panelled with the 246 different types of wood she collected around the world, and below an elaborate cornice it took her a year to paint, hang the fruits of her travelling career: 832 oils, flamboyant and gaudy, with not an inch of wall-space between them.

Mysterious and frightening attacks of what Marianne uncertainly referred to as 'nerves' began to assail her after the successful opening of the gallery: instead of feeling invigorated, she would come home from her travels frail and tired, and she made her last journey (to Chile) during the winter of 1884–5. After that she settled in a begardened cottage in deepest Gloucestershire and devoted herself to her *Recollections of a Happy Life*—a life that ended just before the book was done.

❧

ROUTLEDGE, Katherine M. (1866–1935)

[with William Scoresby Routledge]: *With a Prehistoric People: The Akikúyu of British East Africa. Being Some Account of the Method of Life and Mode of Thought found existent amongst a Nation on its First Contact with European civilisation.* London: Edward Arnold, 1910; pp. 392, with 2 photogravures, 135 halftones and sketches, and a map

The Mystery of Easter Island: The Story of an Expedition. London: for the author . . . by Sifton Praed [1919]; pp. 404, with 2 photogravures, 134 halftones and sketches, and 11 maps and plans

Both of Mrs Routledge's books are serious, academic works: she was an amateur anthropologist, well qualified through reading history at Somerville College, Oxford and helping her husband in his scientific researches into the primitive peoples of what is now Kenya and of Easter Island, half-way between the coast of Chile and Tahiti. The Routledges financed their own expeditions, after asking the British Museum where they thought they might most usefully go, and even built their own yacht to sail them with their helpers to the Pacific Ocean. Once at their destination, they spent years combing the local archaeology and folklore for statistics to bring home to academe. Thus, even some eighty years on, *With a Prehistoric People* and *The Mystery of Easter Island* are still eminent in their field.

❧

SCHREIBER, Lady Charlotte (formerly Lady GUEST) (1812–1895)

Lady Charlotte Schreiber's Journals: Confidences of a Collector of Ceramics and Antiques throughout Britain, France, Holland, Belgium, Spain, Portugal, Turkey, Austria, and Germany from the years 1869 to 1885. Edited by her son Montague J. Guest [etc.]. London: John Lane, 1911; 2 vols., pp. 503/542, with 8 colour plates, 1 photogravure, and 98 leaves of halftones

Lady Charlotte's travels were the travels of an obsessive: she was an incurable and insatiable collector, who was likely to stop at nothing in the quest for treasures to add to the groaning show-cases of precious china and porcelain at home in Dorset. She did not begin her ransacking travels until quite late in life, after she had met and married the wealthy ironmaster John Guest and borne him ten children in thirteen years, and made a literary name for herself by publishing a translation of Welsh poetry from *The Mabinogion* in 1849. After Guest's death Charlotte married the children's tutor, one Charles Schreiber; and soon after that, with time, money, and companionship to nurture it, 'China Mania' broke out and flourished. Almost every year between 1869 and 1885 Lady Charlotte was to be found engaged in what she called one of her 'chasses', running some rich ceramic to earth—invariably ahead of the steadily growing competition—anywhere from Constantinople to Cadiz. She travelled by whatever means were to hand at the time, and never tired: the excitement of the hunt and the beauty of its quarry fuelled her considerable energies right through her sixties and early seventies until her voluntary retirement on her husband's death. The 'Schreiber Collection' was then presented to the nation, and Lady Charlotte's name went down in history as Europe's first and foremost authority on her subject.

❧

SMITH, Agnes (Mrs LEWIS) (1843–1926)

Eastern Pilgrims: The Travels of Three Ladies. London: Hurst and Blackett, 1870; pp. 328, with wood-engraved frontispiece and title-page vignette

Glimpses of Greek Life and Scenery. London: Hurst and Blackett, 1884; pp. 352, with 4 wood-engraved plates, a vignette, and a map

Through Cyprus. London: Hurst and Blackett, 1887; pp. 351, with 4 wood-engraved plates, a vignette, and a map

How the Codex was Found: A Narrative of Two Visits to Sinai From Mrs Lewis's Journals 1892–1893. By Margaret Dunlop Gibson. Cambridge: Macmillan and Bowes, 1893; pp. 141, with 2 halftones

In the Shadow of Sinai: A Story of Travel and Research from 1895 to 1897. By Agnes Smith Lewis. Cambridge: Macmillan and Bowes, 1898; pp. 261, with 21 halftone vignettes

Agnes Smith's claim to scholarly fame rested on her discovery in 1891 of the so-called Sinaitic Palimpsest: the earliest known manuscript of the Gospels in Syriac. She and her twin sister Margaret Dunlop Gibson found the treasure during a stay in an Orthodox monastery on Mount Sinai (they knew where to look, being authorities in their own right on ancient biblical manuscripts and having published several learned works on the subject); it was the climax to a travelling career spent in search of history.

The sisters first went to the Middle East after their father's death in 1868 with a friend, partly to occupy their minds, partly to sightsee, and partly to prove to themselves and others that 'any woman of ordinary prudence (without belonging to the class called strong-minded) can find little difficulty in arranging matters for her own convenience.' After their journey to Turkey, Egypt, and the Holy Land the three ladies ventured to Athens and the Peloponnese. For the next trip to Cyprus the twins went alone, and roamed around the island's historic sites (with Mrs SCOTT-STEVENSON's book as their guide) with a modest cavalcade of mules and asses, sleeping sometimes in alarmingly collapsible tents, and at others in rich Greek monasteries.

Both women married Cambridge historians, but their scholarly visits to Mount Sinai did not begin until their husbands' deaths a few years later. There were four journeys made in all, to work on and transcribe the palimpsest, during which the two English widows became the welcome guests of the bedouins with whom they travelled across the Sinai desert, and the monks of the monastery in which they studied. They were fifty-four when they eventually retired to live together in Cambridge; Margaret died twenty-three years later in 1920 and Agnes, paralysed and venerable, in 1926.

STOPES, Marie C. (1880–1958)

A Journal from Japan: A Daily Record of Life as Seen by a Scientist. London: Blackie, 1910; pp. 280, with 8 halftones

Yes, *the* Marie Stopes, the one who became Britain's youngest Doctor of Science at twenty-four, a much-published authority in the somewhat esoteric field of palaeobotany, and at last the high priestess of 'married love'. She travelled to Japan in 1907 to study the islands' fossils, she said; in fact the visit was more of a sporting expedition than a scientific one. Whilst studying in Munich in 1903, Marie had become besotted with a middle-aged Japanese academic; they had met several times since then and now Marie was going to claim him for a husband. When it became clear that the Professor preferred not to be claimed, Marie bravely stayed on and combed the island of Hokkaido—probably still reeling from Isabella BIRD's visit twenty years before—for coal and other promising palaeobotanical nuggets. She stayed for eighteen months in all, writing this rather self-consciously whimsical journal to record her many field trips (for which she daringly dressed in Japanese trousers with an elegant fan in one hand and a hammer in the other) and to distract her attention (for a little while) from love.

◆

THOMAS, Margaret (1843–1929)

A Scamper through Spain and Tangier. London: Hutchinson [1892]; pp. 302, with 18 halftones and vignettes throughout

Two Years in Palestine and Syria. London: John C. Nimmo, 1900; pp. 343, with 16 coloured plates

Denmark, Past and Present. London: Anthony Treherne, 1902; pp. 302, with 13 halftones

From Damascus to Palmyra. By John Kelman. London: A. & C. Black, 1908; pp. 368, with 67 colour plates by Margaret Thomas and a map

It is a pity Margaret Thomas's travel books are so impersonal: she must have been a most interesting woman. Born in England, she emigrated with her parents to Melbourne, Australia in 1852 and there began to develop a lifelong interest in art. She later studied in London, Rome, and Paris and exhibited at the Royal Academy, her best work being based on the travels she took all over Europe and the Middle East, from Denmark to Damascus. She was usually accompanied by a girl-friend (perhaps the same one to whom she dedicated a volume of richly erotic poetry in 1873) and together they spent month after month living on a shoe-string, just wandering and painting. Although her narrative is dry and factual the illustrations are vibrant and obviously evocative: she need not have bothered with the words at all. Miss Thomas settled in England, travelling little during the last twenty years of her life, and died (her work by now out of fashion) in Hertfordshire at the age of eighty-six.

❧

VISSER-HOOFT, Jenny (1888–1939)

Among the Kara-Korum Glaciers in 1925. With contributions by Ph. C. Visser. London: Edward Arnold, 1926; pp. 303, with photogravure frontispiece, 20 half-tones, and 2 maps

Jenny Visser-Hooft was a respected Dutch explorer and mountaineer, variously Honorary Member of the Royal Netherlands Geographical Society and the Dutch Alpine Club, and Vice-President of the Ladies' Alpine Club in London. She learned to climb (in the Alps) a year before her marriage to the doctor and diplomat Philip Visser in 1912; by way of a belated honeymoon they planned a mountaineering and exploration trip to the Caucasus in 1914—to which the outbreak of war soon put a stop.

During the war the Visser-Hoofts worked in a hospital in Petrograd, and afterwards moved to Stockholm, where Philip was appointed Dutch Military Attaché. A grand total of four expeditions to the Karakoram Glaciers soon followed, during which time the couple became first-hand experts on 'one of the most difficult fields of exploration on earth': that great, inaccessible watershed between the Hindu Kush and the Hunza Valley. Jenny took charge of botany and general planning, relishing the long marches through hazardous serac and rotten ice or along blady, unknown mountain ridges, and by the 1930s the Visser-Hoofts and that other climbing partnership the WORKMANS had effectively filled in most of the blank spaces on the Karakoram map.

After their Swedish stint the Visser-Hoofts were posted to Calcutta—the base for two more central-Asian expeditions—and in 1937 moved to Ankara in Turkey, where Jenny died soon after a nostalgic visit to the Swiss Alps, at the age of fifty-one.

7

Quite Safe Here with Jesus

UNTIL the sixteenth century to be a woman, travel, and remain respectable one had to be generally either a queen or a pilgrim. Most of us owe our travelling heritage to the pilgrims, whose journeys led them, in the abbess ETHERIA's words, 'right from the other side of the world' to the major shrines of Christendom. Indeed the very name pilgrim, derived from the Latin *peregrinus*, means stranger— or, by implication, traveller.

Chaucer's Wife of Bath was a past mistress at the art of pilgrimage:

> . . . thries hadde she been at Jerusalem:
> She hadde passed many a straunge strem;
> At Rome she hadde been, and at Boloigne,
> In Galice at Seint-Jame [Compostela], and at Coloigne.
> She koude [knew] muchel of wandrynge by the weye.

Even then (he was writing in about 1387) such worthy women from both the lay and the religious communities had been familiar figures along the pilgrim paths west of Jerusalem for centuries. All of them had found accommodation arranged in special hostels once they arrived at their destination (*if* they arrived—not all were as sturdy as Chaucer's characters); manuscript or printed plans of the local sites would be provided and a guide made available to show them round (for a price); and when they had done their devotions they would return home after a year or more's absence, shriven and assured of heaven's reward—and, if they were anything like Mrs KEMPE, with a definite taste for travel.

The natural successors to the pilgrims were missionaries, still fuelled by faith but with the extra momentum of evangelism to speed their travels. Two rather frenetic Quakers, the Misses EVANS and CHEEVERS, were amongst the first missionary women to write an account of their travels. They were imprisoned by Inquisitors on landing at the island of Malta on their way to Alexandria in 1658, and the testimony they published soon after their eventual escape to England was designed as a salutary story of how belief—even a woman's belief—can overcome the treacheries of the Antichrist. It no doubt served as a cautionary tale for travellers, too: it is a dangerous business, travel, even with God as your guide.

Danger was merely an occupational hazard for the thousands of women sent away to spread the word of the Lord during the late nineteenth century, the 'golden age' of missionary travel. Some never even survived the trek to their destination (a remote settlement in the steamy midst of equatorial Africa, perhaps, or some infidel forbidden city like Lhasa). If they were married, transport was usually arranged for them, although in the case of Annie HORE, expected to trundle to Lake Tanganyika in a Bath chair, that did not make things much easier. Spinsters Mildred CABLE and Eva and Francesca FRENCH had to organize their own vehicles on the 'itinerations' that took them five times across the Gobi Desert: they rigged up a couple of bizarre-looking carts, as dire objects of curiosity to the local inhabitants as the ladies were themselves.

The Misses Cable and French built up a rapport with those inhabitants, but others were not so fortunate. Susie RIJNHART's tale is the saddest: her missionary husband was killed by Tibetan bandits, leaving her to cope in a hostile wilderness on her own, dazed with grief not only for him but for her baby son who had died a few weeks before. Indeed, many—perhaps most—children born to missionaries in the field either failed to survive or had to be sent home for safety: an added tax on the zeal of this extraordinary band of women.

All this said, the books written by them are amongst the most uplifting and exciting of all travel books. Despite the nature of the journey the description of it is rarely sanctimonious; despite the physical and emotional stresses involved and, often, the apparent lack of success in terms of 'converts', there is a spirit of confidence and affectionate humour which, when allied with the missionary's other stock-in-trade, the ability to tell a good story, proves often quite irresistible.

BIRD, Mary Rebecca (1859–1914)

Persian Women and their Creed. London: Church Missionary Society, 1899; pp. 104, with halftones throughout

Mary Bird—'Tiny' to her friends—was a shining example of that selfless and doughty race of English spinsters who chose to devote their lives to the Lord. Sunday-school teachers told her story as an inspiration to the young, and mission directors publicized it to win recruits for the Field.

It was certainly a remarkable story. Miss Bird joined the Church Missionary Society in 1891 at the age of thirty-two, hoping to be sent to Africa but soon finding herself instead on the way to Persia (Iran), where she had been sent by the Society as its first-ever female worker. The arduous journey to Jolfa took a month, and was crowned by a 500-mile camel-ride through a strange country whose exotic and benighted population either spat at or ignored her. Once she had arrived, however, and set herself up as the local quack (her own word), things became easier.

The women were beguiled by this minute, pallid-looking woman's ability to cure their illnesses with a prayer and a bottle of quinine; curiosity soon turned to trust, and trust to love. It was mutual. Mary was not one of the betrousering and 'civilizing' brand of missionary: at home in her Jolfa dispensary and on her extensive travels with Whitey the donkey into the outlying areas, she grew to respect (if not condone) the Muslim traditions of the Persians, and was able to explain the Christian faith with a potent mixture of deference and authority. Her work in Jolfa paved the way for the first (official) female medical missionary to be sent out there in 1897, enabling Mary to come home to care for her ailing mother. Six years later she was back, in Yazd this time, and there she stayed until her death 'in action' (of dysentery) at the age of fifty-five.

❖

BROAD, Lucy

A Woman's Wanderings The World Over. London: Headley Bros. [1909]; pp. 189, with 14 halftones

Miss Broad travelled as an evangelist for the Women's Christian Temperance Union (or WCTU). She took the pledge herself when still a child, and spent her youth at home in a Cornish village working for the Band of Hope and caring for her sick mother. The familiar pattern of the spinster daughter followed when on her mother's death, Lucy was left with freedom, time, and a little money. Like many others—Isabella BIRD and Dervla MURPHY amongst them—she decided to travel.

In 1897 she accompanied an ailing friend to the Riviera for ten months;

the friend recovered and Lucy journied on over the Apennines to Florence *on her bicycle*. In 1898 she sailed for South Africa (the bike aboard) and spent three years in and around Durban, as a Methodist missionary and lecturer for the WCTU. She awarded herself two cycling holidays during this time: one in Madagascar and the other in neighbouring Mauritius. A two-year tour of Australia followed next (with brief asides in Tasmania and New Zealand), and a voyage home via Samoa and Fiji. Finally, a WCTU Congress at Boston in 1906 and another in Tokyo rounded off a proud reckoning of 100,000 miles in ten years, and over a thousand public meetings all over the world.

Miss Broad's life was a singularly happy combination of purpose and pleasure: the excitement of evangelism and preaching was only matched by a 'thirst for speed' and love of 'escapades' ('we ought not to have any I suppose, respectable middle-aged women like ourselves'). These escapades included native attacks on lonely African bicycle rides, getting lost on the brink of Tasmanian precipices at dead of night, and being washed away whilst wading across a swollen tropical river, losing not her life but her brolly, cape, handbag, and (most inconveniently) one boot. Such a zestful missionary must have been hard to resist.

❖

CABLE, (Alice) Mildred (1878–1952) and FRENCH, Francesca Law (1871–1960)

Dispatches from North-West Kansu. London: China Inland Mission, 1925; pp. 73, with 7 halftones, endpaper sketches, and a map

Through Jade Gate and Central Asia: An Account of Journeys in Kansu, Turkestan and the Gobi Desert. London: Constable, 1927; pp. 304, with 13 halftones and a map

The Challenge of Central Asia: A Brief Survey of Tibet and its Hinterlands. London: World Dominion Press, 1929; pp. 136

Something Happened. London: Hodder and Stoughton, 1933; pp. 320, with 6 halftones and a map

[with Evangeline FRENCH]: *A Desert Journal: Letters from Central Asia*. London: Constable, 1934; pp. 261, with 16 halftones and a map

The Gobi Desert. London: Hodder and Stoughton, 1942; pp. 301, with 3 colour plates, 49 halftones, and a map

Journey with a Purpose. London: Hodder and Stoughton, 1950; pp. 192, with 19 halftones and 4 maps

'The Trio', as they called themselves, are probably the best-known twentieth-century missionaries of all. Mildred Cable and Evangeline and

Francesca French all worked for the China Inland Mission (the CIM) before, during, and for forty years after the horrific Boxer Rebellion of 1900. Most of their time was spent travelling—three bespectacled 'foreign devils' together—in the vast desert areas of Chinese Turkestan.

Evangeline was the first French sister to be sent to China (in 1893) and she suffered the worst of the Uprising of the Righteous Fists, escaping to England in 1901 after a 50-day journey from Kie-Hiu to Hankow. She returned later that same year with a junior helper, Mildred Cable. Francesca French joined them after her mother's death in 1909. Their first mission together was to set up a teaching complex at Hwochow in the north-western province of Shanxi; it was there that they heard—and answered—'the call'. The call required them to take up a roving commission to travel beyond the westernmost station of the CIM in Kansu to spread the word of God beyond; they obeyed by setting out together in 1926 on the first of their astonishing evangelical journeys.

During the next fifteen years they crossed the Gobi desert five times, travelling often at night and always either on foot or on their two carts, christened the *Gobi Express* and the *Flying Turki*. They dressed in Chinese gowns, carried supplies of Bibles and a portable harmonium, and left the moot question of their survival to the Lord. Survive they did, to write eagerly and humorously of their nomadic lives as the most sympathetic and gentle of proselytizers.

Although they retired from the Field in 1941, the ladies carried on travelling, visiting missions in New Zealand, Australia, and India on behalf of the British and Foreign Bible Society in 1947, and in South America three years later. Mildred was 72 by this time, Francesca 79, and Evangeline 81: it was their final journey together.

❖

CRAWFORD, E(mily) May (d. 1927)

A Record of Medical Missionary Work and Travel in British East Africa. London: Church Missionary Society, 1913; pp. 176, with coloured frontispiece, 23 halftones, and a map

It was a missionary's lecture in Richmond, Surrey, that moved May Grimes to offer her life to the South Africa General Mission in 1889. She was a poet and composer of hymns—well known and respected in Victorian Church circles—but feeling paper praise was not enough, she determined to further 'the British Empire's Evangelistic Spread' abroad, and travel. She first went to Pondoland (Transkei) in 1893; after her marriage to Thomas Crawford, a Canadian doctor with the Church Missionary Society and admirer of her poetry, in 1904, she ventured further north with him to spend the next eight years travelling and working amongst the Kikuyu and Embu tribes of Mount Kenya. Dr Crawford established local hospitals, sometimes treating up to four hundred patients a day, and May taught the children. They occasionally went on 'medical itinerations' to the wilder tribes of the regions beyond the Tana River, where May would often be the first white woman ever seen. She loved it all, but sadly her own illness forced the Crawfords to resign from the CMS a year after this travel account was published, and they were never able to return to Africa again.

❖

DAVIES, Hannah

Among Hills and Valleys in Western China: Incidents of Missionary Work [etc.]. With an Introduction by Mrs Isabella Bishop. London: S. W. Partridge, 1901; pp. 326, with 49 halftones and a map

I include this account because it is one of the best examples of a journey fuelled entirely by evangelism. Most missionaries might confess to curiosity of one sort or another, but Miss Davies (of the China Inland Mission) seemed to have none—only an impelling and consuming desire to spread God's truth. Isabella Bishop (née BIRD) prefaced the book in her capacity as mission-hospital builder rather than traveller, and amidst the zeal of Miss Davies' account it is not easy to appreciate the scale of her own travels. She left England in 1893, but due to the rumblings of the Japanese war and local rioting in China, she did not arrive at her remote station in Sichuan until February 1896, frequently being forced to retrace her steps and broaching the fantastic Yangtze gorges in a ridiculously frail-looking river boat three times. Once there, there were the regular and gruelling 'itinerations' to be faced, which involved travelling as far afield as possible and back in a week, reaching as many heathen ears as she could. The Boxer Rebellion in 1900 drove Miss Davies home, in common with hundreds of those missionaries lucky enough to survive—Evangeline FRENCH amongst them—and that last journey was perhaps the most difficult of all. Missionaries such as this might be travellers by default, but are often as sorely tried and tested as any real explorer.

❖

ETHERIA (or Egeria) the Blessed

The Pilgrimage of St Silvia of Aquitania to the Holy Places c.385 A.D. Translated with an Introduction and Notes by John H. Bernard. London: The Palestine Pilgrims' Text Society, 1891; pp. 150, with 2 maps and a plan

Egeria's Travels. Newly Translated with supporting documents and notes by John Wilkinson. London: SPCK, 1971; pp. 320, with maps and plans throughout

Just over a century ago, a copy of a description sent home by a fourth-century pilgrim from the Holy Land was discovered in an Arezzo monastery. The copy, made by a seventh-century monk named Valerius, was a fragment of twelve pages bound in at the back of another work in the monastery library, chronicling just four months of what its author mentioned was a three-year pilgrimage. It started and finished tantalizingly in mid-sentence, was written in execrable Latin (faithfully reproduced by the scholar Valerius) and, what is more, was by a woman. She was a Blessed Lady, said Valerius, from the western shores of Europe (probably an abbess from Spain or south-western France), and he had copied the account she had kept for her Sisters to show what heroism, energy, and devotion she showed in making her remarkable journey. He didn't name her; she has been supposed in turn to be the fourth-century ascetic St Silvia (although there is nothing at all ascetic about her attitude), the daughter of Emperor Theodosius the Great of Constantinople, and finally the abbess Etheria, or Egeria, whom she is likely to remain.

Her journey, undertaken between 381 and 384, does indeed seem to be remarkable, though the pilgrims' track from Europe to the East was even then a well-worn one. Personable as the description is, Etheria never mentions danger or discomfort, even though she can tell a bishop she meets that she is 'right from the other end of the earth'. She carried the Bible as her only guide, and appears to have visited all the requisite holy places, from the tomb of Job to Rachel's well, and Moses' burning bush to the pillar of salt that was once Lot's wife (even she, admits Etheria, cannot pretend she actually *saw* the salt, but the place where it had been was carefully pointed out to her by an enterprising guide).

Etheria's particular interest involved the wanderings of the Israelites into Egypt, to which end she mentions travelling twice into Egypt, both times checking her geography against the Book of Exodus. We learn much of her itinerary by inference or passing mention in the extant account; but there is a full description of an ascent she made of Mount Sinai. Pilgrims (those who were up to it) were led on foot round the several peaks of the mountain until they reached the base of the summit peak; then they must climb straight upwards, instead of spiralling, to the church at the top, where they would be rewarded with the Eucharist and the most breathtaking view of their lives. Etheria revels in it all.

Not all of the fragment we have deals with Etheria's travels—part of it gives the first extant description of certain rites of the Christian Church in Jerusalem. Here is the first mention—noticed and puzzled over by Etheria—of the use of incense in Christian worship, of the Festival of the

Purification of the Blessed Virgin Mary, and of Palm Sunday, as well as a close description of all the offices and celebrations of Holy Week and Easter.

Etheria's importance can hardly be overestimated, not just as the first woman travel writer; she should be the patron saint of them all.

❧

EVANS, Katharine (d. 1692) and CHEEVERS, Sarah

This is a Short Relation Of some of the Cruel Sufferings (For the Truth's Sake) of Katharine Evans and Sarah Cheevers, In the Inquisition in the Isle of Malta [etc.]. London: for Robert Wilson, 1662; quarto, pp. 104

In 1647 one George Fox, a Leicestershire preacher, founded what he called the society of 'The Friends of Truth' (nicknamed soon afterwards the Quakers), a small religious body dedicated to listening to the word of God within themselves, with none of the trappings of 'steeple-houses' and the like, and to spreading their plain and impassioned beliefs to as many people as they could reach.

Two of the new faith's earliest overseas evangelists were the English-women Katharine Evans and Sarah Cheevers, who declare in their remarkable story that as soon as God asked them in 1658 to travel to Alexandria (an unheard-of journey for two respectable lone women to make in the mid-seventeenth century), they immediately scurried to Plymouth and secured their passage to the lands of infidelity. After a desperately rough crossing of thirty-one days they reached the Italian port of Leghorn, where they found places on a Dutch vessel bound for Cyprus and Egypt by way of Malta. They arrived at Valetta on a Saturday evening, and since there could be no question of setting sail on a Sunday, the two women gathered up their printed Quaker homilies for distribution on shore and made for the British Consul's house. At this time Malta was clenched firmly in the teeth of the Jesuit Inquisition and riddled with religious mistrust: the strange English-women were noticed handing out distinctly un-Catholic tracts, and when they refused to bow to the high altar of the convent chapel they visited, it was decided to haul them in as the devil's disciples.

They were kept at the embarrassed Consul's house for three months pending 'examination', and then sent down to the Lord Inquisitor's dungeons. There the Jesuits separated their captured witches (for so they were called—how else could their crazy plans to travel towards the Holy Land be explained? They weren't men; they were not even pilgrims: they *must* be witches). Each was told the other had confessed her evil and apostatized, but neither gave in, chanting prayers and hymns and visions, refusing food, and wearing out the gaolers with their prodigious spiritual energy. They were eventually and grudgingly released three years after their arrival in Valetta (after intervention from England) and bundled home with a company of Knights of Malta bound for London on a King's frigate, more vigorous and mature in their belief than ever and vowing to carry on with their mission wherever God wished, as soon as He gave the word.

EYTON-JONES, Theodora

Under Eastern Roofs. London: Wright and Brown [1931]; pp. 300, with 49 halftones and endpaper map

This is a charming and affectionate account of most unusual sojourn: an English girl's stay in a Greek Orthodox monastery in Bethlehem. It came about as a result of the Orthodox patriarchs' and prelates' visit to the Bishop of London in 1895: 'Theo' was the daughter of one of the bishop's clergy and had been enlisted to pour the tea at a Fulham Palace garden party. She was enchanted by the kind and dignified 'wise men of the East', and determined to study their Church more closely *in situ.* The opportunity arose five years later. Now a charity and church worker, Theo was invited to visit the Archbishop of the Jordan; the chance of being ambassadress not just for 'English womanhood' but the Anglican Church as well was too good, she thought, to miss. She travelled alone to the Archbishop's monastery in the Holy Land, and after several weeks there retraced her steps as far as Egypt for a spot of sightseeing before sailing home by way of Smyrna and Athens. The whole journey seems to have been an idyll of innocence and sympathy; the book is quilted with ecclesiastical prefaces and epilogues all praising its author (newly created Knight of the Holy Sepulchre—a singular honour for an English girl!) and acknowledging her gentle contribution to East–West Christian unity.

FISHER, Ruth B. (1875–1959)

On the Borders of Pigmy Land. London: Marshall Bros. [1905]; pp. 215, with 32 halftones

Twilight Tales of the Black Baganda. London: Marshall Bros. [1911]; pp. 198, with 24 halftones

The Church Missionary Society sent Miss Ruth Hurditch out to Uganda in 1900. She was just twenty-four, and spent much of the sixteen-week journey from England to Toro suffering from acute homesickness. Once arrived, however, there was too much to do to pine for the 'old country': her life assumed the gruelling round of teaching, nursing, visiting, and 'tramping' common to all young CMS recruits, as long as they kept their health. She describes the setting up of the Toro mission, including the building of a cathedral in which she and fellow missionary Mr Fisher were the first to marry in 1902; she plots their 700-mile honeymoon through the four kingdoms of Uganda, and the various evangelistic 'tramps' she took—sometimes for weeks at a time—on board a Muscat donkey or, more surprisingly, a bicycle. The gentle pygmies were her favourite converts, and the 'Black Baganda' tribesmen of Bunyoro, where she and Fisher settled in 1904, her favourite story-tellers. Ruth's purpose in writing was as much to tell these stories of indigenous Ugandan culture as to proselytize: her books are unusually free of missionary propaganda, being 'the result of an insistent endeavour to make the country . . . yield up its own secrets, and to reveal the story of its peoples and their beliefs, before the white man trespassed on their domains'.

HASELL, (Frances Hatton) Eva (1886–1974)

Across the Prairie: A 3,000 mile tour by two Englishwomen on Behalf of Religious Education [etc.]. London: SPCK, 1922; pp. 115, with 18 halftones and a map

Through Western Canada in a Caravan. London: Society for the Propagation of the Gospel in Foreign Parts, 1925; pp. 254, with 24 halftones

Canyons, Cans and Caravans. London: SPCK, 1930; pp. 320, with 26 halftones and an endpaper map

These three books chronicle the amazing travels of Eva Hasell, one of the Lake District's most determined daughters. Eva grew up in Cumbria and was trained as a missionary at St Christopher's College in Blackheath, London; it was there that she first learned about the desperate dearth of religious education there was in the prairie provinces of western Canada. During the First World War she learned to drive as a VAD and then made her first trip to the degenerate expanses of Saskatchewan, Manitoba, and Alberta. Her idea was to drive across the prairies—all thousands of miles of them—in a specially built caravan named after a saint and loaded with books and tracts and other missionary knick-knacks: if the wide open spaces of Canada were too wide open for parishioners to visit their church regularly, then the church would just have to visit them instead.

Every year found Eva out in her 'vicar taxi' (frequently mistaken for the booze waggon or a Black Maria) puttering across the prairies with a lady companion and stopping every now and then to hold one of the Sunday-school sessions which soon became the highlight of the local season. And during the few months she spent in England each winter, she would first regale her lecture audiences with anecdotes of travel—what to do when St Vincent gets stuck crossing the railway line just as the Calgary Express is due, perhaps, or how best to welcome a visiting grizzly bear at breakfast time—and then, catching them unawares, she would squeeze them of every penny they had. The money was not just for the caravans and their equipment (by now there were several), but for the hundreds of permanent Sunday schools Eva set up in Western Canada, and for miscellaneous expenses like road-mending, which had often to be done as one went along, or paying for hospital treatment when one of the numerous accidents that befell the motorized missionaries occurred.

Eva retired—reluctantly—to her family home near Penrith after the Second World War. She was awarded the MBE for her services to the people of western Canada and the honorary degree of Doctor of Divinity; and even now, I am told, there are people out in the back of some Canadian beyond who still remember the excitement of the arrival of 'the Van' each summer with Eva and her companion inside: 'the heavenly twins', they called them.

HORE, Annie Boyle (b. 1853)

To Lake Tanganyika in a Bath Chair. London: Sampson Low, Marston, *et al.*, 1886; pp. 217, with 1 photograph, 1 halftone, and 2 maps

This must surely be the most intriguing title of any travel book. It is a very serious account: Annie Hore was still on the shores of Lake Tanganyika when she wrote it, the sole white woman on a mission station there and teacher of all the local children. It was not primarily as a missionary that she set out, however.

Annie's husband was a surveyor sent to Ujiji to map the shores of Lake Tanganyika and study the ten nations around its coast. He had been there for two years before Annie and their infant son joined him to help set up a mission station on Kavala Island, just off Rwanda. Her journey was to be an experiment: Hore was determined to prove that wheels could be used on the track from the Zanzibar coast inland to the lake. Any wheels would do: 'if he could succeed in getting no other vehicle, he would at least take his wife to Ujiji in a wheelbarrow.' It did not quite come to that: after a number of false starts, Annie set off on her journey into the interior in a wicker Bath chair, complete with hood and wheels, and an emergency bamboo pole for when the going got too rough.

In fact the wheels were totally impracticable, and Annie was carried all the way, suspended from the pole in sedan-chair fashion. This was uncomfortable enough: to make things worse her baby suffered dangerously from fever, and would be carried nowhere else but on his mother's lap. Annie was ill too, but unlike many previous travellers through the Darkest Continent, she survived the journey—the first white woman to do so. Her book is so unassuming and sanguine, it is difficult to appreciate what she is really saying when she casually mentions the 'strange black shapes' along the track that turn out to be the dead bodies of previous caravans, or when she artlessly notes that the 830-mile trip from the sea to Ujiji was accomplished in a record ninety days: it was an astonishing journey.

❖

KEMPE, Margery (*c.* 1373–*c.* 1440)

The Book of Margery Kempe 1436. A Modern Version by W. Butler-Bowdon [etc.]. London: Cape, 1936; pp. 385, with facsimile frontispiece

The Book of Margery Kempe. The Text from the Unique Manuscript . . . Edited with an Introduction and Gloss by Professor Sanford Brown March with . . . Hope Emily Allen [etc.]. London: Early English Text Society, 1940; pp. 441

'Here begynnyth a schort tretys and a comfortabyl'. So Margery Kempe opened the first autobiography in the English language: an exuberant and candid narrative of her life as one of the most colourful of medieval pilgrims. The 124-leaf document she dictated to scribes between 1436 and 1438 was for centuries housed in the Carthusian priory of Mount Grace in North Yorkshire, and not discovered again until the 1930s, in the library of an eminent book-collector; it eventually found its way to the British Library, where it remains. It is an account of Margery's travels (what she called her visits 'to certeyn places for gostly helth') throughout England and to the principal shrines of Christendom: Jerusalem, Rome, and Santiago de Compostela.

Before the mystical revelations that set Margery on the road in 1413, she had been quite happy to live the comfortable life of a mayor's daughter and merchant John Kempe's wife in King's Lynn (then Bishop's Lynn) in Norfolk. But after the birth of her first child, she became frighteningly depressed (she called it madness) and tried to commit suicide. A vision of Christ saved her life, and she promised in return to devote it to Him. It was never easy: although she persuaded her husband to accept her vow of celibacy, somehow thirteen more children seem to have been born and Margery admits that she was sorely tempted once or twice to be unfaithful not only to God but to John too. And although she dutifully trekked to all the holy places of England (Canterbury, Lambeth, Lincoln, and York) she seemed to arouse suspicion rather than inspire piety wherever she went. This was largely due to the bouts of weeping and wailing that accompanied her frequent visions, her unlaymanlike love of discomfort and mortification, and her embarrassing habit of pointing out others' faults, roundly rebuking bishop and beggar alike for any unseemly behaviour. If not a Lollard, they said, she was certainly insane—and she was regularly arrested for one crime or the other.

In the autumn of 1413 Margery left Yarmouth for the Holy Land on the first of her excursions abroad. The details of route and means are sketchy, but she appears to have walked a good deal of the way from Zierikzee on the Dutch coast through Konstanz and Bologna to Venice, where she stayed for thirteen weeks awaiting passage (probably via Jaffa) to Jerusalem. Once there she visited all the sacred places that ETHERIA would have seen a thousand years before, and rode to Bethlehem on an ass's back; she saw Lazarus' grave and what she calls 'Mount Quarentyne' (the Mount of the Temptation, a quarantine being forty days) before travelling back through Damascus and probably Tyre or Beirut to Venice again. From there she followed God's directions to Rome, via Assisi, and stayed there for Christmas 1414 and the following Easter. The voyage home across the North Sea had to be made in an old fishing smack: she had missed the pilgrims' ship, and could neither afford (having given all her money away) nor wait to catch the next one.

Margery's second journey abroad was to Spain, a week's sail from Bristol; after this she stayed in England looking after her own health during an eight-year attack of 'flux', and that of her ageing husband. After his death she was joined in Lynn by her widowed daughter-in-law. She was from Danzig (young Kempe had been a Baltic trader), and before long had decided that life alone in 'Dewchland' was better than life with an embarrassing mother-in-law in Norfolk. She was not to be rid of Margery yet, however: by now this 'elde woman' was as confirmed a traveller as she was a mystic—and she had never been to the Baltic. Fitting in a quick trip to Walsingham, she accompanied daughter-in-law in 1433 to Ipswich and from there to Danzig (Gdansk), where she stayed for five or six weeks. The journey home alone was even more perilous than usual, thanks to the King of Poland's forces then invading parts of northern Europe. Margery's ship got as far as Stralsund on the German coast; its passengers had then to make their way overland to Aachen and eventually Calais.

Margery survived, and even found the strength to fit in short pilgrimages to Canterbury, London, and Sheen before travelling home to Lynn and—in accordance with God's instructions—dictating her travel account for posterity. The account is surprisingly frank about the difficulties Margery's mystic gifts caused her fellow pilgrims. The poor woman was usually left to travel alone: pilgrimages were fashionable affairs for both laymen and religious (such as Chaucer's Wife of Bath or his Prioress) and Margery's *outré* behaviour was definitely bad form. Her near-incessant wailings (what she calls her holy 'cryingys') and well-publicized mystical encounters frightened off her companions and as often as not, admits Margery cheerfully, 'thei wold not go wyth hir for an hundryd pownd'. Still, she managed with the odd understanding guide and charitable friend to live the bizarre life God had told her to and to record it for His glory and—incidentally—her own.

❧

McDOUGALL, Mrs Harriette (c. 1817–1886)

Letters from Sarawak: Addressed to a Child. London: Grant and Griffith, Norwich: T. Priest, 1854 ['Third thousandth' (same year as first edition, none earlier located)]; pp. 144, with 4 wood-engraved plates and 3 vignettes

Sketches of Our Life at Sarawak. London: Society for Promoting Christian Knowledge [1882]; pp. 250, with 4 wood engravings and a hand-coloured map

In Winchester Cathedral is an unobtrusive memorial tablet to Harriette McDougall: 'She first taught Christ to the women of Borneo', it says. Harriette went to Sarawak with her husband in 1847, leaving the son to whom her *Letters* were addressed at home. They were funded by its charismatic Rajah Sir James Brooke to build the country's first church, mission-house, and school. It was a highly romantic destination—imagine the Union Jack flying bravely over a new-found sovereign land of savages—and the McDougalls enjoyed their early years there. Harriette taught at the school and collected a household of half-caste and refugee Chinese orphans; she had children herself, and the novelty of Christianity seemed to be catching on well amongst the Muslim Malays and head-hunting Dyaks of Kuching and thereabouts. In 1850, however, things began to go wrong. Harriette lost three children in fifteen months (not surprisingly falling ill herself) and the Borneo Mission funds were running out. The McDougalls came home to deliver their surviving baby into English safe keeping and find a new sponsor for their mission work; by 1854 they were back in Kuching. During the five years until their next visit home the newly appointed Bishop of Labuan and his wife immersed themselves in the business of evangelism—so much so that they hardly noticed the horrific Chinese insurrection of 1857 until the rebels were chasing them up the Morotabas river from Kuching; they were forced to live in a lifeboat with the few other European inhabitants of the neighbourhood for several days until the situation calmed down.

For the rest of their time in Sarawak, the McDougalls' life was punctuated by recurring illness, the birth and death of more children, the odd episcopal visit as far afield as India and the Philippines, and occasional holidays in a local mountain cottage or with friends in Singapore and

Penang. They finally returned to England in 1866, both of them riddled with malaria, and settled into a quiet and well-deserved Cathedral-close life. Harriette's two books were hugely popular—not just as missionary adventure stories, but as the first easily accessible descriptions of life in Rajah Brooke's new kingdom.

❦

MACPHERSON, Annie (*c.*1824–1904)

Canadian Homes for London Wanderers. London: Morgan, Chase and Scott (for *The Christian* magazine) [1870]; pp. 61, with a woodcut frontispiece

Summer in Canada. London: Morgan and Scott [as above] [1872]; pp. 62

Lowe, Clara M. S.: *God's Answers: A Record of Miss Annie Macpherson's work at the Home of Industry, Spitalfields, London, and in Canada* [includes passages from letters and journals of A.M.]. London: James Nisbet, 1882; pp. 195

Birt, Lilian M.: *The Children's Home-Finder: The Story of Annie Macpherson and Louisa Birt* [with further extracts from A.M.'s letters and journals]. London: James Nisbet, 1913; pp. 262, with 17 halftones

Trained in Glasgow (near her native Campsie) and under Dr Froebel in London, Annie Macpherson looked set for a distinguished career as a teacher until one evening in 1861, when she heard a missionary give a talk on the East End of London. From that day on, Annie felt teaching would not be enough: she must become a missionary too. Not an evangelist, converting the heathen of Africa or China, but a reformer here at home. Three years later the first Home of Industry was opened in an old warehouse in Spitalfields, set up by Annie to house some of the hundreds of 'London Arabs': the orphaned (or as good as orphaned) children of the thieves' quarter scratching a vicious living for themselves by making matchboxes. She founded a Farm Home too, in the countryside outside London, where suitable boys could go to learn the basics of agriculture before leaving their faithless motherland for a new and innocent country: Canada. This was where Annie's mission really lay: in taking in all about 14,000 child emigrants out of squalid poverty in England to the clean and healthy backwoods (if just as poor) of Ontario and Quebec.

From 1870 onwards she made countless summer voyages across the Atlantic, helped by her two sisters and their families. Usually she would sail from Liverpool to Quebec City with as many as 500 children in her care; from there they trekked on to Belleville (home of Susanna MOODIE), or to Galt or Knowlton, where Annie had set up 'Distributing Homes', and then the *real* travelling would begin. She would take a detachment of orphans with her—perhaps twenty, perhaps a hundred—and tour the surrounding country by train or waggon, sending telegrams (if practicable) before her to the villages and townships on the way, inviting everyone to an Important Meeting. As soon as they arrived at their day's destination, the meeting would be held and as many children placed with kind or needy local families as possible. Annie would travel on until all the children had been placed—and then go back for another lot. A rash business, we might think, stealing children from their homeland and forcing them to travel often in the most appalling conditions only to give them away to strangers from the back of beyond. But contemporary accounts, and the testimonies of the children themselves, speak of Annie as a Scottish saint, a kind-hearted innocent capable, with her unquestioning faith, of inspiring nothing but good in others.

❧

MOIR, Jane F.

A Lady's Letters from Central Africa: A Journey from Mandala, Shiré Highlands to Ujiji, Lake Tanganyika, and Back. Glasgow: James Maclehose, 1891; 12mo, pp. 91, with 3 halftones, 3 sketches, and 2 maps

Mr and Mrs Fred Moir were cheerful entrepreneurs, inspired by their compatriot David Livingstone to leave their Scottish home for Darkest Africa. They were the founders of the African Lakes Company, which doubled as a trading outfit and anti-slavery mission based near Blantyre in what is now Malawi.

Much of their time in Africa was spent on the move, on trading expeditions, and Jane rarely countenanced Fred travelling anywhere without her. This book tells of a trip to Ujiji on the north-east coast of Tanganyika, a 250-mile marathon by steamer and on foot which, although it took nearly a month, sounds from Jane's description like a Sunday-school outing. It was, she says, 'the very pleasantest journey that two mortals could have'.

All went well once they arrived: the Moirs bought ninety elephant tusks at Ujiji and looked forward to their return voyage along the lake. But then things began to go wrong. They managed to miss the Tanganyika steamer, and so had to rely on a dubious-looking native dhow which was promptly beached by a gale on to the shores of the Attongwe tribe. The Attongwe lived up to their treacherous reputation by first trying to drug the Moirs and then shooting at them. Somehow the party managed to escape (Jane afterwards found several bullet-holes in her hat), only to be faced with a smallpox outbreak on board the dhow some days later. But Jane put her trust in the Lord and, four months and four days after setting out, everyone miraculously arrived back at Blantyre, where she remained until her book was published, keeping home for her husband and calmly dispensing tam-o'-shanters and the word of God amongst delighted natives.

MONTEFIORE, Lady Judith (d. 1862)

[Anon.]: *Private Journal of a Visit to Egypt and Palestine by way of Italy and the Mediterranean.* London: Joseph Rickerby ['Not Published', i.e. for private circulation only], 1836; pp. 322

[Anon.]: *Notes from the Private Journal of a Visit to Egypt and Palestine by way of Italy and the Mediterranean.* London: Joseph Rickerby [as above], 1844; pp. 410, with a folding table

Lady Judith was the wife of one of the nineteenth century's great phil-anthropists, Sir Moses Montefiore. Their journeys to Palestine (these accounts describe two of four made by the author) were not just Jewish pilgrimages, but reconnaissance trips for Sir Moses to collect information for his mission: the emancipation and well-being of Jews world-wide.

Each journey lasted nine months. The first was made between 1827 and 1828, when the Montefiores travelled to Naples on the overland route via Mount Cenis and Milan, sailing then via Sicily and Malta (where they stayed a while with friends) to Alexandria. They digressed up the Nile to Cairo and back before sailing on to Jaffa (wearing Turkish costume to baffle the corsairs) and riding at last to Jerusalem. The route differed on their second journey (1838–9): this time they toured Belgium and Germany before making for Marseilles, and skirted the Riviera on the way to Sicily, Malta, and finally Beirut. Sir Moses documented the visits in official papers and reports, while Lady Judith wrote conventional journals—the first accounts in English by a Jewish woman traveller. As well as describing the zeal of Sir Moses in managing to find synagogues for Saturdays in the most unlikely places, and movingly acknowledging the joys of travelling as a Jew in the Promised Land, she notes some delightfully common-place details: the best souvenirs to collect for gifts, for instance, which include straw hats from Nice and curious lava ornaments from Sicily (nothing changes!).

PATON, Maggie Whitecross (d. 1905)

Letters and Sketches from The New Hebrides. Edited by her Brother-in-Law, Revd. Jas. Paton. London: Hodder and Stoughton, 1894; pp. 382, with photogravure frontis-piece, 23 halftones, and a map

These *Letters and Sketches* were published by public demand after the appearance (as chapter 9 of her husband's 1889 *Autobiography*) of some brief reminiscences of Pacific-island life written by a young Scots girl, Maggie Paton. Maggie was the second wife of pioneer missionary John G. Paton, who devoted his life (and his first wife's) to the Presbyterian Church in the New Hebrides and, later on, in Australia. He took his new bride out to Aniwa, neighbouring Tanna in the archipelago between the Solomons and New Caledonia in 1865, and there they stayed, weathering inhospitable natives, repeated illness, the birth of six children (and the death of two more), until 1881.

Maggie would be the first to admit she was not a dyed-in-the-wool missionary. She was too prone to what she called 'risibles', or disastrously dissolving into giggles at moments of high ceremony, and dreaded the thought of her children being 'called' to suffer the same hardships as she and John had suffered. Her account of life on Aniwa is by turns high-spirited and despairing—the latter particularly when one by one her children are sent for their own safety to live with relatives in Australia; it is less saccharine than many a missionary's tale and one of the most often reprinted of them all.

❧

PRINGLE, M. A.

Towards the Mountains of the Moon: A Journey in East Africa. Edinburgh: Wm. Blackwood, 1884; pp. 386

The 'Mountains of the Moon'—what are now known as the Ruwenzori in Uganda—were reported by explorer John Speke in 1863 to cradle the near-mythical source of the Nile. There could hardly be a more romantic destination for travellers during the late nineteenth century, and when Mrs Pringle was offered the chance to accompany her husband there in 1880, she jumped at it.

Actually, the Pringles' destination was about as far from the Mountains of the Moon as Manchester from Mont Blanc, but still, it was close enough. It was no pleasure trip: they were travelling as volunteers for the Church of Scotland, keeping company with an elderly Doctor of Divinity sent to investigate rumours of serious unrest at the church's Mission Station at Blantyre. Their journey took them through the Red Sea to Aden and on to Zanzibar; after luncheon with the bishop there they sailed on to Mozambique, and up the Zambesi and Shiré rivers (on steamers swarming with drunken natives and hungry-looking rats) to Blantyre. Mrs Pringle says nothing of her husband's work during the six weeks they stayed at Blantyre; instead she relates the local difficulties faced by a white woman in 'the Interior', where she was beset by a 'dreadful depression' and forced to guard not so much against fierce tribesmen as against their wives, who followed her about trying to tear off her clothes. It was, we gather, a thoroughly debilitating experience.

On their way home, the couple misguidedly visited Delagoa Bay ('even horses cannot live here', gasped Mrs Pringle—she should have met Rose MONTEIRO!) thereby missing the steamer home. Desperate to leave east Africa, they hailed the first vessel they saw and eventually arrived in Scotland, feversome but relieved. So much for romance.

◆

RIJNHART, Susie Carson (1868–1908)

With the Tibetans in Tent and Temple: Narrative of Four Years' Residence on the Tibetan Border, and of a Journey into the Far Interior. Edinburgh: Oliphant, Anderson and Ferrier, 1901; pp. 406, with 13 halftones and a map

This must surely be the most pathetic traveller's tale of all. Susie Carson, a Canadian doctor, was married to the Dutch missionary Petrus Rijnhart and in 1895 the newly wed couple travelled to the village of Lusar, near the Kumbum monastery on the Sino-Tibetan border, to open a dispensary there and preach the word of God. The first part of Susie's book tells of their time there and at nearby Tankar, where they learned to love and trust the native people—Chinese, Tibetan, and Mongolian—and to understand something of their faith and customs. For their own part, the Rijnharts were regarded as magicians, able to cure and repair the impossible, and story-tellers *extraordinaires*.

During the spring of 1898 Susie and Petrus, now joined by baby Charlie, the apple of their eye, were called to journey into 'the Far Interior' of Tibet. Like Annie TAYLOR before them, they felt it their duty to the people of that benighted country to acquaint them with Jesus and save their heathen souls. Well-used to travel and full of confidence, they set out towards the sacred and forbidden city of Lhasa. Had they reached their goal, they would have been the first Westerners to visit Lhasa since the Jesuit missionaries Huc and Gabet in 1846. But they did not reach it.

All went as well as could be expected in such inhospitable terrain until about three months into the journey, when the little family were within two hundred miles of the capital. In quick succession two of their servants ran away, five of their ponies were stolen, and then, as Susie says, 'the darkest day in our history arose': baby Charlie, 'the very joy of our life, the only human thing that made life and labour sweet amid the desolation and isolation of Tibet', suddenly and quite simply stopped breathing and died. His stunned parents could do nothing but bury their son in a medicine box and leave him behind them in the bleak Tibetan earth. A few days later their dispirited struggle towards Lhasa was halted by officials who, instead of killing them, allowed them to turn back towards China—which they did, blindly pressing on through the growing winter blizzards, until they had almost reached the monastery of Tashi Gompa, some five hundred miles from the border. Finding themselves lost a day or two's trek from Tashi Gompa, they camped by a river opposite a small tented settlement: perhaps Petrus could find help there. He left Susie alone while he crossed the river and disappeared from view towards the tents, waving and shouting something she could not quite catch as he went. She was not worried when her husband had not yet returned by next morning: he was obviously organizing a guiding party to get them to the monastery. But as the day wore on and yet another night passed with still no sign, the desperate truth began to dawn.

Petrus Rijnhart was never seen again. With astonishing courage, Susie managed the two-month journey back to China alone, numb and ill and terrified. She followed the river to another camp, where friendly Lamas provided her with food and a yak, and staggered on with a variety of escorts (some honest and some out to kill) until she reached the mission station at Ta-chien-lu (K'ang Ting). And even that is not the finish of her incredible story: she went home to Ontario in 1899, married another missionary, a Mr Moyes, and bravely returned to the Tibetan borderlands with him. She had another child, too: a well-deserved happy ending, you might think. Except that Susie herself died just three weeks after its birth.

❖

STENHOUSE, Fanny (1829–1904)

Exposé of Polygamy in Utah: A Lady's Life Among the Mormons. A Record of Personal Experience as one of the Wives of a Mormon Elder during a Period of more than Twenty Years. New York: American News Co., 1872 [first published in London by George Routledge, 1873, as *A Lady's Life Among the Mormons*]; pp. 221, with 9 wood-engraved plates

'Tell It All': The Story of a Life's Experience in Mormonism. An Autobiography [etc.]. With Introductory Preface by Mrs Harriet Beecher Stowe. Hartford: A. D. Worthington [by subscription], 1874 [first published in London by Sampson Low, 1880, as *An Englishwoman in Utah*]; pp. 623, with 2 steel-engraved plates and 25 wood engravings

Fanny 'Tell It All' Stenhouse became a household name on both sides of the Atlantic when edition after edition of her remarkable story was published throughout the 1870s and 1880s. Most were illustrated with discreetly salacious wood engravings, showing a comely collection of maidens glaring at one another with the caption 'I could tear you to pieces', or a stout gentleman introducing the same comely collection to a friend who asks 'Are these *all* you have brought?' For Fanny was not the only Mrs Stenhouse in her husband's keeping: she was married to a Mormon.

Fanny, born in the Channel Islands, was on the point of marrying a respectable Frenchman she had met whilst working as a governess in Paris when, on a visit home in 1849, she was swept off her feet by a Mormon evangelist. She converted to his religion and married him within the year. Together they travelled to Switzerland as missionaries, living on next to nothing and suffering horribly from cold and hunger: the Lord provided only intermittently. In 1855 they moved on to New York, where Stenhouse edited a Mormon newspaper and Fanny did menial work until the Call came to Salt Lake City, the Mormon's headquarters in Utah. This involved a weary and dangerous summer trek across the plains along a route on which hundreds of last winter's emigrants had perished, and what was waiting for Fanny at the end of it was even more horrific. She had heard rumours of authorized polygamy amongst the Mormon leader Brigham Young's followers at Salt Lake City (the justification being that only married women were eligible for heaven), but had refused to believe them until her own brainwashed husband announced he was taking another wife—and another one after that.

Fanny wrote her account of the humiliation and oppressions of the next fifteen years in the hope it might trigger off some sort of reformation. Its publication after she and her husband had eventually broken away from the Church and settled in Utah as monogamous 'Gentiles' did lead to several criminal charges against some of the Elders, and enlisted much feminine sympathy for Mormon women. But I suspect most read it—in Britain at least—as a lickerish cautionary tale of what might happen to an innocent abroad . . .

❧

TAYLOR, Annie Royle (b. 1855)

The Origin of the Tibetan Pioneer Mission, Together with Some Facts about Tibet [including 'My Experiences in Tibet']. London: Morgan and Scott [1894] [reprinted in enlarged form as *Pioneering in Tibet*]; pp. 20, with a wood engraving and a map

Carey, William: *Travel and Adventure in Tibet: Including the Diary of Miss Annie R. Taylor's Remarkable Journey from Tau-Chau to Ta-Chien-Lu through the Heart of the Forbidden Land.* London: Hodder and Stoughton, 1902; pp. 285, with 75 halftones [the diary: pp. 173–285]

Little Annie Taylor from Cheshire was the first European woman ever to enter the Forbidden Land of Tibet, and came closer to its sacred capital of Lhasa than any European at all since the Jesuit missionaries Huc and Gabet in 1846. She was a missionary herself, called to China in 1884 convinced that God meant her eventually for the most inhospitable country in the world. She spent the next seven years on Tibet's Chinese and Sikkimese borders preparing to march to Lhasa and save its soul. She learned the language and the habits of Tibetan natives and travellers whilst living amongst lamas and refugees (unlike Alexandra DAVID-NEEL, Annie had no interest in their beliefs—'Poor things, they know no better; no one has ever told them about Jesus') and as soon as she was confident of her disguise as a pilgrim nun, it was just a question of waiting for God's word. That came in September 1892, whilst Annie was at Tao-chou: a Chinese called Noga was about to leave for Lhasa with his wife, and agreed to act as Annie's guide. With her faithful servant Pontso (a Tibetan convert to Christianity) and two more hired men, she stole across the forbidden frontier at daybreak and her dangerous crusade began.

The journey was terrible. Winter was approaching and the bandits of the 'Tea Road' to Lhasa were getting desperate—but Annie would be safe: 'I am English and do not fear for my life.' The rivers she had to cross on inflated bullock-skins were swollen and ringing with ice ('the Lord will take care of us') and her horse soon grew so tired that it literally lay down and died beneath her ('it looks as though I could not go on without a good horse. God will provide me with one'). One of her servants died on the way, another turned back to China, the food began to run out and the yaks were reduced to eating the travellers' clothes—yet despite everything, Annie's faith never wavered: 'He has sent me on this journey, and I am his little woman.' Come Christmas, she solemnly boiled a cherished plum-pudding even though she was too ill from altitude sickness to eat much of it, and every day her calm refrain was 'Quite safe here with Jesus'. Annie might well have made it to Lhasa had not Noga betrayed her to the officials there; as it was, she and Pontso were turned back within just three days of their goal.

The whole harrowing journey of 1,300 miles took exactly seven months and ten days to complete. Annie was not beaten, though. She eventually settled in Yatung, a trading and mission station just over the Tibetan border with Sikkim. She had lectured in England since her return from the journey and set up the 'Tibetan Pioneer Mission' there; now she was content to live the rest of her days in useful obscurity. She left 'Lhasa Villa', her rusty shack in Yatung, for England some time between 1907 and 1909, but when and where she died we do not know.

8

Journeys into Print

THE profession of travel writer for women is a relatively recent one. It is true that the publication of one's carefully reworked travel-diary, full of contrived spontaneity, was at one stage considered the *sine qua non* of literary high fashion. But the ladies who wrote them were hardly professionals. Early nineteenth-century salons were full of débutantes and dowagers eager not just to tell one another of their latest Grand Tour, but to immortalize it between the gilt-embossed covers of a book, usually printed 'for private circulation only' (i.e. for a limited number of subscribers who paid, along with the author, for the privilege), and prefaced with the modest and obliging insistence that it was written only under duress.

Such women did not expect to make money from their work—quite the contrary. Others, however, were more ambitious. They were probably authors already, like Lady Sydney MORGAN and Julia PARDOE, both writing in the early 1800s. Travel was as legitimate a subject as fiction, after all, and becoming increasingly popular a pastime amongst female readers: why not turn the previously unproductive time spent away from home sightseeing or recovering from illness into profit? So popular were Lady Morgan's controversial books on France and Italy, and the maudlin Miss Pardoe's on Turkey, that now their authors are best-remembered for those alone.

Neither woman, however, would have been able to subsist entirely on her travel writing. Each relied on the attraction of novelty, and had her own trade mark (a brazenly libellous attitude to foreigners in Lady Morgan's case, and in Julia Pardoe's the wonderful engravings she used to illustrate her books); once the novelty wore off, they were forgotten. It was left to Louisa COSTELLO and her peers to emerge as the first women actually to earn a living by describing their travels.

They tended to alternate books and journeys, building up an avid audience the while and managing to satisfy the armchair traveller with their cheerful anecdotes as well as encouraging the more active with alluring descriptions and discreet practical advice.

Not all the professionals confined themselves to narrative travel books, although those skilled in the art, like Fanny TROLLOPE or Ethel MANNIN, enjoyed the widest audiences. Mariana STARKE made a career out of 'helpful hints', pre-dating both Baedeker and Murray with a series of volumes 'written for the Use and Particular Information of Travellers'. I have only included guidebooks here with a suitable pedigree: Mariana's evolved out of her first-hand account *Letters from Italy*, published in 1800, and others, like Ellen TAYLOR's *Madeira* or Emma ROBERTS's *East-India Voyager*, were based on the author's description of her own journey to the place concerned. In fact sometimes there is hardly any difference between a guidebook and a narrative at all.

There is a fine line to be drawn too, of course, between the writer who travels, and the traveller who writes. I would call Jan MORRIS, for example, a writer first and foremost, even though she has probably travelled further than anyone else I have mentioned. But Freya STARK, the most prolific author in the book, is essentially a traveller. Perhaps it is all to do with the art of travel. They are all artists of one sort or another, the women in this chapter; for some the consummation of that art is the written book, while for others it is the journey itself.

BIGLAND, Eileen (1898–1970)

Laughing Odyssey [Russia]. London: Hodder and Stoughton, 1937; pp. 307, with 5 colour plates and a map

The Lake of the Royal Crocodiles [Zambia and north-east Zimbabwe]. London: Hodder and Stoughton, 1939; pp. 300, with 8 line drawings and 2 maps

Into China. London: Collins, 1940; pp. 315, with 16 halftones and a map

Journey into Egypt. London: Jarrolds [1948]; pp. 192, with 20 halftones and an endpaper map

Russia Has Two Faces. London: Odhams, 1960; pp. 240, with 19 halftones and 2 maps

Mrs Bigland trained as a dancer, worked as a journalist and publisher's reader (she it was who 'discovered' Eric Ambler), and produced a string of what she called slushy novels. But her real vocation was travel writing. In her biography *Awakening to Danger* (1946) she explained her motives for travel. Like Martha Gellhorn, she chose to visit unquiet places at unquiet times; she went where the complacent British public would rather not go, and wrote what they would rather not hear.

In Russia and China, for example, she travelled as a passionate evangelist for Communism at a time when Fascism was sickening Europe for war. She took the train through Germany and Poland to Moscow, and then worked her way slowly through the Ukraine—spending some time on a collective farm there—to the Black Sea and the Caucasus. This was in 1936; later that same year Eileen travelled to Spain as a journalist at the outbreak of the Civil War. Her next journey was to Shiwa Ngandu in Zambia, where she stayed in the summer of 1938, gathering material for a travel book and a political study of black Africans *(Pattern in Black and White*, 1940).

The journey to China during the Sino-Japanese war was another pilgrimage in search of the Communist ideal. She entered the country by the Burma–Yunnan highway and travelled up to Chongquing, there sleeping in a hospital bath because the hotels had been bombed; for her keep at the hospital she acted as an escort for those wounded soldiers who needed to get back to their families, and accompanied one of the dying men as far as Hanoi. After the war, which she spent lecturing on Russia and China, Mrs Bigland visited Egypt and Sudan. Her last travel book was about a return visit to Russia in 1958.

CAMERON, Charlotte (d. 1946).

A Woman's Winter in South America. London: Stanley Paul [1911]; pp. 292, with 35 halftones and a map

A Woman's Winter in Africa: A 26,000 mile Journey. London: Stanley Paul, 1913; pp. 403, with illustrations throughout and 2 maps

A Cheechako in Alaska and Yukon. London: T. Fisher Unwin, 1920; pp. 292, with 36 halftones and a map

Two Years in Southern Seas. London: T. Fisher Unwin, 1923; pp. 315, with 44 halftones and 2 maps

1 Eliza Bradley, newly shipwrecked, being kidnapped by Barbary Arabs

2 Canadian missionary Dr Susie Rijnhart in her Tibetan home

3 A home from home aboard Lady Anna Brassey's schooner
Sunbeam

4 The Hon. Impulsia Gushington tasting the joys of Egyptian travel

5*a* May French Sheldon's splendid African palanquin . . .

5*b* . . . and an uninvited guest

7 Emmeline Lott's uncompromising uniform as governess in a royal harem

6 Charlotte Canning embarking on an Indian sketching expedition

9 Mary Kingsley, elegant explorer of 'fish and fetish' in West Africa

8 A typically glamorous study of Amy Johnson, the 'darling of the skies'

11 Dame Freya Stark pony-trekking in Nepal at the age of seventy-seven

10 The entomologist Evelyn Cheesman dressed for action in New Guinea

12*a* and *b* Forceful Fanny Bullock Workman accepting a little
help in her conquest of the Himalaya

Lady M_y W_r-t-l_y M_nt-g-e
The Female Traveller
In the Turkish Dress.

Let Men who glory in their better sense,
Read, hear, and learn Humility from hence;
No more let them Superior Wisdom boast,
They can but equal M-nt-g-e at most.

14 Lady Mary Wortley Montagu, the most
celebrated wayward woman of them all

The Indian Alps
And How We Crossed Them

13 The alluring cover to Nina Mazuchelli's classic of
Himalayan travel

Wanderings in South-Eastern Seas. London: T. Fisher Unwin, 1924; pp. 269, with 45 halftones and a map

Mexico in Revolution: An Account of an Englishwoman's Experiences and Adventures in the Land of Revolution [etc.]. London: Seeley, Service, 1925; pp. 278, with 24 halftones

'Travel has been my comrade, Adventure my inspiration, Accomplishment my recompense.' This motto, and the titles of her books, would suggest that Mrs Cameron must be one of the most serious and intrepid of women travellers. In fact she should be the patron saint of cruise liners; although she covered a good 250,000 miles between 1910 and 1925, almost all of them were by courtesy of some shipping line, steamer company, or local railway—a remarkable feat in itself, I suppose. Her style is remarkable, too: florid and gushing, no doubt designed to allure those other ladies of leisure and means who had only to get to their nearest Thomas Cook office to be a 'cheechako' (tenderfoot), wanderer, or 'woman alone' themselves. Mrs Cameron also wrote novels based on her panoramic cruises and whilst in Delhi in 1911 represented *The Ladies' Pictorial* and several other periodicals as a foreign correspondent. She was one of a new breed of 'globe-trotteress'—not of the pioneering independent type like Isabella BIRD or Miss NORTH, but of dedicated sightseers, determined to get to the end of a beaten track, but never off it.

COSTELLO, Louisa Stuart (1799–1870)

A Summer amongst the Bocages and the Vines. London: R. Bentley, 1840; 2 vols., pp. 411/358, with 4 lithotints and 2 wood engravings

A Pilgrimage to Auvergne, from Picardy to Le Velay [etc.]. London: R. Bentley, 1842; 2 vols., pp. 360/362, with 4 lithotints

Béarn and The Pyrenees: A Legendary Tour to the Country of Henri Quatre. London: R. Bentley, 1844; 2 vols., pp. 428/391, with 2 lithotints and 32 wood engravings

A Tour to and from Venice, by the Vaudois and the Tyrol. London: John Ollivier, 1846; pp. 453, with 3 lithotints

Louisa Costello combined three considerable talents to become the first professional lady travel writer: she was an accomplished artist, a successful author (poet, novelist, and historian), and an experienced and imaginative tourist.

Her father's death in 1815 left the family in a precarious financial position, and to help out Louisa moved to Paris, where she could best paint and market the miniature portraits in which she excelled. On her return home to England in 1820 she turned to writing, and after each of her subsequent tours to the Continent published an illustrated account, well informed, easy to read, and cheerfully inviting. She did tourism a great service in the early days of excursion travel by recommending simple itineraries with plenty of historical and topographical interest, and by pointing out that if such tours were only suitable now for 'travellers whose time is their own', soon 'just a few idle days' would carry the curious right across Europe. She was right. It wasn't long before Mr Cook's and Mr Gaze's clients were bustling all over France and beyond, armed with their copies of 'Mr Murray'—and Miss Costello's good advice.

❦

EASTLAKE, Lady Elizabeth (*née* RIGBY) (1809–1893)

[Anon.]: *A Residence on the Shores of the Baltic. Described in a Series of Letters*. London: John Murray, 1841 [second edition published the following year as *Letters from the Shores of the Baltic*]; 2 vols., pp. 293/286, with 20 etchings

[Anon.]: 'Lady Travellers', *Quarterly Review*, 151 (1845), 98–137.

Journals and Correspondence of Lady Eastlake. Edited by her Nephew Charles Eastlake Smith. London, John Murray, 1895; 2 vols., pp. 326/331, with 17 halftones

'I have no doubt that she is the cleverest female writer now in England.' So said the venerable John Murray of Miss Rigby on the first publication of *A Residence . . . on the Baltic*. It is the book that made her reputation, although after her death in 1893 she was best remembered not as a travel writer but as novelist, critic, translator, and art historian. The Baltic book shows signs of all these talents; in describing a visit to her married sister in Reval (now Tallinn in the USSR) it is unusually well crafted, sensitive, intellectual, and finely illustrated with the author's own sketches.

Unlike most of her predecessors, Elizabeth Rigby had little interest in the mechanics of the journey itself, although it was hardly commonplace for a twenty-nine-year-old woman to be travelling in 1838 in subzero temperatures amongst the wolves and 'stiffening ice' of Estonia with just a Russian manservant for company; it is the place itself that interests her, and we get a full description of ill-frequented Estonia and the north Baltic coast. More familiar territory is analysed instead—like Imperial Russia, 'where the learned man wastes his time, the patriot breaks his heart, and the rogue prospers'.

It is a shame that Elizabeth Rigby did not write further travel accounts: she went to Reval twice more, and after her marriage to Charles Lock Eastlake, RA in 1852 visited the Continent with him in pursuit of Art almost every year until his death in 1865. She did bring up the subject once more, however, in a lengthy article in Murray's *Quarterly Review* in 1845, giving us the only critical contemporary view we have of the early Victorian Lady Traveller—already a race apart—by one of their own kind.

EDWARDS, Matilda Betham (1836–1919)

A Winter with the Swallows [Algeria]. London: Hurst & Blackett, 1867; pp. 286, with 2 wood engravings

Through Spain to the Sahara. London: Hurst & Blackett, 1868; pp. 317, with 2 wood engravings

Holiday Letters from Athens, Cairo, and Weimar. London: Strahan, 1873; pp. 247

A Year in Western France. London: Longmans, Green, 1877; pp. 346, with 1 wood engraving

Holidays in Eastern France. London: Hurst & Blackett, 1879; pp. 325, with 2 wood engravings

East of Paris: Sketches in the Gâtinais, Bourbonnais and Champagne [etc.]. London: Hurst & Blackett, 1902; pp. 275, with 4 colour plates and 2 halftones

Unfrequented France: By River, Mead and Town; London: Chapman & Hall, 1910; pp. 204, with 32 halftones

In the Heart of the Vosges and Other Sketches by a Devious Traveller. London: Chapman & Hall, 1911; pp. 327, with 16 halftones

In French Africa: Scenes and Memories [etc.]. London: Chapman & Hall, 1912; pp. 324, with 16 halftones

This looks a prodigious enough output—but it is fractional compared with the rest of Miss Betham Edwards's work. She was a largely self-educated Suffolk girl—the cousin of Amelia Blandford EDWARDS—whose first verses were published by Charles Dickens in *All the Year Round* when she was twenty-one. She went on to produce sixty years' worth of novels, children's histories, and these nine books, based mostly on her travelling life in France.

What makes the travelogues notable (apart from their quantity) is their fresh approach to an apparently hackneyed subject. But Miss Edwards was a socialist at heart; she prided herself in travelling without hotels or guidebooks, and always in rural, unfrequented areas. Whenever possible she stayed in French peasant homes (and was eagerly invited to do so), and wrote the accounts of her journeys not so much to encourage others to visit as to explain the 'real' French people to the short-sighted English tourist. She aimed to do the same in Algeria, or French West Africa. Certainly there was none of the customary *esprit de corps* between her and any other English tourists she might meet abroad. She mentions only one such encounter, on board a P. & O. steamer to Alexandria, when the 'so-called gentlemen' of the First Class saloons rampaged around the ship drunk each night, and chased the galley's sheep until its legs broke. The French recognized her mission, and appointed Miss Edwards Officier de l'Instruction Publique de France for her evangelistic efforts. I think her English readers somewhat missed the point—she was reviewed as a bright and sunshiny teller of rural tales—which she was, but more besides.

❖

ELLIOT, Frances (Minto)

Diary of an Idle Woman in Italy. London: Chapman and Hall, 1871; 2 vols., pp. 316/329

Diary of an Idle Woman in Sicily. London: Richard Bentley, 1881; 2 vols., pp. 251/208

Diary of an Idle Woman in Spain. London: F. V. White, 1884; 2 vols., pp. 298/258

Diary of an Idle Woman in Constantinople. London: John Murray, 1893; pp. 425, with a map and 4 wood engravings

These books are hilarious. They were obviously written for other 'idle women', bored and moneyed ladies, to lure them abroad and guide them when they got there. Mrs Elliot went nowhere that it was not easy by now to follow, and identified perfectly with an audience of comfort-loving upper-class women of a derisive turn of mind. She was herself an excellent vanguard for them—the books are carefully detailed, the journeys well organized, and her style pacey and fashionable offhand. What makes them so funny now (and perhaps she had her tongue in her cheek even then?) is the strain of proud British trenchancy that runs through them all. Mrs Elliot is like Mrs CALDERWOOD in this respect—nothing is worth seeing, but how satisfying it is proving so! Rome is a 'third-rate modern City', and ·St Peter's 'sorely disappointing'; there are too many tourists(!) and bad hotels in Sicily; Madrid is ugly and corrupt, the people of Valencia 'the stupidest people in the world', and Malaga 'detestable'; and as for Constantinople, it is 'depressed', 'degraded', desolate', and 'degenerate'. Mrs Elliot did write very well, and her books were popular, but I bet they were never translated.

❖

ERSKINE, Mrs Steuart (Beatrice Caroline) (d. 1948)

Trans-Jordan. London: E. Benn, 1924; pp. 126, with 48 halftones and a map

The Vanished Cities of Arabia. London: Hutchinson, 1925; pp. 324, with 40 sketches (8 in colour) by M. Benton Fletcher

Vanished Cities of Northern Africa. London: Hutchinson, 1927; pp. 284, with 40 sketches (8 in colour) by M. Benton Fletcher and a map

Palestine of the Arabs. London: Harrap, 1935; pp. 256, with 38 halftones

Mrs Erskine travelled in order to write her books, not (as is so common) the other way round. Her work is mostly of a historical or purely descriptive nature—just letterpress to accompany the illustrations provided by her collaborator Major Benton Fletcher—the modest coffee-table books of their day. She was no devotee of the art of 'roughing it', and hired a car or lorry wherever possible, just occasionally resorting to horseback. Still, she was an enthusiastic student of the Middle East, publishing an authorized account of King Faisal of Iraq in 1933 and becoming somewhat of an authority on the archaeological sites ('Vanished Cities') of Palestine, Tunisia, and Algeria. The two books she published on Naples and Madrid, not listed above, are not travel but picture books, for which she just wrote the lengthy captions.

❧

GAUNT, Mary (1872–1942)

Alone in West Africa. London: T. Werner Laurie [1912]; pp. 404, with photogravure frontispiece and 97 halftones

A Woman in China. London: T. Werner Laurie [1914]; pp. 390, with 130 halftones

A Broken Journey: Wanderings from the Hoang-Ho to the Island of Saghalien and the Upper Reaches of the Amur River. London: T. Werner Laurie [1919]; pp. 295, with 61 halftones

Where the Twain Meet [Jamaica]. London: John Murray, 1922; pp. 335, with 10 halftones

Reflection—In Jamaica. London: Ernest Benn, 1932; pp. 258, with 4 halftones

Mary Gaunt's journeys were all business trips, in a way. She travelled to gather material for the ripping yarns of adventure that made her name. She had grown up in Australia, where her English father was Governor of the Victoria Goldfields, nourished by tales of brave prospectors and of her grandfather's travels to China in the service of the Honourable East India Company.

Her own travelling career began in 1900, by which time she had taken her degree at Melbourne University, had her first novel published, and married a doctor. He died tragically soon, and Mary bravely sailed to England as a young widow and aspirant writer in search of inspiration. Edwardian London life proved lacking, however, and, deciding she would lose nothing (and perhaps gain material for more novels), she surrendered to a childhood dream and booked a passage to Dakar.

The travel books she wrote about this expedition to West Africa and subsequent ones to China, Turkestan, Siberia, and the Caribbean are just as exciting as the reams of 'Tales of Savage Lands' she produced. She was down-to-earth, sympathetic, observant, tenacious, and adventurous; fiercely independent but never above laughing at herself (after all, she was not exactly *petite*, nor in her first flush of youth; that is why she got on so well with the Chinese, she said—they respect bulk and age!). She felt she should earn her material, so never complained at the discomforts of being swung in a hammock along the shores of the Gold Coast by truculent and inebriated natives. She even preserved her equanimity whilst white-water canoeing on the Volta, and never felt in ruder health than when travelling. She was like Mary KINGSLEY in her disapproval of missionary methods and in questioning the Imperial Spirit held sacred by pre-Great-War Britain, and like Miss Kingsley too in that her books are an entertainment and an education—just as the art of travel should be.

GRIMSHAW, (Ethel) Beatrice (1871–1953)

In the Strange South Seas. London: Hutchinson, 1907; pp. 381, with 56 halftones

From Fiji to the Cannibal Islands. London: Eveleigh Nash, 1907; pp. 356, with 108 halftones

The New New Guinea [Papua]. London: Hutchinson, 1910; pp. 322, with 45 halftones and a map

Adventures in Papua with the Catholic Mission. Melbourne: Australian Catholic Mission Society, 1912; pp. 62, with 3 halftones.

Isles of Adventure. London: Herbert Jenkins, 1930; pp. 307, with 30 halftones

Beatrice Grimshaw wrote over forty books based on her travels in the South Seas. Most of them were romances or adventure stories—like Mary GAUNT she collected her subject-matter as she went along, and regularly sent manuscripts home to pay for her independence. She had always wanted to go to the south Pacific; life at home in Dublin was simply not challenging enough for her, even though by the time she was thirty she had become a respected journalist and champion cyclist, holding the women's world 24-hour distance record. She moved to London, set herself up as a free-lance journalist offering press coverage to shipping companies in return for her passage, and left for Tahiti in 1906.

Beatrice travelled amongst the islands of the south Pacific and Indonesia almost incessantly for the next thirty years, settling (periodically) in Papua

New Guinea in a series of houses she built herself. Here the challenges were endless. This rather fluffy-looking, wide-eyed woman, with a revolver tied round her waist and wearing a decorous (but detachable) riding-habit positively encouraged adventure. Once she found herself joining a party of missionaries searching for a colleague captured by cannibals (they found most of him), and as early as 1909 she was diving for pleasure in the Torres straits. She ran a coffee plantation in Papua for several years, went diamond prospecting, and almost incidentally became the first white woman to navigate the deadly Sepik and Fly rivers. In 1939 she retired to Australia—still writing—and died there at the age of eighty-two.

❧

MACAULAY, Rose (1881–1958)

Fabled Shore: From the Pyrenees to Portugal. London: Hamish Hamilton, 1949; pp. 200, with 31 halftones and an endpaper map

The Towers of Trebizond, perhaps the best known of all Rose Macaulay's novels, is often mistaken for a genuine first-hand travel account. I wish it were—it would be even more marvellous were it true—but in fact Dame Rose only indulged herself in one real travel book (not counting various archaeological rambles and social satires set abroad). And 'indulged' is the word. She was infamous amongst her friends in Britain, in America, and in Europe as a desperately bad driver, and yet she insisted on driving whenever she could: she adored speed, and could never quite bring herself to believe in danger. In 1929 this renowned middle-aged English authoress took herself off in an Essex four-seater from New York to California to Texas, and she was still at it some twenty years later when she decided to try a land cruise along the coast of Spain from Port Bou to Cape St Vincent. She hardly saw a soul on the roads, she said (they had probably heard the scream of the motor and fled for their lives), and so vivid were her rapturous and well-informed descriptions of the route that soon everyone who *was* anyone (i.e. who had a car) was clamouring along the coast to see for themselves. *Fabled Shore* started the tourist affair with Spain that has lasted ever since: it is partly responsible for the way the coast is today. Which must make it one of the most successful—or most disastrous—travel books ever published.

❧

MANNIN, Ethel (Edith) (1900–1984)

Forever Wandering. London: Jarrolds, 1934; pp. 352, with a frontispiece

South to Samarkand. London: Jarrolds, 1936; pp. 355, with 19 halftones and 2 maps

German Journey. London: Jarrolds [1948]; pp. 168, with 21 halftones

Jungle Journey [India]. London: Jarrolds [1950]; pp. 256, with 41 halftones and an endpaper map

Moroccan Mosaic. London: Jarrolds, 1953; pp. 248, with 19 halftones and endpaper maps

Land of the Crested Lion: A Journey through Modern Burma. London: Jarrolds, 1955; pp. 256, with 23 halftones and a map

The Country of the Sea: Some Wanderings in Brittany. London: Jarrolds, 1957; pp. 224, with 18 halftones and a map

The Flowery Sword: Travels in Japan. London: Hutchinson, 1960; pp. 288, with 32 halftones and a map

A Lance for the Arabs: A Middle East Journey. London: Hutchinson, 1963; pp. 319, with 29 halftones and a map

Aspects of Egypt: Some Travels in the United Arab Republic. London: Hutchinson, 1964; pp. 264, with 26 halftones and a map

The Lovely Land: The Hashemite Kingdom of Jordan. London: Hutchinson, 1965; pp. 203, with 24 halftones

An American Journey; London: Hutchinson, 1967; pp. 212, with an endpaper map

An Italian Journey. London: Hutchinson, 1974; pp. 191, with colour frontispiece, 30 halftones and an endpaper map

At the age of thirty-two, when she had long left school, long become bored with her secretarial job in an editorial office, and long been a carefully cultivated feature of that era of 'bright young things' between the two world wars, Ethel Mannin made a vow. She would publish two books a year, one fiction and one non-fiction, and become one of the country's best and most prolific authoresses. She may not have been the best, but never mind: 95 books were churned out before her retirement in the late 1970s, and of those 95, some 13 were devoted to travel.

Miss Mannin was rather self-consciously glamorous: it behoved a brilliant young lady like her to float gracefully around the world getting herself into elegant scrapes and being surprisingly brave and plucky. One gets the impression that she was more interested in having *been* to a particular place than in exploring it while she was there, and in fashioning her mild adventures for the delectation of her eager readers. Once home again, however, she never stinted what she saw as her duty as a spokeswoman for the more benighted peoples of the globe: she was a keen (right-wing) political observer, a committed pacifist, and as passionate about justice as she was about her reputation. Foreign governments in search of a freshly scrubbed image recognized the sanitizing effect of a visit by Ethel Mannin and would invite her to stay with them awhile—Egypt, Jordan, and

Iraq, for example—and sure enough, a book would appear within the year, eminently readable, conscientiously polite, and as popular as ever.

❖

MARRYAT, Florence (later CHURCH, Mrs Ross and LEAN, Mrs Francis) (1838–1899)

'Gup': Sketches of Anglo-Indian Life and Character. Reprinted from *Temple Bar.* London: Richard Bentley, 1868; pp. 284

Tom Tiddler's Ground [America and Canada]. London: Swan Sonnenschein, Lourey & Co., 1886; pp. 212

Florence Marryat, daughter of author Captain Frederick, was well known in Victorian literary circles, for her persistence if not her talent. She wrote over seventy novels, appeared on the London stage as actress and opera-singer from time to time, and travelled the provinces and abroad giving picturesque lectures and monologues.

She seems to have been a forceful woman, with a well-developed ego and a sense of humour bordering on the facetious. Unfortunately her travel books show these qualities off to an uncomfortable advantage. *'Gup'* was written after she had married an officer of the Madras Staff Corps at sixteen and travelled with him to Bangalore and Burma to suffer 'seven years passed in exile' amongst witless memsahibs (who are either 'gay, religious or inane', each as bad as the other) and disgusting natives (who deserve all the cruelty they can bear). The only moments of Anglo-Indian life Florence enjoyed were those spent hunting in the beautiful Nilgiri Hills—or, as she put it, knocking tigers' brains out.

'Gup' means gossip, and gossip is the stated subject of Florence's book on America, too. She went there to perform her 'musical and dramatic monologue' *Love Letters* and to harry her American publishers for more royalties, travelling from New York to Boston, to Belleville and Toronto, Halifax, Pittsburgh, and Chicago. The gossip this time is of American ladies' laughable manners and fashion sense, and the hilarious New York niggers. But then she talks of middle-class English boors with just as much contempt: a saving grace. Perhaps she had just reread *'Gup'*.

❖

MILES, Beryl (b. 1919)

The Stars My Blanket [Australia]. London: John Murray, 1954; pp. 235, with 28 halftones and a map

Islands of Contrast: Adventures in New Zealand. London: John Murray, 1955; pp. 200, with 17 halftones and a map

Attic in Luxembourg. London: John Murray, 1956; pp. 263, with 35 halftones and a map

Candles in Denmark. London: John Murray, 1958; pp. 235, with 36 halftones and a map

Spirit of Mexico. London: John Murray, 1961; pp. 208, with 33 halftones and a map

Miss Miles's travels began soon after the realization in her 'neat Worthing office' one day that she was vegetating, in a rut, going nowhere—and the next thing she knew, she was on her way in a Ford freighter from Sydney to Darwin on a six-month aboriginal-cave-painting expedition. For the next ten years she hardly stopped. Always picking up opportunity and friends along the way, she proceeded to New Zealand for a spot of glacier-climbing before covering Luxembourg on a bicycle and exploring the hinterlands of Denmark, finally coming to rest after a vibrant trip to Mexico. Her books are breezy and bright, but nothing particular happens in them: she is just a terribly nice English girl with a taste for foreign life and a talent for describing it—about as balanced a woman traveller as you can get.

❖

MILLER, Lady Anna (1741–1781)

Letters from Italy, Describing the Manners, Customs, Antiquities, Paintings, &c. of that Country, in the Years MDCCLXX and MDCCLXXI, To a Friend residing in France. By an English Woman. London: for Edward and Charles Dilly, 1776; 3 vols., pp. 468/416/351

Lady Anna Miller is best known as the good wife of Bath who, with the amused support of such luminaries as Dr Johnson, Mrs Thrale, and Horace Walpole, set up what she called a Literary Society there. The society flourished: all the ladies of that most fashionable resort crowded to Batheaston House to recite their poetry to each other and discuss the latest *bon ton* and tittle-tattle, and every so often would publish the best of their 'novelties and poetical amusements'. She was too easy a target for satire, considering herself the last thing in intellectual elegance and artistic sensibility when in reality she was (as Fanny Burney gleefully confided) 'round, plump and coarse-looking', 'an ordinary woman in very common life'. The *Letters*, written to her mother who was baby-sitting the Miller children in Paris, were designed to promote the reputation of Lady Anna as public arbiter of taste, and apart from a few rhapsodic passages on the 'glacieres and chrystals' of the Alps, concentrate largely on the art collections visited on an eight-month circuit of Italy's major cities. They are not an easy read.

❖

MILLS, Lady Dorothy (1889–1959)

The Road to Timbuktu. London: Duckworth, 1924; pp. 262, with 14 halftones and a map

Beyond the Bosphorus. London: Duckworth, 1926; pp. 224, with 35 halftones

Through Liberia. London: Duckworth, 1926; pp. 240, with 38 halftones and a map

The Golden Land: A Record of Travel in West Africa. London: Duckworth, 1929; pp. 212, with 40 halftones and a map

The Country of the Orinoco. London: Hutchinson, 1931; pp. 288, with 34 halftones and a map

Lady Dorothy was a glamorous society gal, daughter of the Earl of Orford, who married in 1915, was divorced two years before her silver wedding anniversary, and only then embarked on a career as a travel writer. Although never quite shaking off her former image—touring Istanbul in a chauffeur-driven Ford, for example, or paddling up Venezuelan waterways in an utterly chic cloche hat—she nevertheless earned herself a reputation as one of the most daring and adventurous of all those crowds of ladies travelling between the two world wars. In 1923 she became the first Englishwoman to enter Timbuktu after a three-month journey by way of the Sahara and the Niger; three years later she organized an expedition 'painfully but joyfully trodden' through Liberia before broaching what was then Portuguese Guinea (now Guinea-Bissau) in 1929. Her last fling took her up the Orinoco as far as Sanariapo on the Venezuelan border with Colombia—it was supposed by her friends to be an elaborately staged swan-song which she could not possibly be expected to survive. Of course she emerged from the depths of South America as fashionably spruce as if she'd just returned from the Riviera. And with consummate grace, she chose this point in her career to retire.

MORGAN (née OWENSON), Lady Sydney (*c.* 1783–1859)

France. London: for Henry Colburn, 1817; quarto, pp. 252 (pp. i-clxxix contain appendices by Sir T. C. Morgan)

Italy. London: for Henry Colburn, 1821; 2 vols., quarto, pp. 357/484

France in 1829–30. London: Saunders and Otley, 1830; 2 vols., pp. 527/559, with a mezzotint portrait

What Lady Morgan lacked in talent, she made up for in opportunism. She was the author of several rather awful romances (*The Wild Irish Girl* (1806) is probably best known) and, for a while, a professional traveller. Her first trip abroad (except to sail from her native Ireland to England) was to Paris in 1816; it was a thoroughly successful business venture. Politically, it was hardly the most propitious time to lambast the French monarchy and celebrate the spirit of Napoleon and the Revolution. But that is exactly what she did.

France caused uproar on both sides of the Channel when it was published: already somewhat of a literary and society darling, Lady Morgan now became notorious. The critics called the book 'a sort of rhapsody, written without order, method, or knowledge of the subject', full of 'ignorance, flippancy and bad taste'; a flurry of patriotic pamphlets were published with titles like *France as it is, not Lady Morgan's France*, and Sydney relished it all. Making the most of his best seller, publisher Henry Colburn immediately commissioned a similar book on Italy, and Lady Morgan happily obliged with an opinionated work based on a two-year tour of its major cities. But *Italy* was less controversial: like a restaurant expecting an influential bon viveur, the country had prepared for Lady Morgan and put on its best show for her, giving balls in her honour, offering her lakeside villas, and arranging hand-picked society soirées. By the time *France in 1829–30* was published, Lady Morgan was no longer fashionable. So she did not bother to travel again.

MORRIS, Jan (formerly James) (b. 1926).

Coast to Coast [USA]. London: Faber, 1956; pp. 271, with 61 halftones and a map

Sultan in Oman. London: Faber, 1957; pp. 165, with colour frontispiece, 23 halftones, and 2 maps

The Market of Seleukia [Middle East]. London: Faber, 1957; pp. 337, with 9 half-tones and 7 maps

Coronation Everest. London: Faber, 1958; pp. 146, with 8 halftones and 3 maps

South African Winter. London: Faber, 1958; pp. 196, with 6 halftones and a map

Venice. London: Faber, 1960; pp. 337, with 30 halftones, a line drawing, and 3 maps (including 1 on endpapers)

Cities [essays]. London: Faber, 1963; pp. 375, with 29 halftones

The Presence of Spain. London: Faber, 1964; quarto, pp. 119, with photographs throughout, and a map

The Great Port [New York]. New York: Harcourt, Brace and World [1969: published the following year in London by Faber]; pp. 223, with 20 halftones and endpaper map

Places [essays]. London: Faber, 1972; pp. 186, with 15 halftones

Travels [essays]. London: Faber, 1976; pp. 155, with 6 sketches

Destinations [essays]. Oxford: University Press, 1980; pp. 242, with 10 halftones

Journeys. Oxford: University Press, 1984; pp. 173

Last Letters from Hav. London: Viking, 1985; pp. 176, with line-drawn vignettes and a map

Hong Kong Xianggang. London: Viking, 1988; pp. 320, with 27 halftones and 4 maps

Jan Morris has only officially been a woman since what she calls her 'gender rôle change' operation in 1972. But, as she explains in her autobiography *Conundrum* (1974), even in her early childhood she was aware that she should always have been what she eventually became: 'I was three or perhaps four years old when I realised that I had been born into the wrong body'. It is arguable that her almost insatiable thirst for travel was all part of the yearning for a destination she felt within herself—certainly, she spent her life as a young journalist with *The Times* and later as a free-lance writer notching up an incredible number of miles, until by the early 1980s she was able to claim that she had visited every major city in the world—and many more besides.

Her first major assignment for *The Times* was like a fairy-tale: as James Morris, the promising young special correspondent, she was dispatched with Edmund Hillary's party in 1953 for Mount Everest, and was the first to break the news of its conquest—with sublime timing—on Queen Elizabeth's Coronation Day. It made Morris's name, and when her books on Arabia and the Middle East were published in the years that followed, there was an eager audience already waiting to welcome them. And the books have been coming ever since: histories, studies, portraits, and essays.

Each of the first-hand travel accounts I have listed here has a discernible political flavour, but the detached commentary in all of them is tempered by an unashamed sense of affection for the 'Places', 'Cities', and 'Destinations' they describe: for many years, Morris felt more at home when she was travelling than when she reached base in England or Wales. Of course, she was not the first to feel so—most of the great lady travellers of the golden age in the nineteenth century confessed the same. In a way, as a man, Jan was just as unliberated at home as they had been, explaining that she travelled 'because I liked it, and to earn a living, and . . . only lately recognised that incessant wandering as an outer expression of my inner journey'.

PARDOE, Miss (Julia) (1806–1862)

Traits and Traditions of Portugal, collected during a Residence in that Country. London: Saunders and Otley, 1833; 2 vols., pp. 308/338

The City of the Sultan; and Domestic Manners of the Turks, in 1836. London: Henry Colburn, 1837; 2 vols., pp. 514/500, with 16 lithographs and 2 vignettes

The Beauties of the Bosphorus; Illustrated in a series of Views of Constantinople and its environs, from original drawings by W. H. Bartlett. London: for the Proprietors, by George Virtue, 1938 [originally issued in parts]; quarto, pp. 164, with steel-engraved vignette, portrait, 78 plates, and a map

The River and the Desart: or, Recollections of the Rhône and the Chartreuse. London: Henry Colburn, 1838; 2 vols., pp. 264/261, with 2 lithographs

The City of the Magyar, or Hungary and her Institutions in 1839–40. London: George Virtue, 1840; 3 vols., pp. 322/328/431, with 7 steel-engraved plates and a hand-coloured facsimile leaf

Yorkshire girl Julia Pardoe's first book was published when she was only fourteen. It was a volume of poetry which soon went into a second edition: an auspicious start to a remarkably prolific career as an author. Her first travel book was written in 1833 after a fifteen-month stay (escaping tuberculosis) in Portugal, and all the rest were based on the tour she took to Turkey with her father two years later in 1835.

They travelled through France (visiting the monasteries—or *Desarts*—of the Grand Chartreuse en route), then sailed from Marseilles to Constantinople, returning after six months via the Black Sea and the Danube through Hungary to Austria, Germany, and home. Such colourful destinations offered rich pickings for the historical novelist and romantic storyteller Miss Pardoe turned out to be. But in her travel books she wrote about each place—even exotic Constantinople—in a strangely dry and over-analytical manner. We learn about local politics and social habits from a pen so zealously impartial that it cancels itself out. She avoided personal impressions ('I cannot, as a woman, presume to suppose that any weight can possibly be attached to my particular sentiments') and steered clear of local colour ('I am conscious that more than one lady-reader will lay down my volume without regret, when she discovers how matter-of-fact are many of its contents')—both the stuff of which real travel literature is made. Not surprisingly, the only one of these books to achieve any long-lasting success was *The Beauties of the Bosphorus*, with Bartlett's sublime steel engravings to lift the spirit; the rest were promptly forgotten.

❧

ROBERTS, Emma (*c.* 1794–1840)

Scenes and Characteristics of Hindostan, with Sketches of Anglo-Indian Society. London: Wm. H. Allen, 1835; 3 vols., 12mo, pp. 323/310/323

The East-India Voyager, or Ten Minutes Advice to the Outward Bound. London: J. Madden, 1839; pp. 263

Notes of an Overland Journey through France and Egypt to Bombay. London: Wm. H.
Allen [published posthumously], 1841; pp. 333 (pp. xi–xxvii contain a 'Memoir of
the Author')

Emma Roberts was a sort of journeyman writer. On her first visit to India in
1828 she published a volume of poetry (Calcutta, 1830); she edited the
Oriental Observer newspaper there and contributed to countless periodicals
and annuals. She wrote the text to several collections of steel-engraved
plates of India and submitted academic papers to learned journals: anyone
would think she loved the place.

In fact India—or Anglo-India, since none other appears to exist—treated
her with considerable contempt. She went there reluctantly a year after her
mother died: her sister and brother-in-law were to be posted to Bengal, and
there was little else a poor spinster in her mid-thirties could do but accept
the invitation to accompany them. She would join the displaced population
of maiden-lady 'companions' in the Raj: a race apart. Emma hated it: 'There
cannot be a more wretched situation', she wrote later, 'than that of a young
woman in India who has been induced to follow the fortunes of her married
sister under the delusive expectation that she will exchange the privations
attached to limited means in England for the far-famed luxuries of the East.'
She trailed along to Agra, Cawnpore, and Etawa, with the occasional
holiday tour on a ratty, smelly *budgerow* up the Ganges or squashed into a
stifling palanquin, until her sister died in 1831. Then she tried supporting
herself in Calcutta as a writer for a year, before ill health sent her home again
to England.

By 1838 Emma had recovered physically and financially enough to travel
back to India under commission from the *Asiatic Journal* to report on the
novel Overland Route via Alexandria and Suez. She planned to stay for a
year or so collecting material for another book but again her constitution let
her down and she died, in Poona, in the autumn of 1840. Her philosophy
was always to make the best of a bad lot: for the armchair traveller she wrote
romantic poetry and glamorous letterpress, whilst for the practical she
produced the best vade-mecum available, and a ruthlessly objective and
cautionary tale of Anglo-Indian society.

ROMER, Mrs Isabella Frances (d. 1852)

The Rhone, The Darro, and the Guadalquivir; A Summer Ramble in 1842. London:
Richard Bentley, 1843; 2 vols., pp. 416/428, with 4 lithotints (2 hand-coloured) and
7 woodcut vignettes

A Pilgrimage to The Temples and Tombs of Egypt, Nubia, and Palestine, in 1845–6.

London: Richard Bentley, 1846; pp. 375/390, with 5 lithotints (1 hand-coloured) and 8 woodcut vignettes

Isabella Romer was an incurable narrator. She suffered from a disorder familiar to many travel writers: 'no sooner was I on the move than I found the *cacoethes scribendi* creeping upon me; and, unable long to resist the contagion, I began most vehemently to deface all the blank paper within my reach.' A cultured woman, who published a popular novel in 1841, a series of 'light tales' set abroad in 1849, and a learned biography of the Duchess of Angoulême ten years later, Mrs Romer was no stranger to travelling. Italy had been her second home for years, she says, and France was boringly familiar both to her and, she assumes, her readers.

She was in the habit of setting out across the Channel with no fixed destination in mind and leaving the rest to whim and serendipity. Her first travel account describes a colourful tour from Marseilles to Barcelona, through southern Spain to Gibraltar, Tetouan (on the north African coast), and Malta. The second, more serious, follows the 'petticoated pilgrim' up the Nile from Cairo (where she met the celebrated Mrs POOLE) to Thebes and from Beirut to Jerusalem and back. And she never stopped writing, whether trapped in an overturned carriage amongst the 'savage inhabitants' of Alicante, sobbing with compassion at a Malaga bullfight, sipping sherbert in a Turkish harem in Egypt, or swaying sickeningly in a Palestinian palanquin made for two. Anything and everything that happened to her on the move was legitimate subject-matter, and with cheerful equanimity she recorded it all.

❖

SEWELL, Elizabeth Missing (1815–1906)

[Anon.]: *A Journal kept during a Summer Tour, for the Children of a Village School.* London: Longman, Brown, Green and Longmans, 1852; 3 parts, pp. 124/193/203, with 2 maps

[Anon.]: *Impressions of Rome, Florence, and Turin.* London: Longman, Green, Longmans and Roberts, 1862; pp. 330

Mrs Sewell was the author of some forty books: novels (including the famous *Amy Herbert*), histories and theological works, and these two undemanding travel accounts. The *Journal* is one of the few first-hand travel narratives written expressly for children. Those children were pupils of Miss Sewell's own private school on the Isle of Wight, and were able to follow, in three instalments, their mistress's route from Ostend to Lake

Constance and over the Simplon Pass to Genoa. The style she uses is not *too* edifying, thank goodness; with a gentle sense of humour she avoids the stuffiness of a long-distance history lesson (something many contemporary lady travel writers for adults could not manage!). Even though she called herself 'an inexperienced traveller', Miss Sewell does appear to have spent several summers abroad. In 1862 she chronicled a holiday to Italy 'at the request of private friends', and in her autobiography (published posthumously in 1907) she mentions a trip to Dresden in 1870: pleasant journeys, pleasant books.

❖

SHAW, Flora Louisa (Lady LUGARD) (1852–1929)

[Anon.]: *Letters from South Africa by* The Times *Special Correspondent*. Reprinted from *The Times* of July, August, September, and October 1892. London: Macmillan, 1893; pp. 116

[Anon.]: *Letters from Queensland by* The Times *Special Correspondent*. Reprinted from *The Times* of December 1892, January and February 1893. London: Macmillan, 1893; pp. 110

A Tropical Dependency: An Outline of the Ancient History of the Western Soudan with an Account of the Modern Settlement of Northern Nigeria. London: James Nisbet, 1905; pp. 508, with a map

Flora Shaw, one of the most influential foreign correspondents of her time, was born the daughter of a military man in Ireland, the third of fourteen children. Until she was nearly thirty most of her time and energy were devoted to the younger members of the family (her mother died when Flora was eighteen) and only when they had grown up and her father remarried was she free to live her own life. Her twin passions by this time were writing (she published the successful children's novel *Castle Blair* in 1878) and the British Empire. Our colonies abroad, she thought, held the key to solving the problems of poverty and over-population at home: they needed more publicity.

Between 1886 and 1888 Flora earned a modest living writing objective articles for the *Manchester Guardian* and the *Pall Mall Gazette* based on her travels as a lady companion in Gibraltar, Morocco, and Egypt. Their journalistic calibre was so high that on her return, she was chosen as the *Guardian*'s accredited correspondent at an international congress in Brussels on the suppression of the African slave trade (the only woman reporter there) and in 1892 was sent by *The Times* to South Africa. Her dispatches on the parlous political and economic state there were published in book form at the special request of 'the most prominent public men in South Africa', who considered Miss Shaw their most balanced spokesman: her name was made.

Just a year later Flora became *The Times'* Colonial Editor, and was hurried off to Australia as a reward to write about what she regarded as Britain's most vigorous outpost. On the way home she visited New Zealand, Hawaii, the United States, and Canada—to which last place she returned in 1898 for a series of articles on the Klondike Gold-rush. She left *The Times* in 1901, just before her marriage to the diplomat Lord Lugard; with him she travelled and lived in Nigeria and Hong Kong, no longer as a journalist, but as a respected 'colonial wife'. She died a widow at home in Surrey at the age of seventy-six.

Those are the bare bones (it is difficult not to be statistical where Miss Shaw, analyst *extraordinaire*, is concerned). To flesh out the picture we have contemporary accounts of the dauntless lady as vivacious and surprisingly attractive, 'slim and graceful, undeniably beautiful, with a very fine profile and striking eyes'; she was a traveller of 'unusual pluck and nerve' and a political commentator without female equal, 'so much more interested in the thing she was discussing than in herself, that men . . . forgot she was a woman and talked to her as freely as to another man'. In short, she was an honorary male in all but appearance and in that, as her *Times* obituary put it, she was 'a thoroughly womanly woman': the perfect imperial combination.

❧

STARKE, Mariana (?1762–1838)

Letters from Italy, between the years 1792 and 1798, containing A View of the Revolutions in that country, from the Capture of Nice . . . to the Expulsion of Pius VI . . . pointing out The matchless Works of Art which still embellish Pisa, Florence, Siena, Rome . . . with Instructions For the Use of Invalids and Families Who may not choose to incur the Expence attendant upon travelling with a Courier. London: for R. Phillips, 1800; 2 vols., pp. 383/409, with a map

Travels on the Continent: written for the Use and Particular Information of Travellers [title changed in 1824 to *Information and Directions for Travellers in the Continent*]. London: John Murray, 1820; pp. 545 + 300 [appendix]

Mariana Starke was born in Surrey, the daughter of a future governor of Fort St George at Madras, and spent her youth in India. She first made a name for herself back in London as a dramatist (two of her plays, *The Sword of Peace* and *The Widow of Malabar*, were based on India), before retiring in 1791 to accompany a sick friend to Italy. Then, nine years later, she published her first travel book, and her real career began. Miss Starke's Companion Guides—reissued in various forms over and over again—

formed the basis on which her publisher John Murray styled his famous Guides from 1836 onwards; she even pre-dated the ubiquitous Herr Baedeker in providing the tourist with, quite simply, everything he or she might need to know.

In the first book she confined herself to Rome and Tuscany, and in the second (which absorbs much of the first, being written after a reconnaissance visit in 1817–19) she expanded her coverage to France, Switzerland, Germany, Hungary, the Netherlands, Spain, and Portugal, with notes on Sweden, Denmark, Norway, and Russia for good measure. Both include exhaustive details of expenses, itineraries, boarding-houses, food, and luggage requirements, regularly revised (either through first-hand knowledge or meticulous research) by Mariana herself. 'Pirate' editions began to appear on the bookstalls of the European capitals and soon the 'Starke' became an indispensible part of any serious tourist's paraphernalia.

Not only did Mariana point out the highlights of travel, and explain its mechanics, but she also gave pithy warning of its perils. Some places were surprisingly unhealthy ('we saw at Nice no instance of recovery from pulmonary complaint'), some dirty (sulphuric acid is recommended as a sure-fire water-purifier) and some plain dangerous: only recently were street lamps being installed in large towns to put a stop to 'that dreadful practise of assassination'. Nevertheless, Mariana herself managed to survive the pitfalls for a good seventy-five years, dying on the way back from a trip to Milan in 1838 of nothing more exciting than old age.

❧

STONE, Olivia M.

Norway in June . . . Accompanied by a Sketch Map, A Table of Expenses, And A List of Articles Indispensible to the Traveller in Norway. London: Marcus Ward, 1882; pp. 448, with 39 wood engravings and a map

Tenerife and its Six Satellites: Or, The Canary Islands Past and Present. London: Marcus Ward, 1887; pp. 477/459, with 72 wood engravings and 9 maps

A rather snide remark in *The Atheneum*, in a review of Mrs Stone's book on the Canary Islands, called the author a 'pioneer of picnic travel': it sums her up very well. Her aim in writing these hefty tomes was to provide touring gentlefolk with everything they needed to know in order to spend as comfortable and picturesque a time as possible in what were in the 1880s still rather outlandish holiday destinations. She and her husband personally covered every inch of the strictly regimented itineraries she recommends to her readers and impeccably researched the local history. She pads out the narrative with hordes of useful facts—how much luggage a Norwegian carriole can carry, for example, or how long a boarding-house bed tends to be—and even goes so far as to analyse the local mineral waters: what more could the 'picnic traveller' require? Just one thing: a guidebook perhaps a third the size and weight of Mrs Stone's.

⬧

STRUTT, Elizabeth

[Anon.]: *A Spinster's Tour in France, The States of Genoa, &c. during the year 1827*. London: for Longman, Rees, Orme *et al.*, 1828; 12mo, pp. 427

[Anon.]: *Six Weeks on the Loire with a Peep into La Vendée*. London: W. Simpkin and R. Marshall, 1833; pp. 408, with 4 etched plates

Domestic Residence in Switzerland. London: T. C. Newby, 1842; 2 vols., 12mo, pp. 282/288, with hand-coloured lithographed frontispiece and title-page vignette to each volume

Good Mrs Strutt was apparently one of the early nineteenth century's most prolific ladies of letters, writing countless novels between 1806 and 1860 (most sunk without trace) and producing these three travel accounts in what was presumably her prime. They are rather unexciting guidebooks in narrative form, encrusted with the odd flourish of romantic poetry and aimed at a decidedly feminine audience.

The first sets out to prove that spinsters do not have to limit their pleasures to the drawing-room: a little preparation, a sanguine heart, and Mrs Strutt's book are all that are needed before 'the "Lady Traveller" . . . may begin her peregrinations'. The second, written when the author was a spinster no more, still favours the unaccompanied female in its hotel recommendations and carefully points out the most edifying and uplifting scenes for impressionable hearts along the gentle stretch of the Loire from Tours to Nantes. By the time Mrs Strutt reached Switzerland she was confident enough a Lady Traveller herself to plan out a whole itinerary for her readers (concentrating on the Vaud canton), and to regale them with the received philosophy of the Englishwoman abroad: 'The great end of travelling, I was going to say the only thing that can excuse it, ought to be the transplantation to *home* of all that we may see abroad, really deserving of imitation.'

There is no published record of Mrs Strutt's life after Switzerland— except for the notice that in 1863 she was granted a small government pension 'in consideration of her straitened circumstances at a great age and after fifty-eight years of contributions to literature': rather a depressing end.

⬧

TAYLOR, Ellen M.

Madeira: Its Scenery, and How to See It. With Letters of a Year's Residence, and Lists of the Trees, Flowers, Ferns, and Seaweeds. London: Stanford, 1882; pp. 261, with wood engraved frontispiece, a plan, and a map

During the Victorian period Madeira reached its peak as an easily accessible haven for moneyed invalids who might spend a year or more basking in its temperate and fragrant airs. Ellen Taylor went out there with a sick friend in 1880, and on her return produced this exhaustive guidebook and short narrative (the *Letters of a Year's Residence*) with just such visitors in mind. Madeira was not supposed to be adventurous: one lived in an hotel or in a hired *quinta*, and rationed one's excitement to the odd nature ramble in a portable hammock or a stroll along the level path of a *levada*, or waterway; both the island and its guidebook were dedicated to the passive pursuit of health.

❧

TROLLOPE, Frances (1780–1863)

Domestic Manners of the Americans. London: for Whittaker, Treacher, 1832; 2 vols., pp. 336/271, with 24 lithographed plates

Belgium and Western Germany in 1833; including visits to Baden-Baden, Wiesbaden, Cassel, Hanover, The Harz Mountains [etc.]. London: John Murray, 1834; 2 vols., pp. 329/298

Paris and the Parisians in 1835. London: R. Bentley, 1836; 2 vols., pp. 418/412, with etched half-titles and 12 etched plates

Vienna and the Austrians; with some account of a journey through Swabia, Bavaria, the Tyrol, and the Salzbourg [etc.]. London: R. Bentley, 1838; 2 vols., pp. 388/419, with etched half-titles and 12 etched plates

A Visit to Italy. London: R. Bentley, 1842; 2 vols., pp. 402/396

The excellent Mrs Trollope—Anthony's mother—is many people's favourite lady traveller. She it was who scandalized the smug and vulgar Americans with a spirited exposé of their 'Domestic Manners' in 1832; who wrote admiringly of the wily Germans and fondly of the indulgent French, analysed the stuffy Austrians, and fell in love with the irresistible Italians; and all with such style and confidence.

Her story is a beguiling one. She married Casaubon-like Thomas Anthony Trollope, an earnest barrister in 1809; and as their family grew (seven children in nine years) its fortunes fell, until one day in the autumn of 1827 Fanny decided the time had come to go treasure-hunting. She took up an invitation from a friend, Frances Wright, who had set up a colony for the rehabilitation of Negro slaves in Tennessee. The colony, called Nashoba, attracted Fanny: she imagined it as a sort of Utopia, full of philanthropic (and financial) promise. Taking some children with her, she arrived in New Orleans two months after setting out and proceeded up the Mississippi to Nashoba, near Memphis, only to be met by scenes of utter desolation. The land was acrid and sad, the slaves miserable, and Miss Wright less than welcoming. None deterred, Fanny moved on to Cincinnati to establish a centre for local 'refinement'—a seemly sort of bazaar offering kiosks, games rooms, concerts, a library, and other such delights designed to educate the dull Ohio worthies. The project failed miserably, and as a last resort to raise money, Fanny (by now joined by her desiccated husband) decided she would have to write a book. Inspired by a recent publication of Captain Basil Hall (in which he openly criticized the Americans) and her own well-developed sense of caricature, she travelled from Baltimore in the south to Niagara in the north preparing the work which, at the Americans' expense, became an instant best seller both here and in incredulous New York. It was the first of her many books (mostly novels), and the best known of them all.

Fanny found that she enjoyed travel for its own sake ('There certainly is in the blood of our race a very decided propensity to locomotion'); making sure she had secured commissions for books before she went, she made prolonged forays with various members of her family throughout the Continent, healthily laughing at the Briton abroad and the foreigner at home in equal measure. After her first rapturous visit to Italy in 1841 she made Florence her home, and there she died—by now a celebrated and rather befuddled English Personality—at the age of eighty-three.

TWEEDIE, Mrs Alec (Ethel Brilliana) (d. 1940)

A Girl's Ride in Iceland. London: Griffith Farran, Okeden and Welsh, 1889; pp. 166, with 18 sketches and a map

A Winter Jaunt to Norway: With Accounts of Nansen, Ibsen, Björnson, Brandes, and Many Others [etc.]. London: Bliss, Sands and Foster, 1894; pp. 316, with 26 halftones

Through Finland in Carts. London: A. and C. Black, 1897; pp. 366, with 18 halftones and a map

Mexico As I Saw It. London: Hurst and Blackett, 1901; pp. 472, with 2 wood-engraved plates, 115 halftones, and a map

Sunny Sicily: Its Rustics and Its Ruins. London: Hutchinson, 1904; pp. 392, with 13 halftones and a map

America As I Saw It: Or, America Revisited. London: Hutchinson, 1913; pp. 395, with 16 sketches and 36 halftones

Mainly East. London: Hutchinson [1923]; pp. 320, with colour frontispiece, 52 halftones, and a map

An Adventurous Journey (Russia–Siberia–China). London: Hutchinson [1926]; pp. 397, with 66 halftones, 4 wood engravings, and 2 maps

Tight Corners of My Adventurous Life. London: Hutchinson [1933]; pp. 315, with 29 halftones and 4 wood engravings

My Legacy Cruise (The Peak Year of my Life). London: Hutchinson [1936]; pp. 319, with 4 colour plates and 66 halftones

'Will you take a little jaunt with me, unknown friend—you see, I take it for granted you are a friend. So let us go a jaunt together . . . '

It is easy to tell that Mrs Alec Tweedie had a style (thank goodness) all her own. She was somewhat of an Impulsia GUSHINGTON, a mistress of the melodramatic (while the journeys she described were often comparatively tame), and one of those unctuously humble women who proclaim their modesty so often that it is liable to total eclipse. That 'little jaunt' she was just talking about, for example, was actually a two-years' accumulation of 50,000 miles. But if Ethel gushed, at least she did it ingenuously. She saw herself quite simply as a pillar of Edwardian womanhood whose duty it was to enchant others as much as she enchanted herself.

The travel books Ethel produced, following hard on a plethora of journalistic articles on everything from whether or not women should ride astride to how one should cope when the rudder of one's liner snaps in mid-Atlantic, and accompanied by biographies, autobiographies, and books on etiquette and anti-Communism, were justly popular at first. They were written to keep Ethel's head above water during the devastating years when her father and husband died and both sons were killed. While the journeys they describe were rarely as adventurous as those of her peers, the way in which she described them was so heavily perfumed with gay, egotistic enthusiasm that it is difficult not to succumb. As she got older, however, her self-appointed position as British Ladies' Oracle became embarrassingly pronounced, with book titles like *My Table-Cloths: A Few Reminiscences* verging on the surreal. The melodramatic quality of her travel writing began to smack of bathos and it was only Ethel's unerring sense of admiration for herself that saved her from literary ridicule. And I suspect her work suffers in retrospection: she cannot have been alone in considering herself one of the country's most fashionable travel writers, and she certainly covered a good few thousand miles more around the world than most women of her time. Clichés were made for writers like Ethel: she was one of the old school, a game old bird, and a law unto herself.

9

Ornaments of Empire

THERE are no mere Women here. They are Ladies all (many literally so), the favoured daughters of Albion sent out into a naughty world to teach it how to behave. It should be stressed that the world, in this case, means those parts shaded nice and pink on the atlas: the glorious British Empire. It was hardly likely to cramp their style, confining themselves like that: the Union Jack billowed bravely over one quarter of the earth's surface at imperialism's height at the end of the nineteenth century, which was when most of the books in this chapter were published.

It is not surprising, then, that many of these books have a markedly parochial air about them. At worst, it is a sign of the bigotry being British tended to engender in certain susceptible breasts. Zélie COLVILE's *Round the Black Man's Garden* is a prime example, the garden being Africa and the Black Man, who thinks it is his, a laughable little fellow. At best, it means that the writer is not only sympathetic to her subject, but is trying to learn from it herself. So Julia MAITLAND's commentary on India refreshed an audience brought up on the assumption that all natives are savages, by telling how much she admired some of the Hindu women she met (and, incidentally, how much she despised some of the missionaries bent blindly on converting them into bewildered little Christians). Flora STEEL was a kindred spirit—she even published one of her books in Urdu, unwilling to talk about the local people behind their backs.

Annie BRASSEY, the heroine of many a jolly seafaring tale, hardly ever left Britain at all. Her yacht *Sunbeam* was her own mobile empire, and she the popular empress on a continuous sort of progress, or state visit, to anchorages all over the world. Some of the diplomatic or army wives dotted about the globe managed the same illusion. Thus there were little Englands to be found in the hill stations of Burma, the Hottentotty African Cape—even the unhealthy climes of Sierra Leone. Standards were kept up with the help of clubs, afternoon levées, nightly theatrical or musical performances, and, of course, the writing every Sunday of letters home soon to find themselves in print.

These diplomatic appendages (several of whom have strayed into chapters elsewhere) were the most obvious ornaments of empire. It was their duty to be British at all times, which to one of them at least proved terribly taxing: 'Please to remember', wrote Emily EDEN from India, 'that I shall return a wornout woman.'

Not all of them minded so much. Lady DUFFERIN tackled the rather limited life she was allowed to lead with great gusto: there was always travel to take one's mind off the petty etiquette and conventions of Victorian diplomacy. She went as far afield as possible on a variety of expeditions during official residences in India, Canada, Russia, and Turkey. Mrs CLEMENTI, whose husband acted as Colonial Secretary in Georgetown, Guyana, spent her leave literally beating a path to Mount Roraima (8,625 ft.)—and not only to it, but up it. Right up it. Lady Florence DONALDSON's holiday was even more original: she and her husband decided to ramble about the Himalaya taking pictures with their new cameras for six weeks. They usually managed to dress for dinner, of course, no matter how far their snow-laden tent might be from the nearest British outpost: Florence never forgot she was a Lady.

How eccentric! Yet it was not really. Members of the British Empire of a certain class were entitled to do the most extraordinary things abroad. Eccentricity was expected of them.

BARKLY, Fanny A.

Among Boers and Basutos: The Story of our Life on the Frontier [etc.]. London: Remington, 1893; pp. 270

From the Tropics to the North Sea, including Sketches of Colonial Life; Five Years in the Seychelles . . . an Interlude at the Falklands . . . followed by Promotion to Heligoland, the

Gem of the North Sea. London: Roxburghe Press [1897]; pp. 252, with 7 halftones, 2 sketches, and a map

If ever there was a suitable candidate for success as a Victorian travel writer it is Fanny Barkly. She was the daughter of a Bishop of Mauritius—a nicely exotic upbringing for a start; then she married into one of the best-established diplomatic families in the empire and found herself keeping house (for the first time in her life) in Thaba-Bosiu, otherwise known as 'Advance Post, Cannibal Valley' during the Basuto Rebellion of 1879. She managed an astonishing escape from the siege of Mafeteng (now in Lesotho) that same year by stowing away in a wagon bound for the Orange Free State and living on locusts, only to endure even worse hardships later when the Seychelles were forcibly cut off from the rest of the world after a smallpox epidemic and the Barkly family, then resident on Mahé, were reduced to eating their pet tortoise to survive. And later still, after her husband had returned from a brief spell in the Falklands, Fanny witnessed Heligoland's last year as a British colony before its cession to Germany in 1888. So even for a traveller, her life was unusually rich. Then why has no one heard of her? For the simple reason that Mrs Barkly was *not* a writer, and sadly, for all the exciting tales she has to tell, these two travel books are resoundingly dull.

❦

BARNARD, Lady Anne (1750–1825)

South Africa a Century Ago: Letters Written from the Cape of Good Hope 1797–1801. Edited by W. H. Wilkins. London: Smith, Elder, 1901; pp. 316, with a portrait frontispiece

Lady Anne Barnard and the Cape of Good Hope 1797–1802. Edited by Dorothea Fairbridge. Oxford: Clarendon Press, 1924; pp. 344, with coloured portrait and 59 halftones

These two books comprise the letters and journals of the wife of Britain's first Colonial Secretary to the Cape of Good Hope. Their Scottish author was a certain Miss Lindsay who caused considerable social excitement by marrying at the grand old age of forty-two the dashing thirty-year-old son of the Bishop of Limerick, Andrew Barnard. Together they travelled out to the 'lately captured' Cape in 1797 to join the governor Lord Macartney in his first administration there.

As chronicles of travel and life abroad, the letters and journals are fresh and appreciative, and quite unmannered. Anne was entranced by the 'Kaffes' (Kaffirs) whose tribes reminded her (rather wistfully) of the Scottish clans, and their homelands inspired a series of detailed and delicate sketches. She enjoyed travelling, making frequent expeditions with Barnard up Table Mountain or into what was darkly known as 'the interior'. She never mentions the hazards of such journeys, only the pleasures.

Lady Anne was on one of her frequent solo visits home to Britain when news came of Barnard's untimely death in an accident up-country in 1807; she never travelled again.

◆

BECHER, Augusta Emily (1830–1909)

Personal Reminiscences in India . . . 1830–1888. Edited by H. G. Rawlinson. London: Constable, 1930; pp. 230, with a frontispiece

Augusta Becher (née Princep) was actually born in transit, on board the *Duke of Lancaster* off the Cape. Much of her life was spent in transit too. When they began their wedding voyage to India she and Bengal Army officer Septimus Becher expected to settle in Lucknow, and it is there that their first child was born. But within a year they were ordered to the fashionable hill station of Simla, where 'Sep' was to be Assistant Adjutant-General. This meant a harrowing journey by 'equirotal' (a box on wheels pulled by sixteen coolies) towards the Himalayas, during which the baby died—and no sooner had they settled in Simla than the Mutiny broke out. Sep was up-country at the time, and as an unaccompanied woman (like Kate BARTRUM) Augusta was forced to leave home to make the 33-day trek through the searing heat to the coast. The passage home was no better: two people died aboard the notorious old steamer *Southampton* where, for lack of space, one of the bodies had to be stored in the piano. A year later Augusta returned to Calcutta to care for Sep, who was ill; each of them was to make the passage twice again before eventually settling to the quiet life they had planned fourteen years ago, in Barrackpore.

◆

BRASSEY, Lady Anna (Annie) (1839–1887)

A Voyage in the Sunbeam: *Our Home in the Ocean for Eleven Months*. London: Longmans, 1878; pp. 504, with 9 wood-engraved plates, vignettes throughout, and 8 maps and charts

Sunshine and Storm in the East: Or, Cruises to Cyprus and Constantinople. London: Longmans, 1880; pp. 448, with 9 wood-engraved plates, vignettes throughout, and 2 maps

Tahiti. With Thirty-one autotype illustrations . . . by Colonel Stuart-Wortley. London: Sampson Low, 1882; pp. 68

In the Trades, the Tropics, and the Roaring Forties. London: Longmans, 1885; pp. 532, with wood-engraved vignettes throughout and 10 maps and charts

The Last Voyage, to India and Australia, in the Sunbeam. London: Longmans, 1889 [published posthumously]; pp. 490, with 20 lithotint plates and a title-page vignette, wood-engraved vignettes throughout, and 2 maps

The Victorian public welcomed Annie Brassey's books with an enthusiasm reserved nowadays for episodes of a soap opera. She was rich, carefree, idyllically happy with a handsome, distinguished husband and bonny children, and spent her life enjoying genial adventures at sea on what seemed like a non-stop holiday. Actually, Lord Thomas Brassey did discharge the odd duty as Admiralty Secretary and diplomat here and there on their travels, but these were never allowed to intrude on the sailors' pleasure. The adventures encountered aboard their beloved schooner *Sunbeam* were usually quite manageable: it is true the deck once caught fire, but there were about thirty-five people on board at the time to extinguish it; once or twice they were able to come to the gallant rescue of a shipwrecked crew less fortunate than themselves, but hardly ever did anyone fall off the *Sunbeam*.

There is no doubt that Annie Brassey admired such contemporaries as Isabella BIRD and Marianne NORTH, but she probably thought their exertions a little *de trop*. The world we see through her eyes is better prepared than theirs, and more distant and refined, since she had a portable little England to return to every night. By rights, she should have led a comfortable old age, regaling countless eager grandchildren with tales of a Lady's life at sea; as it was, Annie Brassey's merry story was cut miserably short when she died suddenly of malaria on the way home from India and Australia in 1897 and was buried at sea at the age of forty-eight.

❖

CANNING, Charlotte (1817–1861)

Hare, Augustus: *The Story of Two Noble Lives: Being Memorials of Charlotte, Countess Canning, and Louisa* [her sister], *Marchioness of Waterford.* London: George Allen, 1893; 3 vols., pp. 381/489/495, with 29 photogravures, 1 steel- and 6 wood-engraved plates, and vignettes throughout

A Glimpse of the Burning Plain: Leaves from the India Journals of Charlotte Canning. Edited by Charles Allen. London: Michael Joseph, 1986; pp. 170, illustrated throughout in colour and black and white, with endpaper map

Charlotte Stuart de Rothesay—vaguely related to Lady Mary Wortley MONTAGU—was appointed in 1842 to the arduous but prestigious post of

Lady of the Bedchamber to Queen Victoria. It meant that she must accompany the Queen wherever Her Highness might go, whether it be to Balmoral or Constantinople, and by the time of her retirement in 1855, she had blossomed from an assured young woman of elegant pedigree into a well-travelled and utterly cultured Lady. So she was perfectly qualified for her next occupation as the Governor-General (and later first Viceroy) of India's wife, and when she and Lord Charles arrived at their sumptuous lodgings in Calcutta she should have had every confidence in her ability to negotiate the frothy feminine waters of Anglo-Indian diplomacy.

In fact, the first lady of the Raj was miserable. She was of such high rank that no one dared speak to her, and she was expected to do nothing but regally sit. 'I never knew what idleness was before', she said; nor had she ever known loneliness like the loneliness of the Governor's Residence in Calcutta. Only in the fresher and more relaxed climes of Barrackpore, where the household repaired in the summer heat, did she manage to enjoy herself in rambling and sketching, or in the cool and sweet-aired Nilgiri Hills. Even travel, the anodyne of many a bored Englishwoman abroad, offered little relief for Charlotte. The two Grand Tours in which she was involved were exhausting, laborious affairs. The first was made to the north of Allahabad immediately after the Indian Mutiny—which she spent, incidentally, in Calcutta, sending regular reports home in letters to the Queen and comforting the refugees from Lucknow and Delhi with great devotion. It comprised a slow, 24-mile procession of camels, bullocks, and men, and there was little leisure for anything but diplomatic posturing. And the second across to Jabalpur wellnigh killed her.

Photographs of Charlotte Canning show far more graphically than her diaries and letters do what a change India wrought in her. She arrived radiant and expectant; six years later, worn down by loneliness, the preoccupation of her husband, the stifling rigidity of Raj society, and, of course, the climate (which she met like all her kind in a vice-like bodice of stiff black silk), she had turned into a gaunt and frail old woman. She died of fever in Calcutta in 1861, and is buried in her garden at Barrackpore.

CHURCH, Mary

[Anon.]: *Sierra Leone, or, The Liberated Africans, in a Series of Letters from a Young Lady to her Sister in 1833 and 1834*. London: Longman, 1835; pp. 49

Miss Church was one of the several European observers resident in Sierra Leone during the early years of its establishment as a colony for 'The Abolition of the Slave Trade, by means of the propagation of Christianity and the advance of Commerce and Civilisation'. Her letters home from

Freetown are revealing both of the country itself and the rather uneasy social arrangements there: although she would not actually want to *shake hands* with a Liberated African herself, she can sympathize with them, she says, suffering indignities as they do even under Britain's halcyon protection. She enjoys the look of the land, the cosmetic cleanliness of Freetown and neatness of the native costumes, and soon gets used to the nightly cacophony (worse than London, she says) of frogs, dogs, monkeys, poultry, and psalm-singing. Mary Church even travels within Sierra Leone itself, frequently and daringly taking rides alone, and sometimes sailing out with friends to the islands just offshore: like most of the British women abroad at this time, she is happy to look, but not to touch.

CLACY, Mrs Charles (Ellen)

A Lady's Visit to the Gold Diggings of Australia, in 1852–53: Written on the Spot. London: Hurst and Blackett, 1853; pp. 302, with engraved frontispiece

Lights and Shadows of Australian Life. London: Hurst and Blackett, 1854; 2 vols., pp. 304/310

Ellen Clacy wrote many books, some of them under the pseudonym of 'Cycla', and most moral tales of one sort or another. These are the only two concerning her six-month stay in Australia, *Lights and Shadows* being a collection of anecdotes 'founded on facts' about the emigrants, old colonials, and gold-digging opportunists she met there, and the first book taken from her own journal. *A Lady's Visit* is an unsentimental and unpompous account (private feelings being 'only so much twaddle') of life in the creeks and gullies around Bendigo, Victoria, and their curious population, assorted and nomadic, in its pursuit of rich pickings.

Ellen had gone to Australia with her brother, Frank, inspired by fulsome newspaper descriptions of the new 'auriferous regions'; Frank proved a tolerably successful prospector and stayed on, and perhaps it is surprising at first that Ellen decided not to join him. She agrees with Mrs CHISHOLM that emigration can be a profitable and fortunate vocation, and spends a chapter on advising who should do it. Women should, especially: after warning against the hazards of bush life (especially at night: 'murder here—murder there—revolvers cracking—blunderbusses bombing'), she says 'the worst risk you run is that of getting married, and finding yourself treated with twenty times the respect and consideration you may meet with in England.' And Ellen should know: the voyage home was her honeymoon.

CLEMENTI, Mrs Cecil (Marie Penelope Rose)

Through British Guiana to the Summit of Roraima. London: T. Fisher Unwin, 1920; pp. 236, with 14 halftones and a map

The Colonial Secretary and his wife had been living in Georgetown for two years when in December 1915 they decided to take a holiday. Georgetown, in what is now Guyana, sits at the mouth of the Demerara river; they planned a journey that would take them, their guide, and porters up that river as far as the village of Wismar, then up the Essequebo and Potaro rivers (with short overland stretches by rail and dogcart) to the unmapped hills on the borders of British Guiana, Venezuela, and Brazil. Marie Clementi's objective account describes the 43-day journey, during which she helped cut what became a permanent trade route from the hills to the sea, and climbed through jungle and savannah to make the first white female ascent of Mount Roraima (8,625 ft.).

❧

COLVILE, (Lady) Zélie

Round the Black Man's Garden. Edinburgh: Blackwood, 1893; pp. 344, with 32 halftones, 20 vignettes, and 2 maps

The title says it all, really. Her husband had suggested substituting 'Gentleman of Colour' for 'Black Man'—but of course they are *not* gentlemen, said Zélie, and Sir Harry had to agree; he had certainly never seen one at *his* club, anyway.

The voyage the Colviles undertook looks very impressive on paper. They visited several of the Red Sea ports on their way from Suez to Durban (taking in Madagascar on the way) and then turned inland to Pretoria and Cape Town before completing the circuit to Brindisi (their starting-point) by steaming up to Cameroon and Sierra Leone and touching on the Canary Islands. A fully organized expedition with ninety-eight porters ferried Zélie across Madagascar in a *filanza*, a sort of dining-chair on poles, in true pioneering style. But for all this 'serious' travel, she seems to have remained remarkably unimpressed by the exotic charms of Africa, except to note that a Cape menu is likely to offer for a banquet 'Big fish cold', 'All dem sweet mouth', and 'stink butter' (salmon, dessert, and cheese), and that the natives are often to be found hanging their brothers from trees with hooks in the heels. All this she accepts with aloof equanimity, and the book closes with the Colviles at home once again, 'dining quietly in our little house in Chapel Street, hardly able to realise that we had ever left it.'

❧

COLVILLE, Mrs Arthur (Olivia)

A Thousand Miles in a Machilla: Travel and Sport in Nyasaland, Argoniland, and Rhodesia, with Some Account of the Resources of those Countries [etc.]. London: Walter Scott, 1911; pp. 131, with 50 halftones and a map

Olivia Colville—née Spenser-Churchill—trod much the same ground (or at least her hammock-carrying porters did) as Helen CADDICK before her. Mrs Colville's attitude to the land and its people was completely different, however. She accompanied her sportsman husband to East and British Central Africa (now absorbed into Malawi, Mozambique, Zambia, and Zimbabwe) and in true imperial tradition considered every inch of African soil her own, rather resenting the vain, obtrusive natives there. Swinging along in her double-poled *machila* on a circular route from Chinde to Kopa, across to Broken Hill (Kabwe) in the west, and back round to Bulawayo and Beira (Sofala), we can imagine her barking instructions and being terribly 'brave' and tiresome. She was very much a passive traveller, taken along for the ride with no object of her own but to have *been* there—and it was, of course, a long and admirable journey for her to make as a 'non-traveller'. It is just a pity her account of it is so unpalatable.

◆

DONALDSON, Lady Florence Annesley

Lepcha Land: Or, Six Weeks in the Sikhim Himalayas. London: Sampson Low, 1900; pp. 213, with 24 halftones, vignettes throughout, and a map

The journey from Calcutta (which is where the Donaldsons were stationed) to the Tibet–Sikkim border involved more changes of climate and terrain than they had bargained for. They chose 'Lepcha Land' for their summer 'photography holiday' for its cool freshness after a steamy spring and had no idea that they would be able to snap tigery jungles, desert-like plains, and snow-blocked passes all during six short weeks. With only a handful of coolies and a single servant, things sometimes became a little difficult—one can imagine how awkward it must have been trying to dress for dinner on some God-forsaken slope in a tent buckled by snow—but Lady Florence was admirably unperturbed. Between visits to isolated officers' messes, she enjoyed the primitiveness of it all. There is one marvellous episode when on meeting some merchants on the road from Lhasa to Kalimpong she was faced with the ticklish social dilemma of how to tell a Tibetan one doesn't think his yak-fat tea is *quite* the thing . . . The tour finished in suitable style, however. Making for Darjeeling, Lady Florence spotted a train (of all things) and they leapt aboard, coolies and all, arriving 'just in time to climb up to Woodlands Hotel . . . to join the table-d'hôte dinner'.

❖

DUFFERIN, Lady Harriot (Blackwood) (1843–1936)

Our Viceregal Life in India: Selections from my Journal 1884–1888. London: John Murray, 1889; 2 vols., pp. 344/346, with a portrait plate and a map

My Canadian Journal 1872–8 . . . Extracts from my Letters Home written while Lord Dufferin was Governor-General. London: John Murray, 1891; pp. 422, with 8 wood engravings and a map

My Russian and Turkish Journals. London: John Murray, 1916; pp. 350, with 12 halftones

Lady Harriot Dufferin was a gracious and appreciative woman. While her husband Lord Frederick wrote volumes on the political consequences of his diplomatic positions in Ottawa, St Petersburg, Constantinople, and India, she steered well clear of business and wrote instead of almost everything else. She was an Irish girl with few of the trammelled imperial narrow-mindedness one might expect of a viceroy's wife. In Canada, indulging in yachting trips or fishing on the St Lawrence, she was noted for her 'plain and sensible' clothes and liking for the muddier corners of the Dominion; in Russia she enjoyed a bear-hunt with the best of them, and she learned the complexities of cricket in Constantinople. She founded a Trust for medical aid to native women in India, and was diligent in visiting mission schools and hospitals there. And in all the countries she visited, she travelled whenever she could, with an ever-growing brood of little Dufferins behind her. Of course, it was part of her official duty to travel, but Lady Harriot was the perfect woman for the job.

❖

EDEN, The Honourable Emily (1797–1869)

Portraits of the People and Princes of India. London: J. Dickinson, 1844; pp. 24, with 24 lithographed plates and a lithographed list of plates

'Up the Country': Letters written to her Sister from the Upper Provinces of India. London: R. Bentley, 1866; 2 vols., pp. 302/263

Letters from India [etc.]. Edited by her Niece Eleanor Eden. London: R. Bentley, 1872; 2 vols., pp. 352/297

Miss Eden's Letters. Edited by her Great-Niece Violet Dickinson. London: Macmillan, 1919; pp. 414, with 4 portraits

EDEN, The Honourable Frances (1801–1849)

Tigers, Durbars and Kings: Fanny Eden's Indian Journals 1837–1838. Transcribed and Edited by Janet Dunbar. London: John Murray, 1988; pp. 202, with halftones throughout and a map

Emily Eden's book '*Up the Country*' is well known. It is a wittily entertaining account of the eighteen months she spent in the upper provinces of India, beginning with a journey from Calcutta to Simla in 1837, as first lady of the Raj. It was an enormous undertaking—a six-month epic that began with steamers and barges from Calcutta up the Ganges to Benares (Varanasi), followed by a succession of carriage, palanquin, sedan-chair, horse, and elephant rides through Allahabad and Delhi to the hill station, with frequent diplomatic stops along the way.

Emily and her sister Fanny had accompanied their brother Lord Auckland to Calcutta on his appointment as governor-general in 1835. They were the only unmarried sisters of a family of fourteen, and had lived with their eldest brother for several years; that they should not join him on his most prestigious posting was never even contemplated. Not that Emily wanted to go—she was a busy and admired society hostess in intellectual Whig circles, and a beloved—if rather eccentric—maiden aunt. She endured her six years in India as an exile, but, true to her breeding, made the best of it.

Fanny Eden—some of whose letters are included in Violet Dickinson's edition above—was a far better traveller than Emily. Fanny enjoyed the long voyage out, whilst Emily and her spaniel Chance scarcely surfaced, and Fanny once joined a tiger-hunting expedition (jauntily described in her journals) while it was the most Emily could do to stagger around Calcutta. A robust sense of humour stopped Emily from being maudlin, however, and indulging her mild prejudices and writing these marvellous letters home kept her occupied. In fact these letters probably took up more time than they should have done. Although they dutifully visited local hospitals and schools, and did their share of rather tiresome official entertaining, neither of the Misses Eden was exactly a paragon of feminine English refinement. In fact Miss FANE declared on meeting them that 'both are great talkers . . . both ugly, and both s—k like polecats!' A kind heart and appreciative eye won Emily respect, however, not least from the notorious old maharajah Ranjit Singh, whom she met in Kashmir and who proceeded to shower her with rather embarrassing gifts of jewels. She in turn was almost won over by India (although constantly counting the days to her return home).

What seems to have impressed Emily most on her journey 'Up Country' was not the poverty and political disarray of India—although she was not at liberty to publish her political views—but the material splendour of its princes. She painted a series of portraits of them—great, rich watercolours—which were published on her return to England and much admired. She also published two novels, *The Semi-detached House* in 1859 and *The Semi-attached Couple* in 1860. These were popular, but it is *'Up the Country'* that has been reprinted most, as a classic of the literature of the Raj.

❖

ELLIS, Beth

An English Girl's First Impressions of Burmah. Wigan: R. Platt, London: Simpkins, Marshall, *et al*, 1899; pp. 248

This is one of the funniest travel books ever written. Beth Ellis went on to write a handful of novels, but this was her first literary attempt and must be her best. She was invited to winter in Burma by her sister, whose husband managed a remote hill station she calls Remyo. Beth's journey to Remyo was broken at Rangoon (full of stout European ladies busily and inexplicably writing 'chits' from dawn till dusk); then followed a long train journey to Mandalay (full of lean Americans endlessly singing 'On the Road to Man-da-la-ay . . . ') and a trek up-country by bullock cart and horse. She had never ridden a horse before (I thought *all* lady travellers were horsewomen!) and at first could only make the beast go backwards. Sensibly, she tried to turn it round so that at least they could travel in the right direction; at last it bolted, jolting Miss Ellis's outsize sola topi over her face, where it stayed for much of the journey. Once at Remyo, she scandalized both European and Burmese by producing a bicycle and proceeding to patrol the jungle on it. She once joined a shoot, but disgraced herself by spending most of it up a tree, wielding her brolly uneasily and feeling sorry for the tiger. Her descriptions of such excursions and of life in a down-market hill station are irresistibly merry—but she is not always quite so flippant. Miss Ellis grew to respect her fellow Europeans; she would miss their empire-building spirit. She would miss the peeling splendour of Remyo too, with its impossibly out-of-tune piano (staunchly ignored) and nine-hole golf-course (mostly bunkers). Most of all she would miss the Burmese themselves: 'such jolly people and such thorough gentlefolk'.

❖

FALCONBRIDGE, A(nna) M(aria)

Two Voyages to Sierra Leone, During the Years 1791–2–3, In a Series of Letters [etc.]. London: for the Author, 1794; 12mo, pp. 279 (and appendix, pp. i–iv)

This is the first account by an Englishwoman of West Africa. In 1790 one Dr Falconbridge, an ardent Abolitionist, was commissioned by the St

George's Bay (later the Sierra Leone) Company to help establish a British colony of liberated slaves. This involved making two voyages to the 'white man's grave' of Sierra Leone: the first to gather together what scattered settlers were already in the area and help negotiate the purchase of land; the second to help actively establish a town for these and for new 'free blacks' imported from Nova Scotia. He must have been rather surprised when his new wife, a young Bristolian lady, announced she was coming too.

Falconbridge was an idealist, an enthusiast; his wife reckoned she saw things as they really were. On her first visit to Sierra Leone she was appalled to find that prostitutes (of whom few had survived) had been press-ganged into sailing out from England to provide services and stock for the settlement, and that the altruistic Europeans out there helping them were each furnished with their own black mistress. The management of the new colony was haphazard—Falconbridge himself was appointed Commercial Agent on their second voyage, a position for which he was singularly ill fitted, and his financial incompetence eventually resulted in dismissal. He died soon afterwards from fever or (as Anna Maria thought more likely) from drink. Anna writes that she was much relieved by his death—he was such an irritable, unkind husband and had become rather burdensome to her. Obviously: within a month she had married again, remarking proudly that although three-quarters of the Europeans who had travelled out to the settlement in 1792 had died, she herself had never felt better.

With no patron to please—in fact with a grudge against the Sierra Leone Company who declined to pay her the widow's pension they had promised—Anna Maria's account of the setting up of Freetown is remarkably revealing. Nor does she confine her spirited observations to the settlement. None of her four voyages to and from Sierra Leone was exactly smooth, and by accident or design she visited the West African island of Bance and the Gambia, the Cape Verdes and the Azores, and even Jamaica (on a circuitous route home in 1793 'on slave business'), commenting succinctly on them all.

❖

FALKLAND, Amelia CARY, Viscountess (1803–1858)

Chow-Chow: Being Selections from a Journal kept in India, Egypt and Syria. London: Hurst and Blackett, 1857; 2 vols., pp. 326/287, with hand-coloured frontispieces and title-page vignettes

This is one of many records of Anglo-Indian life seen from Government House, but more leisurely than most. The Falklands' residence in Bombay from 1848–1853 witnessed a political calm before the storm of the Mutiny, and the Viscountess Amelia's account is full of the niceties (and absurdities) of social and diplomatic life in a country she feared was becoming 'too civilised' too soon. She actively enjoyed the inevitable spring trek to the hills (except at night when her palanquin tended to list perilously during the bearers' frequent arguments) and made several lone expeditions within the Deccan high country once there. Travelling appealed so much that the Falklands decided to make an extended tour of their journey home to England, and this *Chow-Chow* (i.e. oriental mishmash) closes with a tireless journal of their sightseeing marathons *en route* in Egypt and the Holy Land.

❖

KING, Agnes Gardner

Islands Far Away [Fiji]. London: Sifton, Praed, 1920; pp. 256, with 5 halftones, 62 line-drawn sketches, and 2 maps

Artist, eccentric, and general Character Agnes Gardner King adored Fiji. Or, more correctly, she adored Fijians. As soon as she arrived with her lady companion (who knew the islands and recommended them as a tonic to Agnes' jaded health) she was eager to meet the natives 'and I realised from the first that we should be in sympathy and get on well together.' During her stay in Suva and amongst the remote villages of Vanua Levu and Viti Levu, Agnes threw herself whole-heartedly into Fijian life, becoming a much-valued sort of aunt in the process, and the affectionately honoured guest at many a local feast. In order to visit a particularly well-tucked-away community she would leap into a dug-out canoe and shoot a few rapids with as much equanimity as we imagine her strolling to church at home in Wroxham. She trekked through mountains, tackled meals of whole-baked shark, and sketched her spirited illustrations in the most unlikely circumstances—and all in the true tradition of Britain's travelling spinsterhood: with curiosity, humour, and delight. And the links of friendship she forged with the Fijians are still strong, I am told, to this day.

❖

LARYMORE, Constance

A Resident's Wife in Nigeria. London: Routledge, 1908; pp. 306, with photogravure portrait and 41 halftones

The most interesting part of this account is headed 'Household Hints' and resembles a sort of tropical Mrs Beeton. Constance Larymore spent five years in Africa, based first in Sierra Leone and then in the provinces of northern Nigeria around Kabba and Kano. She writes with great verve of long journeys on foot, horseback, or in a steel canoe as the sole woman member of various boundary-delimitation expeditions, and seems thoroughly to have enjoyed her time as a pioneer Colonial Wife there. By trial and error, she says, she discovered the best way to run a household on the march in the middle of African nowhere. In India, where she and her husband had previously been stationed, there was at least the possibility of inheriting furniture and other useful goods (and tips) from one's predecessors; here in Nigeria one had to start from scratch. She gives advice on interior decoration (always put up pictures of waterfalls or snow scenes—it keeps one cool), on the management of servants, on gardening and poultry-keeping, on camp life (do not forget the mincing machine) and on what to wear. *Always* wear corsets, she warns: 'to leave off wearing them at any time for the sake of coolness is a huge mistake: there is nothing so fatiguing as to lose one's ordinary support': salutary stuff.

LLOYD, Susette Harriet

Sketches of Bermuda. London: Cochrane, 1835; pp. 258, with 3 aquatints by the author and a map

Bermuda, 'the smallest of all our West India possessions', is just the place for young ladies of a romantic nature. Susette Lloyd spent a thoroughly delightful eighteen months there, as guest of the Archdeacon and his Lady. It was partly a holiday, during which time she would perch in a cedar tree reciting Shakespeare's *Tempest*, sit on some tropic knoll strumming her guitar, or row around the coast in search of the picturesque. The rest of the time she would fulfil her duty as a visitor from the motherland, ardent to help realize Bermuda's claim to be 'the foremost [British colony] to extirpate, without delay, every trace of slavery'. She helped set up a scheme of Sunday schools and joined all the local societies for the welfare of the emancipated Negro and convict population. Miss Lloyd was also well aware of her duties as author of one of the earliest exclusive accounts of Bermuda, travelling its length and slender breadth to provide her *Sketches* with accurate detail.

LUSHINGTON, Mrs Charles (Sarah) (d. 1839)

Narrative of a Journey from Calcutta to Europe, by Way of Egypt, in the years 1827 and 1828. London: John Murray, 1829; pp. 280

This is a useful account—with notes on route, itinerary, and the necessities of travel—for ladies intending to brave the 'Overland Route' from India to

England. Mrs Lushington wrote it to show that one's quality of life whilst travelling need not suffer too much, even when steaming up the 'imperfectly surveyed' Red Sea, or crossing the Syrian desert in a camel litter. Indeed the eleven-month journey was actually *enjoyable*—on the whole. Every evening in the desert, for instance, the Lushingtons and their party, which included Lord Elphinstone, late Governor of Bombay, would settle down to a light supper of 'roast turkey, ham, fowls, mutton in various shapes, curry, rice and potatoes, damson tart and a pudding; madeira, claret, sherry, port and . . . beer; . . . almonds and raisins, water-melons, pumplenose [grapefruit] . . . and a plumcake'. There were thrilling 'mummy-openings' to attend in Thebes, balls in Cairo, and soirées in Naples and Paris. In fact the experience, said Mrs Lushington, was decidedly not to be missed.

❧

MACKENZIE, Mrs Colin (Helen)

Life in the Mission, The Camp, and the Zenáná: Or, Six Years in India. London: Richard Bentley, 1853; 3 vols., pp. 359/307/336, with 3 lithotints and 2 maps

Helen Mackenzie was the second wife of Brigadier Colin Mackenzie, veteran of the first Afghan War of 1838–1842 and hero of the Siege of Kabul. She sailed out to Calcutta with the brigadier in 1846, and spent most of the next six years in the Punjab at Ludhiana, 'one of the ugliest stations in the country'.

In Ludhiana the Mackenzies lived in an American mission station, and Helen's chief interest throughout her stay—good Scots stalwart of the Free Kirk as she was—was the state of Presbyterianism amongst the *Zenáná*, or women's communities of India. She was also a military enthusiast, fascinated by the tactics and strategy of soldiering (seen at first hand during the Punjabi uprisings of 1848) and often to be seen inspecting 'The Lines' on her favourite elephant. It is a sign of her general stoutness that she did not succumb to the slightest fever until the end of her second year, and that when she did eventually become seriously ill in 1851, she still insisted on leading her elderly mother (who had come to visit) on a sedan-chair tour of the jungle—and enjoyed it.

❧

MAITLAND, Julia Charlotte

Letters from Madras, During the Years 1836–1839. By a Lady. London: John Murray, 1843; pp. 300

A witty and sensitive account, this is, of a young Englishwoman's enchantment with India. Julia Maitland sailed there in the summer of 1836; it was a wedding journey, and her husband was on his way to a new appointment as district judge based in the hills above Rajahmundry. With unwonted application, Julia whiled away the voyage with learning Tamil verbs, and could hardly wait to get away from Madras—'England in perspiration' she called it, and full of colonial dinner-parties, 'grand, dull and silent'. The journey to the hills was made in a sumptuous palanquin, and the house awaiting them was delightfully cool and picturesque.

Once settled (with a comparatively modest staff of twenty-seven: 'we wish to be economical') Julia lost little time in learning the local language, Gentoo, and helping her husband to set up a thriving school and library for the Hindus. She had little time for meddling missionaries, and still less for the peculiarly bird-brained breed of European women to be found in India, whom she summarily divided into the Civil Wife and the Officer's Wife. The first are

generally very quiet, rather languid . . . almost always lady-like and *comme-il-faut* . . . rather dull, and give one very hard work in pumping for conversation. They talk of 'the Governor', 'the Presidency', the 'Overland', and 'girls' schools at home', and have always daughters of about thirteen in England for education. The military ladies, on the contrary, are always quite young, pretty, noisy, affected . . . chatter incessantly from the moment they enter the house, twist their curls, shake their bustles . . . They talk about suckling their babies, the disadvantages of scandal, 'the Officers' and 'the Regiment'.

During the airless summer, Julia moved with her growing family down to the sea at Samuldavee, started another school there, and became a local activist for the cause of native education and emancipation (in the wake of what she called the East Indian slave trade in Hindus to Mauritius and other British dependencies).

After two years Judge Maitland was promoted to a circuit based further south at Chittoor, and Julia travelled with him whenever her health and that of her children would allow, full of admiration for the 'real India' they found together. Sadly, however, she was forced to leave both husband and country behind at the end of 1839, to accompany her ailing eldest daughter back to England, and there the *Letters* end.

❖

MANSFIELD, Charlotte (d. 1936)

Via Rhodesia: A Journey through Southern Africa. London: Stanley Paul [1911]; pp. 430, with 144 halftones and 2 maps

It was Charlotte Mansfield's intention to become the first woman to travel from the Cape to Cairo in a single journey. Mary HALL had done the journey, but in two separate stages; Charlotte, financed by the novels she had written and by promised articles to a variety of journals, would start at Cape Town and go on (following the route her hero Cecil Rhodes proposed

for a Cape to Cairo railway) until she emerged at the other end of Africa.
And the book she wrote afterwards would not just be a travel account; it
would change the world. All the middle-class women in England 'with just
sufficient capital not to know what to do with it' would stream to Rhodesia to
take their rightful place as chatelaines of the empire; they would marry
expatriate cattle-farmers or mine-owners or cotton-growers, using the
endless supply of Blacks for their labour and producing a race of fine white
offspring to keep the country pure.

Well. Charlotte began her epic journey in 1909 and got as far as Lake
Tanganyika, travelling by train (when available) and by *machila* (a hammock
slung on poles) with a 49-piece retinue of natives. She gamely underwent
the African traveller's baptismal fire of malaria, jiggers, and heat-exhaus-
tion, and seems to have utterly charmed her caravan with her long red-gold
hair (always worn loose) and her prowess at target-shooting with a revolver.
Wherever she came upon a government official or a cache of missionaries in
the jungle she would bound up to them, start lecturing on the virtues of
emigration (and the futility of religious instruction), and leave in her wake a
collection of dazed and disbelieving Englishmen and adoring natives with
bunches of flowers. The trouble at Lake Tanganyika was sleeping sickness:
there was an epidemic there which Charlotte would have been prepared to
risk but her porters would not. So she turned right and trekked to
Mozambique instead. By the time she reached Chinde she was worn out
with fever and fatigue, and remembers little of the voyage home to England.

Within the year Charlotte was bound for Africa again, this time (practising to the letter what she preaches in the book) she was married to a mining
engineer, and an *emigrée* herself. Why the Mansfields did not settle in Africa
(coming back to England during the Great War) I do not know; Charlotte
was writing her memoirs, which might have explained, when she died.

◆

NUGENT, Lady Maria (1771–1834)

*A Journal of a Voyage to, and Residence in The Island of Jamaica, from 1801 to 1805, and of
subsequent events in England from 1805 to 1811.* London: [for private circulation only],
1839; 2 vols., pp. 500/515, with mezzotint portrait, 2 engraved plates, and a hand-
coloured lithograph

*A Journal from the year 1811 till the year 1815, including a Voyage to and Residence in
India, with a Tour to the North Western Parts of the British Possessions in that Country,
under the Bengal Government.* London: [for private circulation only], 1839; 2 vols.,
pp. 428/388, with mezzotint portrait

These two journals show admirably just how difficult it is to be a Lady of
Quality abroad: how hard it is to behave, wherever one may be, just as
though one were in one's own apartments at home. They are chronicles of

the boudoir—traveller's tales by default—written by a thoroughly domestic woman for her family and friends alone.

Maria Skinner was born in New Jersey (still, in 1771, under allegiance to the British Crown) and brought up comfortably in Ireland, where she continued to live after marrying a distinguished English military man, George Nugent, in 1797. Three years later, with successes in the 1798 rebellion and civil war under his belt, Nugent quit Ireland for Jamaica, where he was appointed lieutenant-governor. Undaunted by tales of imminent French invasion and the thought of living with 'darkies', Maria joined him there and immediately began her *Journal* of the Great Adventure.

Actually, there is little of Jamaica in the *Journal* at all. The Great Adventure to Maria was being somewhere different and meeting the challenges of inconvenience; it hardly mattered where that 'somewhere different' was. She did not travel much on the island; apart from the odd early-morning ride to the public baths (side-saddle in nightcap and dressing-gown) and the occasional depressing visit to a sugar-cane estate, most of her time was spent either in the official residence in Spanish Town or on the 'Penn', or farmstead, close by. Here she pondered over her state of health—as any Lady of Quality would—and worried. 'I mean,' she wrote, 'as symptoms arise of any illness, always to mention it; because, if I should die in this country, it will be a satisfaction to those who are interested about me, to know the rise and progress of my illness'. In fact Nugent grew more ill than Maria, while she flourished under the '*agrémens* of a Creole confinement' and bore two babies to take home as souvenirs.

In India, where Nugent was next posted as commander-in-chief in 1811, circumstances were a little different. The children (by now numbering four) had to be left at home this time, and there was no time for worrying: Maria was too busy travelling for that. Her feet hardly touched the ground from first to last (literally, thanks to the ubiquitous palanquin). After the social whirl of Calcutta followed a year's official tour of Bengal and the Punjab, and besides vaguely distinguishing between the splendours of Agra and Delhi, or the charms of Lucknow and Cawnpore, all Maria had the strength to write of was how 'very fatiguing' and 'uncongenial' it all was. But she survived, robust as ever, to enjoy a happy reward of twenty years' domestic tranquillity in England after 1813, and died well beloved and satisfied at sixty-three.

❧

PARKES, Fanny

[Anon.]: *Wanderings of a Pilgrim, in search of The Picturesque, During Four-and-Twenty Years in the East; with Revelations of Life in The Zenana*. London: Pelham Richardson,

1850; 2 vols., pp. 479/523, with 48 lithotints (including 11 hand-coloured and some heightened in gilt), a lithographed panorama of the Himalaya, and a plan

> We are rather oppressed just now by a lady, Mrs Parkes, who insists on belonging to our camp . . . She has a husband who always goes mad in the cold season, so she says it is her duty to herself to leave him and travel about. She has been a beauty and has remains of it, and is abundantly fat and lively. At Benares, where we fell in with her, she informed us she was an independent woman . . .

This was Fanny EDEN's account of the author of what must be one of the most fulsome tomes on Indian travel. Everything about Fanny Parkes seems to have been abundant, not least her ingenuous enthusiasm for all things Indian.

She and her customs-officer husband sailed for Bengal in 1822 in the knowledge that they would be staying there for a minimum of twenty-two years (to qualify them for a pension afterwards)—a prospect Fanny faced with stout equanimity. Immediately upon arrival at Calcutta she began learning Hindustani, gathering a household of fifty-seven servants (being 'quiet people', she says, they would need no more) and exploring on her beloved Arab horse, with Scotch terrier Crab at her heels. In 1826 the Parkeses were posted to Prayag, near Allahabad (where Mr Parkes was put in charge of the local 'pits', or stores, of precious ice imported from America) and from there Fanny staged a series of expeditions, often solo, through Thug country to Lucknow, to Benares (now Varanasi, where she blithely foisted herself on the governor-general's camp and so much 'oppressed' the Edens), and to Agra, Meerut, and Delhi.

Much to the amusement of the other British wives, Fanny became an avid and knowledgeable student of Hindu life and history, and her book is heavy with old Indian proverbs and mythology. She speaks of 'us Indians' and counts 'the native ladies of rank' amongst her dearest friends. Most of her countrywomen wilted in the Indian heat and counted the days until their release; for Fanny, made of sterner stuff, the thought of going home in 1844 was appalling: 'How I love this life in the wilderness!', she cries at the end of the book; 'I shall never be content to vegetate in England.' And elsewhere: 'Roaming about with a good tent and a good Arab, one might be happy for ever in India.'

PENDER, Rose (d. 1932)

No Telegraph: Or, A Trip to our Unconnected Colonies, 1878 [Africa]. London: For Private Circulation: Gilbert and Rivington, 1879; pp. 152

A Lady's Experiences in the Wild West in 1883. London: George Tucker [1888]; pp. 80

Rose and James Pender mixed business with pleasure on what they called their 'missions' to Africa and America. James—later Sir James—was the director of a telegraph cable company with plans to lay a submarine link from Aden to Natal, and so connect 'our unconnected Colonies'; whilst he was busy with his agents on a reconnaissance trip in 1878 Rose rode regally around South Africa visiting ostrich farms and worrying ostentatiously about the Malay immigrants there. Five years later the Penders sailed for a four-month visit to inspect cable business in 'civilised' America and then to venture beyond civilization to the Wild West, where James had an interest in the Indian Territory Cattle Scheme. Their route took them from New York through Washington, across to 'dreary Arizona', Texas, and California; at San Francisco they paused to summon up the necessary courage to turn east for Cheyenne and the wastes of pre-railway Wyoming.

Rose bore the journey through cowboy country very well, living rough under canvas, on ranches, and even in 'bar-infested' towns and still saving enough strength to climb up Pike's Peak with an English couple she happened to meet 'on their way home from China' and to subdue most of the horses, cattle, cowboys, and Indians she encountered with her indomitable British vigour. She was justly proud of her achievement afterwards, quoting an admiring ranch-man who reckoned that 'if all English women were as strong as I was they must be a fine race, as I seemed to be a real "Rustler". This, I believe, is a term of approval.'

◆

POSTANS (later YOUNG), Mrs (Marianne)

Cutch; or, Random Sketches, taken during a residence in one of the Northern Provinces of Western India; interspersed with Legends and Traditions. London: Smith Elder, 1839; pp. 283, with lithographed half-title, 5 hand-coloured lithographs, 1 wood-engraved plate, 6 vignettes, and a map

Western India in 1838. London: Saunders and Otley, 1839; 2 vols., pp. 303-295, with 6 lithotints (2 hand-coloured), and wood-engraved vignettes

Facts and Fictions, Illustrative of Oriental Character [includes notes on Syria, Egypt, and St Helena]. London: Wm. H. Allen, 1844; 3 vols., pp. 287/309/296

Our Camp in Turkey, and The Way to It. By Mrs. Young. London: Richard Bentley, 1854; pp. 313

The Moslem Noble: His Land and People with Some Notices of The Parsees, or Ancient Persians. By Mrs Young. London: Saunders and Otley, 1857; pp. 192, with 9 lithotints (partly hand-coloured) and 8 wood-engraved vignettes

At first glance, this author's beautifully illustrated accounts of life and travel in India are like most others of the period: part guidebook and part adventure story. She wrote them, she said, to popularize the jewel in England's crown to 'the reading public' at home, whom she (somewhat harshly) judged to be 'totally uninterested in India'.

No one took much notice, and all her books have long been out of print. Perhaps they were too uncomfortable to read? Too much of a challenge to a complacent audience? For just beneath the surface of her polished narrative lurk distinctly unfeminine warnings. Warnings that there are millions of ordinary Indians who need real education, not the play schools run by rarefied officers' wives. The squalor of the lower castes is too easily eclipsed by the dazzling wealth of Indian princes. Native soldiers, or sepoys, deserve 'the highest encouragement and consideration' for safety's sake. And British officers, often sent out raw at eighteen, need more time to acclimatize both physically and mentally, else they may turn cruel and reckless.

By the time the Indian Mutiny broke out (I told you so . . .) in 1857, Marianne was well out of India, had married again, and made an expedition on a troop-ship to Scutari and thence to Therapia, Gallipoli, Boulehar, and Smyrna to watch the birth of the Crimean War. In quite what capacity she travelled we do not know, but it seems a characteristically unconventional journey. I suspect she was challenge-hunting again.

❧

RIDDELL, Maria

Voyages to the Madeira, and Leeward Caribbean Isles: with Sketches of the Natural History of these Islands. By Maria R******. Edinburgh: for Peter Hill, 1792; 12mo, pp. 105, with half-titles to each section: 'A Voyage to Madeira 1788', 'A Voyage to St. Christopher's 1788', 'A Tour through Antigua and Barbuda [Lesser Antilles] 1790' and 'Geographical Description and Natural History of Antigua, 1791'

It was not unusual at the end of the eighteenth century for travelling gentlemen like Mr Riddell—diplomats, merchants, landowners, and so on—to take their families with them. What is unusual is that the wife or daughter of such a family should take it into her head to write about it. 'Voyages and Exploration' were still very much male provinces on the library shelves, and authors like Mrs PARKER and Maria Riddell were few and far between. Like Mary Ann Parker, Maria Riddell had no literary aspirations to explain her unwonted appearance in print—just a passionate interest in natural history. She assumes that none of her readers will be much interested in the minutiae of travel, having read all that in previous accounts of the well-sailed road from Madeira to St Kitts; instead she concentrates on the classification of the islands' (and others') flora and fauna. There are occasional references, however, to the Riddell family's insatiable appetite for sightseeing and what sounds now like an incredibly busy and familiar social life—balls in Funchal, afternoon tea in Antigua, and concerts in Basseterre—all hosted, it seems, by fellow Scots.

❖

ROCHE, Harriet A.

On Trek in the Transvaal: Or, Over Berg and Veldt in South Africa. London: Sampson Low, Marston, *et al*, 1878; pp. 367

Now that the Transvaal has been annexed to the British Crown, says Mrs Roche in her introduction, perhaps more of us will visit it for pleasure rather than business—women as well as men. And maybe an account of the visit she made herself to that brave new land in 1875 will encourage them. I doubt it—there can be few less encouraging travel books than this. The journey started well enough, with Harriet happily tagging along in the wake of her husband Alfred, the first Honorary Secretary of the Royal Colonial Institute, on a comfortable voyage to Durban and an interesting (if slightly less comfortable) horse-bus ride to Pieter Maritzburg. But as the Roches and their missionary companion embarked for Eersteling, six hundred miles inland, everything began to degenerate: the weather, the wagon, their health, and even their sterling British stoicism. Towards the end of the two-month outward trek, Harriet was reduced to sleeping on a saddle of mutton for its comfort, and by the close of the return journey, her husband was dead. She had had to organize the final stages of the expedition herself, and take full responsibility for her own survival as well as trying to keep the ailing Alfred alive—and she had only gone for the ride! A brave new land indeed.

❖

SCHAW, Janet (*c.* 1737–1801)

Journal of a Lady of Quality; Being the Narrative of a Journey from Scotland to the West Indies, North Carolina, and Portugal, in the years 1774 to 1776 [etc.]. Edited by Evangeline Walker Andrews. New Haven: Yale University Press, 1921 [enlarged third edition, 1939]; pp. 341, with 7 halftones and map on endpapers

It is clear from her *Journal* that Janet Schaw was a strong-charactered, spirited woman. She must have been, to risk her middle-aged reputation as a Lady of Quality in Edinburgh by taking off across the Atlantic to the steamy West Indies: and she *chose* to go. Janet's brother had just been appointed Searcher of Customs at the Caribbean island of St Christopher (St Kitts), and Janet—just for fun—went with him.

Her account of the voyage (undiscovered until a century and a half afterwards) tells of the six-week crossing to Antigua, enlivened by the crowds of Scots emigrants stuffed into the steerage being 'vastly sick' and the storms that washed the stores overboard from time to time; it describes Janet's short residence (with Scottish 'people of quality', of course) amongst the sugar plantations of Antigua and St Kitts, and follows her north to Brunswick, North Carolina, on a visit to relatives who had settled there. Although she confessed to being a little frightened of the Negroes in Carolina, it was the white Americans (in fact still British at this stage) who really alarmed her. There was 'ill humour' in the air (a masterly understatement: it was the opening year of the War of Independence) and eventually Janet was advised, for her own safety, to leave. She sailed from Brunswick in October 1775 for Lisbon (which she seems to have found just as novel as Antigua!) and after three weeks' sightseeing there left for Glasgow, and home.

❧

SCOTT-STEVENSON, Esmé

Our Home in Cyprus. London: Chapman and Hall, 1879; pp. 332, with 8 wood engravings and a map [copy not seen]

Our Ride through Asia Minor. London: Chapman and Hall, 1881; pp. 400, with a map

On Summer Seas (including the Mediterranean, the Aegean, the Ionian, the Adriatic, and the Euxine, and a voyage down the Danube). London: Chapman and Hall, 1883; pp. 408, with a map

Mrs Scott-Stevenson dedicated all three of her books to her husband, as sort of peace-offerings for not having told him she was publishing them in the first place. Had he known, he would strongly have disapproved: he was a man with an Important Position (officer of the Royal Highlanders and the British Commissioner of Kyrenia in Cyprus) who could not afford a wife who might make a fool of herself. But Esmé's self-confidence was well founded, and the books she wrote on Cyprus (covering the first two years of the island's life under British rule), on the local waterways, and on Armenia and Turkey were for a while standard texts on the area.

Both she and her husband were keen amateur explorers, proudly eschewing the normal (not very well-worn) tourist routes wherever they went and always travelling as self-sufficiently as possible. Esmé seems to have flourished, whether sailing the Ionian Sea in some reeking Greek steamer or 'roughing it' on horseback through the lonely uplands of Armenia; I suspect she spoke for many of her travelling peeresses when she wrote at the close of *Ride through Asia Minor*:

I went to sleep . . . very happy. I felt that I had accomplished a feat that few ladies would undertake; that I had travelled through a country that was almost unknown, and gone through trials and dangers that required great tact and endurance to overcome them . . . I felt, for once in my life, that I had done something really of use.

◆

SIMCOE, Mrs Elizabeth Posthuma (1766–1850)

The Diary of Mrs. John Graves Simcoe, Wife of the First Lieutenant-Governor of the Province of Upper Canada, 1792–6. With Notes and a Biography by J. Ross Robertson. Toronto: Wm. Briggs, 1911; pp. 440, with 237 halftones, maps, and sketches

Elizabeth Simcoe had a classically romantic childhood. Her father died seven months before her birth and her mother in childbed, leaving Elizabeth Posthuma, the Gwillims' only child, in the possession of a distinguished guardian (one Admiral Graves), and a fortune. She married the Admiral's godson in 1782 when she was sixteen and he thirty, and together they settled into luxury with a growing family on the estate Elizabeth bought for them in Devon.

After a brief struggle between the prospects of discomfort and prestige, Simcoe accepted the post of Upper Canada's first lieutenant-governor in 1791, and he and his lady wife set sail with a handful of children and nurses for Quebec, just in time to catch the 1792 summer Season. Elizabeth's diary kept over the five years of their residence in Canada shows just how close the wilderness could come to civilization: it records balls and soirées, picnic trips by canoe, card games in the sumptuous 'canvas house' Simcoe had erected at Toronto and house parties at Castle Frank, their timber Palladian-style villa in the woods. In fact upper-class life in the Canadian raw was soon invested with all the petty jealousies and gossiping of such life at home. At least Elizabeth was able to indulge her hobby of map-making more usefully here than in Devon—several minutely detailed examples have survived, inked on to birch bark—and even the governor's lady was not above the odd excursion to the Indian backwoods.

Elizabeth planned to accompany Simcoe on his next appointment as Commander of the British Forces in India in 1804, but when he died shortly before they were due to set off she elected not to travel again, enjoying instead the privileges of forty-four years' imperious widowhood at home.

◆

STEEL, Flora Annie (1847–1929)

The Garden of Fidelity: Being the Autobiography of Flora Annie Steel [etc.]. London: Macmillan, 1929; pp. 293, with 4 halftones

Flora Steel was a real stalwart of the Raj—one of the few British women who lived their exiled lives in India *usefully*. She arrived in the Punjab from Scotland as a Civil Service bride in 1868, and with the birth of two children (one dead), the repeated ravages of malaria, and having to move house nine times in the first three years, was straight away thrust into the ritual baptism of fire that either killed or sent most memsahibs fainting home. But forceful young Flora was made of sterner stuff: she stayed on for twenty years.

She was a scholarly woman and incurably curious, spending her spare time (and there was plenty of it at first) getting to know the Indian people. She travelled amongst the hills above Lahore (now in Pakistan) collecting folk-tales, many of which she later published, and learning the local lore. As her interest grew, so did her assumed responsibilities: she became the first female school inspector in India, an accomplished nurse and 'doctress'—even a town planner, designing a splendid Municipal Hall for the tiny station of Kasur and fitting it out with a library and numerous meeting-rooms. The definitive *Complete Indian House-keeper and Cook* written by Flora with Grace Gardiner in 1890 (and which soon ran into ten editions) was published, at her insistence, in Urdu 'at the lowest possible price' as well as in English, and she always preferred the dignified presence of Muslim and Hindu to the prattling 'griffin' from home.

Henry Steel retired from India in 1889 and was pleased to go, but Flora was lost. She had spent over half her life there: it was home. She went back five years later, alone, to research her celebrated novel of the Mutiny *On the Face of the Waters* (1896) and again in 1898, this time with husband and daughter. Although she never returned after that last visit, she wrote histories and stories of India for years, and was a traveller—mostly within Europe, although she did manage a lightning tour of New York and Jamaica in 1928—well into her nineties. She was, and through her numerous books will remain, one of the most *sensible* chroniclers of British India we have.

❖

WILSON, Anne Campbell (1855–1921)

After Five Years in India: Or, Life and Work in a Punjab District. London: Blackie, 1895; pp. 312, with 16 halftones

Hints for the First Years of Residence in India. Oxford: Clarendon Press, 1904; pp. 70

Letters from India. Edinburgh: Blackwood, 1911; pp. 417

Scottish Mrs Wilson's first five years in India, from 1889 (a year after her marriage to civil servant James) to 1894, were spent in and around the remote jungle station of Shahpur in the Punjab. It was there, and in the garrison town of Rawalpindi, that she learned the codes and practices of

Anglo-Indian society so handily described in her first two books. They are both cheerful accounts of what pleasures and inconveniences await ordinary novice sahibs and memsahibs on their arrival in India, for once uncluttered by either cynicism or romance. In 1901 the Wilsons moved to Lahore and two years later to Calcutta: both comparatively 'civilized' and blessed with libraries to help Anne research her third book, *A Short Account of the Hindu System of Music* (1904). And as they moved up the ranks of the Indian Civil Service, she and James began to tour not just because they had to, but for pleasure—to the buzzing summer capital of Simla, for example, or amongst the refreshing Kashmiri foothills of the Himalaya. Anne's letters home, published two years after her final return to England, describe it all; they chronicle twenty complacent and unexceptional years—the high summer of the British Raj.

Pay, Pack, and Follow

'Pay, pack and follow.' The terse instructions on the message from Damascus were familiar to Lady Isabel BURTON. She spent her life at the summons of her adored husband, the Arabian scholar-explorer Sir Richard, acting as his agent, secretary, servant, and companion. After years spent paying, packing, and following in his wake, Lady Isabel became something of an explorer herself; were it not for her husband, however, it is doubtful that she would ever have become a traveller at all.

That is what all the women in this chapter have in common: they began travelling by default. For some of them, it was a matter of duty: Mrs Meer Hassan ALI and Emily, Shareefa of WAZAN, went to India and Morocco respectively as the English wives of native husbands (and both, incidentally, wrote their books to help distract them from their unhappiness there): there was no alternative to them but to become strangers abroad. Then there were the diplomats' ladies, whose social obligation it was to be by their menfolk's side. Thus wives, sisters, and daughters who had hardly strayed beyond their own drawing-room before might suddenly find themselves in plague-some eighteenth-century Algiers, or imprisoned in some central-Asian outpost with no other European woman for a thousand miles or more. They were the unlucky ones; the more fortunate revelled in the liberty their second-hand travels offered to them. Mrs GRAY, for example, liked nothing better than to comb the streets of Canton by night in search of something sensational. As Isabella BIRD said, 'travellers are privileged to do the most improper things'.

Diplomats go where they are told. But doctors, scientists, writers, and the like do not *have* to go anywhere, and still less must their spouses go with them. Yet there are books written here by a Mrs Doctor in Vietnam who runs her husband's errands on a tricycle through the jungle; a Mrs Astronomer cheerily pitching camp on the Spartan slopes of the volcanic island of Ascension, and a writer's elderly Scottish mother tripping barefoot through the palm trees of Samoa as though she has lived there all her life. To women such as these, one gets the impression that travel was a welcome challenge— but still one they may not have had the courage or opportunity to discover alone.

Not all the travellers in this chapter felt themselves as equal to that challenge as those I have just mentioned obviously did. Although she was also married to an explorer, Marika HANBURY TENISON was no second Lady Isabel. The weeks she spent struggling in Robin's confident footsteps through the depths of Indonesia nearly killed her, and left her quite contented with a backstage role in his expeditions from then on. Still, she felt she owed it to herself to go with him at some time—not because it was expected of her, but because she was loth to waste the privilege of being able to share in her husband's obsession with travel. For once, she did not want to be left behind: she *had* to go.

They all had to go, these travellers by default. Devotion, duty, obedience, and privilege obliged them to—or, as passionate Lady Isabel would have it: 'A dry crust, privations, pain, danger for him I love . . . There is something in some women that seems born for the knapsack'. Others did not have the inclination to refuse (or the strength, perhaps); and having accepted their circumstances, most reacted with a refreshingly unprejudiced eye. Relieved of the responsibility of their travels, they behaved a little like children on a mystery tour, writing down what strange things they saw and felt (whether or not they enjoyed them) with an objectivity rare amongst travel writers.

ALI, Mrs (B.) Meer Hassan

Observations on the Mussulmauns of India; descriptive of their Manners, Customs, Habits, and Religious Opinions. Made during a twelve years' Residence in their Immediate Society.
London: Parbury and Allen, 1832; 2 vols., pp. 395/427

It is a strangely sad and earnest book, this. Apparently Mrs Meer Hassan Ali (whose maiden name we are not told) met her husband during his six-year visit to England between 1810 and 1816. He was an exotic Indian nobleman, she a committed Christian spinster; both throwing convention to the winds, they married and travelled together to the family home at Lucknow to settle down. She must have been very unhappy, being neither memsahib nor Muslim. Nor was she the sort of wife her husband required, who is supposed to receive her Master 'with undisguised pleasure, although she has just before learned that another member has been added to his well-peopled harem'. There were no children of the marriage, and to occupy her time the author sat quietly taking notes in her father-in-law's house, or travelling (outrageous notion!) to gather material for this book, once even managing to interview King Akbar II at Delhi—and always alone. In 1828 she returned to England after twelve years away, ostensibly on account of failing health. She never went back.

—◆—

ATKINSON, Mrs (Lucy)

Recollections of Tartar Steppes and their Inhabitants. London: John Murray, 1863; pp. 351, with 5 wood-engraved plates

'Being one of a large family, it became my duty at an early period of life, to seek support by my own exertions.' Thus Lucy Atkinson, in common with thousands of other gently bred but impecunious nineteenth-century spinsters, became a governess.

She was engaged by a fashionable St Petersburg family until 1847 when she married the explorer Thomas Whitlam Atkinson, who swept her away for the next six years on his travels from St Petersburg to Moscow, Tomsk, Barnaul, and through the inhospitable Urals. There Lucy found herself 'often left alone with an infant in arms among a semi-savage people', but with true British stoicism, she never complained.

Victorian women writers travelling with (or in the wake of) their husbands at this time often subtitled their work with something like 'an unscientific account of a scientific expedition' (e.g. Isabel GILL and Cara DAVID)—but their artlessness and subjectivity usually make for a far more entertaining book than the 'serious' efforts of their spouses (in this case, Thomas's *Travels in the Upper and Lower Amoor*, 1860). Mrs Atkinson is no exception.

—◆—

BENN, Edith Fraser

An Overland Trek from India by Side-saddle, Camel, and Rail: The Record of a Journey from Baluchistan to Europe. London: Longmans, 1909; pp. 343, with 80 halftones and a map

Major R. A. E. Benn and his brisk wife Edith were at home on furlough in
Clevedon when they got the news. The major was to be transferred from
Quetta in Baluchistan (now in Pakistan) to Seistan (today the Iranian district
of Daryacheh ye Sistan) in south-east Persia (Iran), where he had been
appointed the first British consul. They set off from Seistan's nearest
railway station (which happened to be Quetta) in 1901, and this is a no-
nonsense account of the following journeys to their new home and, two
years later, back to Europe. The first leg from India to Persia involved a
600-mile camel-ride (good for the character, says Edith) with the whole
Quetta household and its dependants for escort and the fox-terrier Tipper
snapping at their heels. The second leg was more difficult. When the time
came to visit England again, the Benns decided to take the 'long' way home,
marching with ponies the 700 miles to Askhabad where the Trans-
Caucasian Railway began. The author does not dwell on hardships (she just
mentions them *en passant*) and it is a mark of her sterling national qualities
that her worst complaints are reserved for the weather.

❖

BROOKE, Lady Margaret (1849–1936)

My Life in Sarawak. London: Methuen, 1913; pp. 320, with 27 halftones and a map

Good Morning and Good Night. London: Constable, 1934; pp. 308, with 10 halftones

When James Brooke, first white Rajah of Sarawak died in 1868, he was
succeeded by his nephew Charles. Already in his forties and eager to secure
a direct line of succession this time, Charles determined to find himself a
wife without delay. His choice fell on a cousin's daughter, who obliged by
accompanying him to Sarawak in 1869 as Margaret Brooke, the Ranee.

The marriage was difficult; Charles was obsessively devoted to his uncle's
little kingdom and was never a tender man. Margaret left him eventually,
but not before she had borne him seven children (of which three sons
survived) and fallen in love with the place and its people herself. She was the
first white woman most of them saw, except for the few that remembered
Harriette McDOUGALL. Not content with the limited role Charles envisaged for her, Margaret tossed off her stiff Edwardian petticoats for the
colourful sarongs of the native women and set up schools and a lending
library for their benefit. She encouraged European visitors, too (amongst
whom was the delighted Marianne NORTH), often guiding them on excursions into the jungles of her adopted country. Once home, the Ranee did her
utmost to ensure her sons married sympathetic wives who would carry on a
tradition of loving guardianship in Sarawak.

❖

BROOKE, Sylvia (1885–1971)

Sylvia of Sarawak: An Autobiography. London: Hutchinson, 1936; pp. 282, with 44 halftones

Queen of the Headhunters. London: Sidgwick and Jackson, 1970; pp. 194, with 25 halftones

Sylvia Brett was earning quite a reputation for herself as a young lady of letters before her travelling life began. She was one of the many female favourites of George Bernard Shaw, and had published several plays and other pieces by the time she met Charles Vyner Brooke, the Rajah Muda (heir) of Sarawak. According to Sylvia, her mother-in-law Margaret BROOKE, despairing of her shy sons ever acquiring suitable wives for themselves, carefully composed a well-bred amateur orchestra of 'willing, desperate virgins' at her London home. The young musicians met once a week, and before long a wife for each of the three boys had been dutifully supplied. Sylvia—the drummer—was chosen by the man who became the third and last white Rajah, and so thrilling did she find the life as 'Queen of the Headhunters' and 'First Lady of Sarawak' that she wrote her autobiography twice. That life included repeated voyages by ship and private yacht to and from that proud corner of Borneo captured so romantically by James Brooke back in 1841; numerous cocktail parties at home and abroad telling tales of the jungle tribes, and writing these two books—along with a popular history of *The Three White Rajahs* (1939)—as proud and personal accounts of the last days of a colourful anachronism in British foreign affairs.

❖

BROUGHTON, Elizabeth

Six Years Residence in Algiers. London: Saunders and Otley, 1839; 12mo, pp. 452, with engraved frontispiece and a plate of music

This book, combining the journal of the British consul-general's wife in Algiers (Mrs Henry Strangford Blanckley) with reminiscences by their daughter Elizabeth, is all about the limited lives led by two English gentlewomen in the heart of Barbary between 1806 and 1812. It is very much a social document, more descriptive of the European abroad than the Algerian at home, and telling (surprisingly tediously, it must be said) of the everyday experiences of a most uncommon life—a life disrupted by earthquakes, involved in the bewilderingly fast-changing fortunes of enemies and allies seeking refuge in the consulate or threatening its inhabitants, and beset with the embarrassments of extravagant Arab gifts (the odd pregnant slave, for example, whom it would be churlish to refuse) and the wearying ramifications of diplomatic etiquette.

BURTON, Lady Isabel (1831–96)

The Inner Life of Syria, Palestine, and the Holy Land: From my Private Journal. London: H. S. King, 1875; 2 vols., pp. 376/340, with 2 portrait photographs, 2 chromo-lithographs, and a map

A.E.I. Arabia Egypt India: A Narrative of Travel. London: W. Mullan, 1879; pp. 488, with 15 wood engravings and 2 maps

'I wish I were a man. If I were, I would be Richard Burton; but, being only a woman, I would be Richard Burton's wife.' Isabel Arundel's life was musky with romance—she *did* marry the charismatic African and Eastern explorer Sir Richard Burton, as she was sure she would the moment they met, and she did live the comparatively 'wild and lawless' life she had dreamed of in her youth.

Burton obsessed her. Her character became the background for his, and during the early years of their marriage all her energies were devoted to satisfying his passion for travel—which of course became her own passion too. Burton's first appointment as a married man was as British consul in steamy Fernando Po, the 'white man's grave' of West Africa; Isabel was only allowed to sail as far as Madeira or Tenerife to see him off before turning back alone. Then in 1865 he was posted to Brazil (thanks to Isabel's petitions to the Foreign Office) and at last she was allowed to join him. But even Brazil was not ideal, for most of Isabel's time had to be spent travelling alone from Santos, where her husband lived, to the healthier climes of São Paolo, and it was not until the coveted Damascus post materialized (again thanks to Isabel) in 1869 that Burton and his wife could travel together. Isabel was ecstatic, and never more so than when galloping along 'en amazone' in Burton's fiery wake across the Syrian desert for days on end. She became as besotted as he had always been with all things Arabian, and grew like him to love the bedouin as her own people. But this romantic idyll was too good to last: various political indiscretions and a marked lack of diplomatic tact lost Burton his consulship, and he was demoted to pedestrian Trieste. During their next furlough the Burtons spent two years exploring India and Egypt, and after that, as the old man weakened, they limited themselves to the Continent for holidays.

After her husband's death in 1890, Isabel made her last journey, taking the body home to Mortlake and building a stone bedouin tent for a mausoleum in the Catholic graveyard there. Quiet at last, she settled down alone until the time came to 'pay, pack and follow' once again.

CALDERON DE LA BARCA, Frances Erskine (1804–1882)

Life in Mexico, during a Residence of Two Years in that Country. London: Chapman and Hall, 1843; pp. 437

This is the earliest and most balanced first-hand account of Mexico to be written by a woman. Its author, Frances Erskine Inglis, was born in Edinburgh but grew up in France and Boston, Massachusetts, where her mother and nine siblings moved after her father's death, and where Frances was to meet the dashing Spanish Minister to the United States, Don Angelo Calderon de la Barca. They married in 1838 and travelled together a year later to Mexico City, via Havana, to take up Calderon de la Barca's prestigious new post as Spanish ambassador. Frances toured Mexico extensively during her two years' stay there; she lived through two revolutions in as many years, and by 1842 was able to give a graphic description of the country's social, political, and topographical landscape—so clear, in fact, that the American army used her book as a guide during the campaign against Mexico in 1847.

Frances' romantic history continued after her husband's return to America and later to Spain: on his death she retired in good Catholic tradition to a convent in France for a few years before becoming governess to the Infanta Isabella during the Spanish royal family's exile in Geneva and later in Madrid. And there she died, a marquesa (surely a singular honour for an Edinburgh girl!) at the age of seventy-eight.

CLIFTON, Violet (1883–1961)

Pilgrims to the Isles of Penance: Orchid Gathering in the East. London: John Long, 1911; pp. 320, with 54 halftones and a map

Islands of Queen Wilhelmina. London: Constable, 1927; pp. 288, with 66 halftones and a map

The Book of Talbot. London: Faber and Faber, 1933; pp. 440, with portrait frontispiece and a map

Mrs Clifton was a woman closely akin to Isabel BURTON. Her husband Talbot was a restless, dissatisfied explorer whom Violet revered and accompanied whenever she was allowed. They met in Peru, where Violet was living with her diplomat father; later they went flower-hunting together (perhaps a little tame for Talbot) in the Andaman and Nicobar Islands, the subject of her first book (whose name, incidentally, was designed to attract the reader who 'whilst he would probably not care for a book the title of which breathed of flowers . . . would undoubtedly favour it did the title

savour of sin'). Celebes (Sulawesi) followed in 1921, and a visit to friend
Gertrude BELL at Baghdad in 1925. The Cliftons' last successful voyage
together was to Java, Sumatra, and Bali, the Indonesian dominions of Dutch
Queen Wilhelmina. Talbot died *en route* to Timbuktu in 1927, and Violet
did not travel again.

❧

DAVID, Mrs Edgeworth (Cara)

*Funafuti, or Three Months on a Coral Island: An Unscientific Account of a Scientific
Expedition.* London: John Murray, 1899; pp. 318, with 18 halftones, 2 vignettes, and
a map

This is a cheerful account, like Isabel GILL's with the same subtitle, of a
wife's tagging along after her eminent husband: uncomplaining, undemanding, and, being 'unscientific', much more revealing of the day-to-day
goings-on of an expedition than most 'scientific' (i.e. husbandly!) accounts
could ever manage. Whilst Professor David of Sydney University busied
himself and his five friends with experiments on the reefs off the eastern
coast of Australia, Mrs David set about becoming a part of the Ellice
(Tuvalu) Island community where they were based. She was happy to live in
a hut amongst the 'dear plump little brownies', to do 'housework' each
morning and 'cleaning up' after the coral-boring scientists in the afternoon,
and in her spare moments to rattle off the Samoan alphabet, make a study of
local folk-songs and tattoos (many of which are reproduced in the book), to
teach in the village school on Funafuti, do a spot of botanizing—and even
bathe with the natives in their lagoons. She seems to have packed into three
months more than many a lady traveller managed in a year, and although she
would never have claimed it, her book's as valuable for that as any scientific
report.

❧

DISBROWE, Anne and Charlotte Anne

Original Letters from Russia 1825–1828. London: Ladies' Printing Press, 1878;
pp. 296

Old Days in Diplomacy: Recollections of a Closed Century. London: Jarrold, 1903;
pp. 320, with 5 photogravures and 2 halftones

Charlotte was the daughter of Colonel Edward and Mrs Anne Disbrowe,
successively stationed in Stuttgart, Stockholm, and the Hague between
1828 and 1851. *Old Days in Diplomacy* is her anecdotal account of the
family's travelling life; it is interesting enough—it could hardly be

otherwise, covering such a fertile quarter century in European politics—but not a patch on her edition of Anne Disbrowe's *Original Letters From Russia*. These letters describe the family's first posting, when the colonel was resoundingly appointed Minister Plenipotentiary and Envoy Extraordinary to the Court of St Petersburg. His wife had the gift of being able not just to correspond through her letters but to converse—they make marvellous reading. Using the comfortable drawing-room jargon of the day, she tells the family and close friends to whom she was writing of the tiresome rigours of court etiquette (especially during the year of mourning after Alexander I's sudden death in 1825), and of the practical effects rather than the political causes of the uprisings that followed. She describes the two journeys she made during her stay, too: one over the frontier to Finland and the other to that heady temple of the beau monde, the Kremlin Palace in Moscow. Charlotte Disbrowe published her mother's letters privately—there were only ninety copies printed by the Ladies' Printing Press ('for the Tuition and Employment of Necessitous Gentlewomen'), which is a shame. They deserve a wider audience.

❧

FANSHAWE, Lady Ann (1625–1680)

Memoirs of Lady Fanshawe . . . To which are added extracts from the Correspondence of Sir Richard Fanshawe. Edited by H. Nicolas. London: Colburn, 1829; pp. 395, with engraved frontispiece portrait

The Memoirs of Ann, Lady Fanshawe, Wife of the Right Honble. Sir Richard Fanshawe . . . Reprinted from the Original Manuscript [etc.]. Edited by H. C. Fanshawe. London: John Lane, 1907; pp. 617, with 4 photogravures, 29 halftones, and family trees

Sir Richard Fanshawe is well known (much by virtue of his wife's *Memoirs*) as the loyal agent and friend of Charles II whose courage and bravery during the Civil War later earned him the ambassadorships of Portugal and Spain. In the best tradition of such heroes, he travelled dangerous distances incognito and by night both at home and abroad; he was imprisoned once by Cromwell in London and narrowly escaped the siege of Cork in 1648. A swashbuckling life, lived close to the edge, and any Royalist wife would have done well just to watch and admire. Ann did more: she chose to live that life herself. If that meant stealing a man's clothing for herself or bribing a guard (so that she could be allowed on deck during a scuffle with a Turkish corsair *en route* for Malaga), or assuming a false name and identity (as she did when travelling alone to meet her husband and Prince Charles in Paris), no matter.

In better times, when Sir Richard was sent in state to Lisbon and Madrid, Ann always accompanied or followed him, and incidentally gives one of the earliest accounts of life and travel in the Iberian peninsula we have by an Englishwoman. Her last journey was her most difficult: Sir Richard died suddenly in Spain in 1666, and all diplomatic arrangements, as well as travelling expenses and enbalming costs, were left to Ann. Highly embarrassed, she managed by selling the family plate, and some of the gifts presented to her by the admiring Spanish royal family, before coping alone with the expedition home.

Ann wrote these *Memoirs* in 1676 for her only surviving son (another burden of her travelling life was that she was constantly pregnant, but only four of her seventeen children survived her). The book would serve, she hoped, as a humble inspiration for future Fanshawes.

❧

FINN, Mrs (Elizabeth Anne) (1825–1921)

Home in the Holy Land: A Tale Illustrating Customs and Incidents in Modern Jerusalem. London: Nisbet, 1866; pp. 520, with a wood-engraved frontispiece

A Third Year in Jerusalem . . . A Sequel to Home in the Holy Land. London: Nisbet, 1869; pp. 341, with a wood-engraved frontispiece

Sunrise over Jerusalem, With other Pen and Pencil Sketches. London: John B. Day, 1873; pp. 40, with 16 lithotint plates

Reminiscences of Mrs Finn, Member of the Royal Asiatic Society [dictated by herself]. London: Marshall, Morgan, Scott [1930]; pp. 256, with 3 halftones

[Mrs Finn wrote one more travel book, *A Month on the Mount of Olives*, which I have been unable to locate, as well as other non-travel books on the Holy Land.]

Elizabeth Finn spent over seventeen years in the Holy Land between 1846 and 1863 as the wife of the British Consul in Jerusalem. James Finn was a sympathetic Hebrew scholar, who spent his tenure working as much for the Jews' benefit as for that of the visiting British pilgrims and tourists in his care. His books and Elizabeth's were designed to explain the Jews and their impossibly beautiful country to the Christian public at home, and to prove their own appreciation of Palestine and its people they bought themselves a 'country house' south of the Dead Sea, complete with gardens and vineyards, and travelled within the Holy Land at every possible opportunity. Once they had arrived back in England (after a three-year journey home!) the Finns dedicated themselves to the cause of Christian and Jewish understanding, raising money, generating publicity, and founding various charities and trusts—among them the Palestine Exploration Fund—which flourished long after their decease.

❧

GILL, Mrs (Lady Isobel)

Six Months in Ascension: An Unscientific Account of a Scientific Expedition. London: John Murray, 1878; pp. 285, with a map

No one would have chosen in June 1877 to visit Ascension for anything but the most pressing scientific purposes. There was no call for any woman to go at all. Even Mrs Gill's first impression of the raw volcanic island was of 'an abomination of desolation', but it did not take long before she and her husband could recognize the 'beauty of perfect ugliness' they found there.

Despite its subtitle, *Six Months in Ascension* is unpretentiously informative on the island's elusive natural history and—more unusually—on the stars above it. Those stars were the reason for the Gills' visit. David Gill (later Sir David, HM Astronomer at the Cape of Good Hope) reckoned Ascension Island the most expedient place to see the forthcoming opposition of Mars, measurement of which would further his research on the distance of the earth from the sun. Isobel was there as his assistant. Their work involved setting up an observatory, moving it to the other side of the island when some bloody-minded clouds set in, and keeping nightly shifts by the telescope. When the stars permitted, the Gills made expeditions to 'the peak', where flowers grew, and to the rocky coasts full of sea-birds and turtles; otherwise they were happy to stay with their poodle Rover in a clinker-ridden camp at 'Mars Bay', Isobel trying to invent new dishes from the limited ingredients available, and David busy triangulating. It was an exhausting six months, and Isobel recognized how ridiculous the whole performance must seem to the handful of Kroo natives on the island (and to the sedentary wives of other astronomers tut-tutting at home in their drawing-rooms)—but it was successful. And, what's more, it was fun.

❧

GRAY, Mrs John Henry

Fourteen Months in Canton. London: Macmillan, 1880; pp. 444, with 20 illustrations

Oh dear, one thinks, another worthy lady's reminiscences of life amongst the coolies in the land of rice and pigtails ... In fact this book is extraordinarily informative and readable. Mrs Gray joined her husband the Archdeacon of Hong Kong when he visited Canton as part of his priestly duties. Once the official business of the day was done, she devoted herself to the most vigorous sightseeing I have ever come across. Day and night she (and sometimes the archdeacon) would comb the streets of Canton and beyond, in rickshaws, sedan chairs, even on foot; they employed one coolie solely to keep them informed of festivals and ceremonies (another, incidentally, just to polish the books in their library, Forth Bridge style!) and she relished all she saw.

Mrs Gray became 'adopted' by a Chinese family, which delighted her enormously—even when it meant enduring meals of hashed dog, stewed black cat, fried rat, and sea slugs; if ever the going got too rough she had an unanswerable method of escape: 'I stood up resolutely and said I must put on my hat.' She was eminently sensible, finding the best place to store her furs in winter to be the pawn shop, for example, and never wasting a moment in languishing *à l'Anglaise*: 'I feel the heat less when employed.'

The Grays' stint in Canton ended rather abruptly when Mrs Gray became seriously ill with some unspecified disease—but not so indisposed to prevent her from commenting on the delights of Singapore and Ceylon on the way home.

❖

HANBURY TENISON, Marika (1938–1982)

For Better, For Worse: To the Brazilian Jungles and Back Again. London: Hutchinson, 1972; pp. 335, with 2 colour photographs, 20 halftones, vignettes, and a map

A Slice of Spice: Travels to the Indonesian Islands. London: Hutchinson, 1974; pp. 331, with 1 colour photograph, 22 halftones, and a map

Before her first trip to South America, Marika Hanbury Tenison spent her days looking after the family farm in Cornwall, being one of the country's most successful cookery writers, and imagining the worst (the lot of most explorers' wives whose husbands, like Robin Hanbury Tenison, disappear for months on end in the most alarming and primitive circumstances). It was partly to dispel these imaginings that she joined Robin in 1971 on a three-month mission to study the Amerindians of the Brazilian jungle and, two years later, on a visit to the tribal communities of Sumatra, Kalimantan (the Indonesian part of Borneo), Sulawesi (Celebes), and the Moluccas. Whether she felt more confident about his career afterwards is debatable—both journeys were frightening, frustrating, and physically punishing (in Indonesia her weight plunged to a mere five stone); she certainly never felt any urge to change her own career. She joined Robin just once more before her premature death at forty-four from cancer, taking the two children to visit him during the Mulu Rainforest project he ran in Sarawak in 1978—but this time she was happy to stay at base camp as the expedition's humble (and absurdly overqualified!) cook.

❖

HARE, Rosalie (b. 1810)

The Voyage of the Caroline *from England to Van Diemen's Land and Batavia in 1827–28* [etc.]. With Chapters on the Early History of Northern Tasmania, Java, Mauritius and St. Helena by Ida Lee. London: Longmans, Green, 1927; pp. 308, with 40 halftones (including maps)

This is an ordinary account, made by the captain's wife, of an ordinary voyage from Hull to Tasmania. It takes up only sixty-nine pages of this super-scholarly (and only) edition and is included here simply because few such accounts exist in print. The *Caroline* was one of several brigs commissioned by the Van Diemen's Land Company to carry livestock, stores, and emigrants to its settlements; this particular round voyage took eighteen months to complete. Rosalie Hare wrote her journal in a careful and uncertain hand; the impressions she noted of life on board the *Caroline* and ashore in the Circular Bay Settlement and (briefly) Java, Mauritius, and St Helena, are the clear and naïve impressions of an untravelled and unsophisticated nineteen-year-old. What happened to her, including an encounter with pirates in mid-Atlantic, repeated bouts of fever, and the birth and death of her first son, may seem sensational to us. But such things could happen to every woman travelling at the time: paradoxically this very 'ordinariness' is what makes Rosalie's simple little account so revealing.

❧

HERBERT, Marie (b. 1941)

The Snow People [north west Greenland]. London: Barrie and Jenkins, 1973; pp. 229, with 27 halftones and maps on endpapers

The Reindeer People [Lapland]. London: Hodder and Stoughton, 1976; pp. 187, with 16 halftones and a map

It is a sweeping generalization, I know, but usually explorers are interested in foreign parts, and explorers' wives in foreign people. This is certainly true in Marie Herbert's case: her books are not on the Arctic or north Norway, but on those who live there.

Before she married the polar explorer Wally Herbert in 1969, Marie had been blithely teaching speech and drama in London with never a thought of travel (despite a childhood spent in Sri Lanka); two years later she was on her first expedition, with her husband and ten-month-old daughter, to Herbert Island off the north-west coast of Greenland. Whilst Wally travelled, Marie and her child lived with the Eskimo, coping with the normal problems of any Arctic housewife: the night that lasts all winter, disguised crevasses and rotten ice on spring sledge-rides, and constantly ravenous huskies (and the odd polar bear) eyeing the baby. But by the time her thirteen months' stay was up, Marie was loath to leave: 'The challenge had tried my mettle ... I had come to grips with life in an environment more hostile than I had ever known, but I had drawn from the people all the warmth I needed.'

Marie's next journey was to Lapland. This time she left her husband and child, and accompanied a Lapp family on their annual spring migration with the reindeer from the mountains to the coast. The journey took two weeks, and furnished her with the material for her second book. She has also written a novel (*Winter of the White Seal*, 1982) set in 1818 on a South Atlantic sealing station, and a general history of *Great Polar Adventurers* (1976)—amongst whom Wally looms large.

HOLDERNESS, Mary

Notes relating to the Manners and Customs of the Crim Tatars; written during a four-years' Residence among that people. London: John Warren, 1821; 12mo. pp. 168, with 3 hand-coloured plates

New Russia. Journey from Riga to the Crimea by way of Kiev; with some account of the Colonization, and the Manners and Customs of the Colonists of New Russia. To which are added, Notes relating to the Crim Tatars. London: Sherwood Jones, 1823; pp. 316, with 5 lithographs (3 hand-coloured)

Mary Holderness belongs to that commendable but frustrating group of women travel writers who stick to their subject. Who she was, or what she was doing in Russia is not known, nor any personal details save that she travelled with her children, and that she was treated with unfailing courtesy and kindness by the Crimean Tatars (or Tartars) with whom she lived. Both books (the first is assimilated into the second) are objective, factual, and thorough. Mrs Holderness sees the country she travelled through in 1816 (including Kiev and recently founded Odessa) and the village of Karagon which was her home until 1820, with a clear and unprejudiced eye, telling us plenty about the land and its people, but nothing about herself.

KEMBLE (sometime BUTLER), Frances Anne (Fanny) (1809–1893)

Journal [east-coast America]. By Mrs. Butler. London: John Murray, [Philadelphia: Carey, Lea and Blanchard,] 1835; 2 vols., pp. 313/287

A Year of Consolation [Italy]. By Mrs. Butler. London: Moxon, [New York: Wiley and Putnam,] 1847; 2 vols., pp. 270/325.

Journal of A Residence on A Georgian Plantation in 1838–39. London: Longman, Green, *et al.*, [New York: Harper,] 1863; pp. 434

Fanny Kemble is of course best known as a sublime actress, and the author of numerous plays, poems, and novels. She came of rich Thespian stock, her father being the actor-manager Charles Kemble, part-owner of the Covent Garden theatre, and her aunt the incomparable Sarah Siddons. In fact Fanny was a reluctant actress; her debut in 1829 was engineered by Kemble as a speculative attempt to save both the theatre and the family from bankruptcy. He gambled on Fanny's native talents as a true Kemble, and won. She was an immediate and sensational success, and by 1832 was enchanting audiences on her first overseas tour as Juliet to her father's Mercutio. The tour opened in New York, and went on to Philadelphia, Baltimore, Washington, and Boston, with a short holiday to Niagara thrown in.

Fanny published an account of her travels in her *Journal*, an uncomfortably gauche work whose indiscretions offended several prominent Americans, and which the English public quite adored. She had been pursued around east-coast America by a number of smitten suitors; in 1834 she succumbed to the most persistent, one Pierce Butler, a wealthy Philadelphia landowner's son. The marriage was not a success—they parted after six years, two daughters, and much unhappiness.

Fanny would never have become a Butler in the first place had she known how the family earned its money: they were slave-owners. Their sugar plantation in Georgia had an average population of seven hundred niggers, boasted Pierce—and Fanny was appalled. She could not help being prejudiced against slavery, she said, 'for I am an Englishwoman in whom the absence of such prejudice would be disgraceful'. She insisted that Butler should take her down to Georgia—a nine-day trial by railway, stagecoach, and steamboat—and spent a desperate winter on the swamps of Butler's and St Simon's Islands at the mouth of the Altamaha River trying to alleviate some of the misery she found there. The passionate *Journal of a Residence* was not published until Butler's death, whereupon (in England at least) its author was hailed as a champion of human rights in the illustrious mould of Harriet Beecher Stowe.

From 1840 onwards, Fanny divided her time between London and Philadelphia, taking a holiday from her lengthy divorce proceedings in 1846—the *Year of Consolation*—through France and Switzerland to Italy. Switzerland caught her imagination and she found herself returning there summer after summer, financed by a successful series of autobiographies (*Records of a Girlhood* (1878), *Records of Later Life* (1882), and *Further Records* (1890)), and by her celebrated Shakespeare readings. She died in London at the age of eighty-three, happily planning another trip to her beloved Alps where she was known by the locals neither as an actress nor an author: just another devoted tourist whom they called 'la dame qui va chantant dans les montagnes'.

◆

KINDERSLEY, Mrs (Jemima)

Letters from the Island of Teneriffe, Brazil, The Cape of Good Hope, and the East Indies. London: J. Nourse, 1777; pp. 301, with copper-engraved frontispiece

The first of these letters is dated June 1764, from Santa Cruz in Tenerife. Jemima Kindersley was on her way to Bengal in company with her army-officer husband, and—as promised to a friend at home—was diligently describing everything she saw *en route*.

In fact we have to wade through a good deal of standard British bigotry before we get to the real descriptions. All we are told about the Canaries, for example, is that they were pocked with the holes and heavings of near-continuous earthquakes. San Salvador on the Brazilian coast was a pernicious place, ruled by a Government of thieves, an insolent Church, and a violent Military. The Cape, on the other hand, was pleasant and prosperous (except for the Hottentots who oiled and put their babies in the sun to bake them brown, and broke their noses to make them flat). But as soon as she reached India, the scales fell from Mrs Kindersley's eyes. The novelty of the country shocked her into actually looking at her surroundings, and describing them objectively. She admits as much, wryly confessing that it is all too easy to submit to 'the true spirit of the Englishwoman, [condemning] whatever is contrary to the customs of my own country'.

By June 1765 the couple had reached Pondicherry on the Coromandel coast. They sailed to Madras and Calcutta, then up the Ganges on a 'budgeroo', or barge-cum-houseboat, to Patna and finally Allahabad, where they settled until March 1768. Mrs Kindersley devotes much time to explaining the differences between Hindu and Muslim, Niab and Nabob, Vizier and Mogul, and carefully records everything from the local zenana (or harem) to the Bengal army's quarters, and all with an air (as a contemporary reviewer said) of 'ease, simplicity and fidelity . . . not the manufacture of a hireling, but the production of a real traveller'.

◆

MACARTNEY, Lady (Catherine Theodora) (*c.*1877–1949)

An English Lady in Chinese Turkestan. London: Ernest Benn, 1931; pp. 236, with 4 halftones and a map

In 1898, young Catherine Macartney made her first-ever journey out of

Scotland. It was her honeymoon (she was only twenty-one) and its destination was what is still one of the most inaccessible towns on earth: Kashgar. Kashgar (now Kashi) is hemmed in by the Tien Shan mountains to the north, the Pamirs and Karakorams to the west and south, and the vast Takla Makan desert to the east. Only the most seasoned travellers attempted the journey there (Catherine did it six times in all), and it was certainly considered no place for ladies. During her first four years as the British Agent's wife at Kashgar, Catherine could not have agreed more. The British Residence, *Chini-Bagh*, was comfortable enough, and Catherine performed her wifely duties with aplomb, producing banquets from a tiny mud-walled kitchen in the middle of nowhere and gracefully entertaining passing officials and explorers—but once the visitors were gone, there was nothing whatever to do. For George Macartney (who later became the consul-general) it was different: he had been stationed in Kashgar for eight years already, and from his point of view, it was not such an exile as one might think. The cloak-and-dagger manœuvres of the Great Game playing between Russia and Britain in central Asia at the turn of the century kept him fully—if clandestinely—occupied. Meanwhile Catherine 'experienced to the utmost what homesickness and loneliness meant'.

Once babies started arriving, however, life in Chinese Turkestan began to blossom. Apart from caring for the three children, Catherine now supervised a burgeoning household of staff (including, incredibly, a Scots nurse, Miss Heath, and Miss Cresswell the governess); she took lessons in Turki and Hindustani; she kept a menagerie of dogs, cats, geese, and gazelles (none of whom were strangers to the drawing-room of *Chini-Bagh*), and gave informal concerts on the portable piano and harmonium. One summer the whole family decided to venture a camping holiday in the Pamirs (excursions were not easy from Kashgar) and all of a sudden, seventeen years had passed, most of them cheerfully spent.

Catherine's last journey home from Kashgar was in 1914, during the first months of the Great War; both she and the children were riddled with fever and the long route home through Scandinavia (to avoid France) seemed to last for ever. Macartney followed three years later, and together they settled into a well-deserved retirement in Jersey—just about as unlike Kashgar as you can get.

❧

PIOZZI, Hester Lynch (formerly Mrs THRALE) (1741–1821)

Observations and Reflections made in the course of a Journey through France, Italy, and Germany. London: for A. Strahan and T. Cadell, 1789; 2 vols., pp. 437/389

The French Journals of Mrs. Thrale and Doctor Johnson. Edited from the manuscripts in the John Rylands Library and in the British Museum with Introduction and Notes by

Moses Tyson and Henry Guppy. Manchester: Manchester University Press, 1932; pp. 274, with 2 halftones and 2 facsimile leaves

Hester Thrale, or Piozzi, was likened in her day to Lady Mary WORTLEY MONTAGU as poet, wit, and intellectual—but she was not as intrepid a traveller as the peeress. Hester only went abroad twice, and then no further than Italy. But those two trips caught the public imagination, both because of the company she kept, and because of the way in which she wrote about them.

The first was in 1775, when she set off with husband Henry Thrale, their eldest daughter, and Hester's dearest friend, Dr Samuel Johnson. It was Johnson's first trip abroad, too, if one does not count the Western Isles (he would, I suppose). The novelty of it all, from the 'deform'd' people of Dieppe, the luxurious countryside, the French prostitutes with crucifixes round their necks (that rankled) right down to the unexpected duckweed in the fountains of the Tuileries, was all highly stimulating; and no sooner had the party returned to England (it was only a two-month tour) than they were planning another journey to Italy.

Sadly, the death of the Thrales' only son prevented the Italian trip, and it was not until Thrale himself had died and Hester remarried in 1784 (after which Johnson never spoke to her again) that she got her second holiday. It was her honeymoon: Gabriel Piozzi, the Thrale children's former music master, took his forty-three-year-old bride through France and Switzerland (which she generously accolades 'the Derbyshire of Europe'), and after two years touring his homeland (during which Hester wrote and had published in England her *Anecdotes of Dr. Johnson*) he brought her back to England via Austria, Germany, and the dismal Netherlands.

The journal Hester kept of the French journey was only discovered in the 1930s, but the Italian journal was published almost immediately the travellers returned to London. Amongst all but the stuffiest readers it was a great success: daringly informal and familiar, full of piquant epigrams to quote at literary *salons*, and irresistibly cheerful.

SHELLEY, Mary Wollstonecraft (1797–1851)

[Anon.]: *History of a Six Weeks' Tour through a Part of France, Switzerland, Germany, and Holland: with Letters Descriptive of a Sail round the Lake of Geneva, and of the Glaciers of Chamouni*. London: T. Hookham and C. J. Ollier, 1817; pp. 183

Rambles in Germany and Italy, in 1840, 1842, and 1843. London: Edward Moxon, 1844; 2 vols., pp. 280/296

Mary Shelley, usually regarded rather dismissively as the tragically devoted widow of Percy Bysshe and unlikely author of *Frankenstein*, was also a travel writer—like her mother Mary WOLLSTONECRAFT. Her first travel book was written in the most romantic circumstances imaginable: it describes the elopement of the sixteen-year-old author with the poet Shelley (five years her senior and already married), tracing their reckless and besotted course across France and the Alps to Lucerne.

The journey there, overland on anything from a third of an ass each (Mary's foster-sister Clara was playing gooseberry) to an 'open voiture' offered by a tender-hearted Swiss, cost £60; the journey back along the cheaply travelled waterways of the River Reuss, the Rhine, and the canals of Holland, only £28. Every guinea is accounted for in the book with precious detail, and the whole thing reads like some joyous picaresque tale. A baby was born a few months after the travellers' return—the first of four children of whom only one, Percy, survived its first few years—and a year or so after that, Mary and Shelley were on the road to Switzerland again for four months' impoverished holiday, described with contributions by Shelley himself in the second part of the *History of a Six Weeks' Tour*.

Shelley married Mary after the death of his first wife in 1816, and two years later the couple travelled to Italy, where they spent the next four years in the heady company of such as Byron, Thomas Love Peacock, and the Leigh Hunts, sauntering between Pisa, Leghorn, Florence, and Rome before finally settling at Lerici on the north-west coast. On Shelley's death in 1822 Mary staggered home to England penniless, and apart from a short trip to Paris in 1828, cut short by her catching smallpox, she stayed there for the next eighteen years, writing to earn a living for herself and her son. She could not afford to travel again until 1840, when she was asked by the newly-affluent young Percy to accompany him and some student friends on two tours of Germany and Italy, the subject of her second and last travel book. It is more instructive than the first (a 'gossiping companion', she called it), but not half so entertaining.

The last few years of Mary Shelley's life were spent editing her husband's letters and planning his biography which remained unwritten at her death in 1851.

❧

SHIPTON, Diana

The Antique Land. London: Hodder and Stoughton, 1950; pp. 219, with 23 halftones, line-drawn vignettes, and an endpaper map

Of the three residents of the British consulate at Kashgar (now Kashi) represented in this book, Diana Shipton is the most recent. She met her husband, the mountaineer Eric Shipton, in India in 1939; she had grown up there and so was used to the vagaries of Asian travel. Even so, she did not face the arduous journey to join Eric at Kashgar until 1946, four years after their marriage. She stayed for eighteen months, desperately trying to guard against 'the rot of frittering' that comes of boredom: unlike Catherine MACARTNEY, she had no children to occupy her (the baby had been left at home) and unlike Catherine's successor Ella SYKES, Diana hated riding and mountain climbing. The book is a record of her gradual absorption into the rarefied life of Chinese Turkestan, the 'antique land'.

In 1948 Diana came home to England, only to return the next year to join Eric in his new post at Kunming. She stayed there for the birth of her second son, and then made the journey home in 1950, trekking right across southern China to Hong Kong with the children whilst Eric made plans for his next expedition to Mount Everest. They were divorced five years later, and while Shipton continued his success as a travel writer until his death in 1977, Diana embarked on a new and different life.

❧

STEVENSON, Margaret Isabella (1829–1897)

From Saranac to the Marquesas and Beyond: Being Letters written by Mrs. M. I. Stevenson during 1887–88, to her sister Jane Whyte Balfour [etc.]. Edited and arranged by Marie Clothilde Balfour. London: Methuen, 1903; pp. 313, with 8 halftones

Letters from Samoa 1891–1895 [etc.]. Edited and arranged by Marie Clothilde Balfour. London: Methuen, 1906; pp. 340, with 12 halftones

Mrs Stevenson—Robert Louis's mother—was a completely unruffleable traveller. Her son seems to have inherited his delicate constitution from her: she first started travelling when sent to Menton for her health. Thereafter she spent seasons on the Continent almost every year, often taking the weakly Louis with her, until the death of her husband in 1887. By that time Louis was married to the indefatigable Fanny Osbourne, and to take Margaret's mind off the bereavement, she was invited to join them on a winter's trip to America. They went partly to see Louis's New York publishers and partly to find somewhere clement to spend the dangerous months between October and March, and eventually settled in a house on the shores of Lake Saranac in the Adirondack Mountains. When spring came, they crossed the United States by train to San Francisco, and then set sail in the yacht *Casco* in search of sunshine and stories, ending up in Hawaii by way of the Marquesas and Tahiti. Not bad for an Edinburgh minister's daughter nearing sixty who had never gone beyond the spas of Europe before.

Margaret's next great journey was to Samoa, to visit Louis in his idyllic home *Vailima* in 1890. During the next five years she visited Australia (several times), New Zealand, Tonga, and other island kingdoms, went home to Scotland once, and settled happily (if intermittently) into the 'strange, irresponsible, half-savage life' of an adoptive Samoan.

Her letters home from all these places are wonderfully laconic and calm. She mentions knitting socks to pass the time usefully whilst cruising amongst the South Sea islands (barefoot and in native dress); to help Louis work in his Pacific paradise she taught herself how to type, and at the age of sixty-two learned to ride, so that she could get to a distant mission church on Sundays. Margaret's travels drew to a precipitate close in 1894 when the sudden death of her son sent her grieving home to Edinburgh; there, in the unkind and unromantic Scottish climate, it did not take her long to die herself.

TOWNLEY, Lady Susan (1868–1953)

My Chinese Note Book. London: Methuen, 1904; pp. 338, with 16 halftones and a map

'Indiscretions' of Lady Susan [western and eastern Europe, China, and America]. London: Thornton Butterworth, 1922; pp 314, with 29 halftones and 2 vignettes

Susan Townley saw herself, I think, as a sort of modern-day Mrs TROL-LOPE. Both described North America as vulgar and snobbish, and both were themselves similarly described by their outraged American hosts. Neither minded very much. On the contrary: it sold their books.

Lady Susan went to the United States in 1905, welcomed as the daughter of the Earl of Albemarle and the wife of a high-ranking embassy official in Washington: it was the latest move in what seemed an endless diplomatic round leading Sir Walter and his wife from Lisbon in 1898 through Berlin, Rome, Peking, and Constantinople, then on to Rio de Janeiro, Bucharest, Persia, and The Hague. Who knows what ambassadorial glory there might have been in store, had it not been for Lady Susan's indulgent 'indiscretions'? For it was undoubtedly due to her unerring talent for offending local dignitaries, upsetting traditional hierarchy, and upstaging visiting VIPs that the embarrassed Foreign Office had to advise Sir Walter to retire in 1919.

Lady Susan took her revenge in print in 1922. After the rather restrained descriptions of Peking in *My Chinese Note Book* she could now afford to let rip and describe her rarefied travelling life amongst the stiff-lipped British abroad, the pesky 'Yellow Peril', the treacherous 'Hun', and those crass, bombastic Yankees.

When the fuss had died down, Lady Susan settled down to an unwontedly quiet and charitable life at home, punctuated by nothing more controversial than the odd letter to *The Times* and fond memories, no doubt, of her proud and flagrant youth.

❧

'TULLY', Miss

Narrative of A Ten Years' Residence at Tripoli in Africa: from the Original Correspondence in the possession of the family of the late Richard Tully, Esq. The British Consul. Comprising Authentic Memoirs and Anecdotes of the Reigning Bashaw. Also, an account of the Domestic Manners of the Moors, Arabs, and Turks. London: for Henry Colburn, 1816; quarto, pp. 370, with 5 hand-coloured aquatints and a map

In the past, because of a statement in the preface to the first edition, this book has been attributed to a certain Miss Tully, sister of the British consul in Tripoli from 1783 to 1793. But later editions call the spinsterly author Tully's sister-*in-law*, which, judging by her native detachment and deference to the consul, seems far more likely. The narrative is composed of the letters she wrote home (whenever the opportunity of 'communicating with Christendom' presented itself) between July 1783 and August 1793; it is a delicious mixture of sensational subject-matter and deadpan delivery.

This well-bred English spinster spent her time in north Africa visiting the royal family, politely congratulating the Bashaw on his fine harem and collection of Christian slaves; sometimes she ventured towards the bazaar— 'it requires some address and resolution . . . to walk full dressed near three quarters of a mile through the streets'—or even, occasionally, into 'the sands', where the plundering Arabs roam and there are 'capital mosques' to be found.

Early in 1785 the plague arrived in Tripoli from Tunis, and kept the Tullys virtually within doors for a year and a half. Hundreds were dying a day at its peak and at one particularly virulent stage, the consul's family was reduced to scavenging left-over ship's biscuits from empty vessels in the harbour and hoarding household wood for its own coffins. All this is recounted by the author with true-blue sanguinity, along with the horrors of civil war in Tripoli and the dangerous neighbouring aggressions between Spain and Algeria, Venice and Tunis; only the Turkish invasion in 1793 which drove out both the Bashaw's family and eventually sent even the Tullys home proved too terrible to describe, and the letters close with tightly pursed lips as the atrocities of Barbary are left behind for England.

❧

VASSAL, Gabrielle M. (d. 1959)

On and Off Duty in Annam. London: Heinemann, 1910; pp. 283, with 65 halftones and a map

In and Around Yunnan Fou. London: Heinemann, 1922; pp. 187, with 43 halftones and a map

Life in French Congo. London: T. Fisher Unwin, 1925; pp. 192, with 49 halftones and a map

We owe her unique view of French territory through resident English eyes to the novelist Gabrielle Vassal's position as wife of a French army doctor. Within weeks of their marriage they were posted to the Pasteur Institute at Nha Trang, on the coast of what was then Annam (now Vietnam). From Nha Trang they travelled inland to the hills (usually by 'tri-car', a strange sort of motorized tricycle) for the sake of health and good game; it was amongst the Annamese mountains that Gabrielle learned how fine a shot she was. Their next post was further north in Tonkin (bordering the Chinese province of Yunnan) where they stayed for a few years before moving to a complete change of scene at Brazzaville, West Africa.

Each of her books rather dryly describes Gabrielle's limited medical duties as her husband's helpmeet, and the hunting expeditions they took together (the 'tri-car' soon grew into a thumping great Ford motor tearing through the jungles of the 'white man's grave'), as well as discussing politics and, for good measure, a smattering of ethnology and local history.

❧

VIGOR, Mrs (formerly WARD and RONDEAU) (1699–1783)
[Anon.]: *Letters from a Lady, who Resided Some Years in Russia, to Her Friend in England. With Historical Notes*. London: J. Dodsley, 1775; pp. 207, with a folding genealogical table

The aptly named Mrs Vigor first went to Russia in February 1730, as the wife of the British consul-general there, Thomas Ward. The sumptuous court of the Empress Anna was based, according to the season, either at 'Petersburgh' (Leningrad) or Moscow; wherever it went, a wash of retainers followed, including the eager Wards. When Thomas died in 1731 of some unnamed disease, his widow stayed on and soon afterwards found herself 'a title, a ribbon, a pompous equipage, and a great estate' in the person of courtier Claudius Rondeau, her second husband. The newly-weds acquired a country retreat not far from St. Petersburg in 1733, and the book closes on their wedded bliss, with a baby, in a 'pretty romantic' land.

The appeal of this book back in London, even though it was not published for another forty years (when Mrs Rondeau had returned and become Mrs Vigor), was in its toothsome repetition of high-born Russian gossip and intrigue—something the only earlier female account by Mrs JUSTICE quite lacked. Its descriptions of a court newly emerged from the empire-building splendours of Peter the Great, and fabled throughout Europe for its glittering extravagances, make it both interesting and significant now. Sadly, there is not much mention of travel at all, except for the odd brief reference to the fur-lined sleighs that whisk one through the empty twilight and the blank faces of the natives, those wretches 'so low and so poor, that they seem to have only the figures of human creatures'.

❖

WAZAN, Emily, Shareefa of (d. 1944)

My Life Story [etc.]. Edited for Mme de Wazan by S. L. Bensusan. Edinburgh: Arnold, 1911; pp. 327, with 20 halftones

Emily Keene was the modestly educated daughter of the governor of Surrey County Gaol. She was a comely, plump young woman whose parents envisaged for her a comfortable provincial marriage and a quiet life: they were stunned when during the winter of 1872 she announced that she was to marry an elderly Muslim she had met, the Grand Shareef of Wazan (now Ouezzane, in Morocco). 'Life would be impossible without him', she said, and in a welter of salacious Press censure and society gossip, she left with him for Barbary in January 1873.

What made this plain, rather stolid Englishwoman do what seemed such a wildly romantic thing is a mystery. But if it *was* because of romance, then she was soon and sadly disabused. She was her sacred husband's third wife; the others were already presenting him with grandchildren and it became clear that Emily was not a welcome influence either domestically or politically in Wazan. By 1884 her elderly husband had turned completely against her, ordering her food to be poisoned (she said) and three years later marrying one of his Arab servants. But Emily stood her ground, tending her three sons' education in France and arranging extensive camping and sporting tours for herself in Spain and Algeria. Once she was no longer a threat to them, the women of Wazan took her into their confidence, and by the time of her death she was known as a universal 'Mamma', a loyal and sympathetic foreigner with none of the Eastern sense of intrigue. And a foreigner she stayed, even though she was the mother of Wazan's next Shareef and has become part of Morocco's history: she never lost her Christian faith, nor ever changed her serviceable Western clothes for the musky drapes of the harem. She retired in her old age to a house in Marshan—still indulging herself in the odd hundred-mile jaunt on horseback from time to time—and died well into her eighties in 1944.

❖

WOLLSTONECRAFT, Mary (1759–1797)

Letters written during a short residence in Sweden, Norway, and Denmark [etc.]. London: for J. Johnson, 1796; pp. [266]

When Mary Godwin and Percy Bysshe SHELLEY eloped to France in 1814, one of the few possessions they took with them was a copy of this book. It was written by Mary's mother, a woman she had never known.

Mary Wollstonecraft is primarily remembered today as the feminist author of *A Vindication of the Rights of Women* (1792), the novel *Mary, A Fiction* (1787), and various tracts on politics and education. She was largely self-educated, rising by turns from the second of a dissolute farmer's seven children to become a lady's companion in Bath, a schoolmistress in Stoke Newington, and a London publisher's reader and translator. She also became one of the leading lights of a band of literary young radicals who moved almost *en bloc* to Paris when revolutionary rumours began circulating in late 1792, and there she lived as a writer and political activist for two years. In Paris she met Gilbert Imlay, and ex-officer in the American Revolutionary Army; they became lovers and their daughter Fanny was born at Le Havre in 1794.

Mary first attempted suicide by swallowing laudanum after following Imlay back to London later that same year. Much of her political ardour had evaporated into infatuation for this one man—and he had tired of her. Perhaps to occupy her—or just to get her out of the way for a while—Imlay suggested after Mary's reluctant recovery that she should undertake a business commission for him. It appears he was involved in illegally running ships against the British blockade in the Baltic and had recently lost a vessel loaded with silver and plate; suspecting that the ship's captain, a Norwegian from Risör, might have something to do with the cargo's disappearance he dispatched Mary to investigate. Wretchedly eager to please him, and remembering a winter voyage she had made to Portugal ten years earlier and enjoyed, she set off with her baby and a miserable maid in mid-June 1795, and travelled for nearly four months in the unfashionable wastes and waters of the north.

Mary's spirit felt at home in the bleak beauty of Scandinavia. The more remote the country became—on the Skagerrak coast of Sweden, for example, or at Tönsberg where she stayed alone for a few weeks—the more desolate she felt. In her book—the perfect pathetic fallacy—the physical landscapes she described become more and more intensely involved in the emotional ones until she returned to London in an unhealthy state of stimulation and exhaustion. Imlay was by this time living with another woman, and so after handing over the money she had recovered for him at Risör, Mary surrendered and threw herself in the Thames.

It is ironic that Mary's own life should so starkly have reflected just the sort of femininity she abhorred. It was the very worst of a woman's lot to be emotionally beholden, like her, to a faithless man (physically too, with the child) until both her pride and her freedom were lost. She was even too weak to kill herself properly. And when the time did eventually come, she died the supreme feminine death in the arms of William Godwin, the man she had married (against both their principles) to save the child she was expecting, Mary, from the stigma of bastardy. In fact, only in her love of travel was she able to achieve the feminist unconventionality on which her reputation rests.

In Camp and Cantonment

THIS is the stuff of real adventure. Tales of hair's-breadth escape from marauding mutineers abound, of frail little nurses facing gunshot and shell to save some poor soldier's life, or of doughty imperial dames taken prisoner by unsuspecting rebels: they are unquiet lives, these writers', all of them. And it is a mark of their bravery, I suppose, that they felt able to relive those days by describing them in print. Some waited half a century or more for peace of mind enough to do it, while others sent their manuscripts home from the very battlefield, dashed off while the full horror of the situation was still upon them.

Perhaps the most affecting stories told by women travellers caught up in wars, sieges, and so on concern the so-called Indian Mutiny of 1857. The memsahibs of British India had until then lived comparatively tranquil, even tedious lives, spent in a stiflingly hot round of resting, supervising the servants, dinner-parties, and discussing, according to Mrs ELWOOD,

> Who danced with whom and who is like to wed,
> And who is hanged, and who is brought to bed.

But everything changed in May 1857, when the sepoy rebellion broke out. A number of issues sparked it off, most attributable to boorish British arrogance and insensitivity and to do with fear over loss of caste and religious suppression on the part of the native soldiers (or sepoys). They mutinied in Meerut on Sunday 10 May, and within days the frenzy of blood-revenge had spread throughout the province of Oudh and beyond. The mutineers drove their fury home in the cruellest way they could by attacking what the British in India held most sacred: their womenfolk. Many memsahibs were massacred in the most appalling circumstances; what happened to the ones who survived is best left to them to tell.

One Mrs DUBERLY, the wife of a regimental paymaster, held the dubious privilege of witnessing both the mutiny and the campaign that immediately preceded it, the Crimean War. The legendary Lady of the Lamp was not the only woman in the neighbourhood of Scutari—there were several officers' ladies there, even an army caterer from Jamaica (a true adventuress if ever there was one), as well as Miss Nightingale's motley flock of amateur and professional nurses. Obviously one did not have to be an army wife or a medical volunteer to find oneself active in the unquiet quarters of the world, as Catherine DAVIES (a governess caught up in the Napoleonic wars in Naples), Lady HODGSON (a diplomat's wife imprisoned in Ashanti during the uprisings of 1901) and Flora SANDES (a sergeant in the Serbian infantry during the First World War) all prove.

The stuff of adventure, as I said, with stories full of derring-do, what Mrs SEACOLE called 'the pomp, pride and circumstance of glorious war', and plangent private tragedy. But that is not all. The books written by these unorthodox eye witnesses are important, as well as entertaining. They commentate on Britain's military past, from the Battle of Waterloo until the Spanish Civil War, from a unique point of view. There is little talk of politics and tactics (the stuff of most military histories); what we are offered instead is a 'domestic', or background chronicle of the fighting, seen through the splintered shutters of some besieged and disease-ridden cantonment, or the squalor of a hospital tent, often with bewildered and compassionate eyes. It is important that such images should not be forgotten.

BARTRUM, Katherine Mary (1834–1866)

[Anon.]: *A Widow's Reminiscences of the Siege of Lucknow.* London: Nisbet, 1858; pp. 102

This must rank with Susie RIJNHART's book as one of the most affecting travel accounts ever written. Its author was by no means a practised writer—

she was just an ordinary young woman living an ordinary Anglo-Indian life with her husband Robert, a doctor in the Bengal Medical Service, and their adored son Bobbie. Until the Mutiny.

They were stationed at Gonda, east of Lucknow, when the first astounding news of a sepoy rebellion began to filter through at the beginning of May 1857: the natives had declared war on the Raj, they had already shot or hacked to death numbers of Europeans in Meerut and Delhi (women and children included), and were reported to be savagely swarming the Presidency of Bengal to flush out the rest. Kate's first reaction on hearing the news was to stay calmly with Robert and Bobbie and wait: at least the family would die together. She was appalled when the British commissioner gave orders that the women and children of Gonda should leave for the Residency at Lucknow where they would stay with refugees from other hill stations until the 'trouble' was over.

The following 80-mile journey on elephant-back was feversome, lonely, and dangerous, and rewarded not by freedom and safety but by all the horrors of the 140-day Siege of Lucknow, described by Kate with awful simplicity. On their arrival at the Begum Kotee, a few yards from the actual Residency building, she and the boy were stowed into a room with thirteen others, the scorching atmosphere already rank with sickness and panic. Their sole possessions were the clothes they had travelled in, what paper, pens, and photographs they had managed to smuggle away, and a few greening copper pans for their cooking and washing (there was not enough fresh water every day to cook *and* wash: they must choose which was most urgent). Rumours regularly seeped through that relief was on the way; when the troops did eventually reach Lucknow on 26 September, only three women and a handful of ailing children were left of Kate's original companions. Still, the first soldiers assured her that Robert was safe and following close behind the vanguard: the thought of him had kept both her and Bobbie alive, she said.

For two days Kate waited at the Bailey Gate, with blissful reunions going on all around her, but Bartrum never came. He had been killed by a sniper as he galloped triumphantly into Lucknow with the first relief force, just minutes away from safety and from Kate. She sank into a sort of daze, and accomplished the following trek to Calcutta, where passages home could be arranged, clinging obsessively to the life and health of her son. But the journey occupied two of India's hottest months, and Bobbie, who had survived the cholera and smallpox of Lucknow, began to grow horribly ill. 'I *cannot* spare him', cried Kate, 'and I do not think that God will take away my little lamb when I have nothing else left.' But the night before they were due to sail for home he died, and Kate's account comes to a shattered, empty end.

BLACKWOOD, Lady Alicia (1818–1913)

A Narrative of Personal Experiences and Impressions During a Residence on the Bosphorus throughout the Crimean War. London: Hatchard, 1881; pp. 318, with lithographed title-page, 10 lithographed plates, and a vignette

Not all Florence NIGHTINGALE's trouble had to do with the hospital wards. She had also to deal with the shoals of worthy English gentlewomen who washed up on the shores of Scutari, eager for glamour and promising the world until they saw—and smelt—what nursing *really* meant. Most of them lost no time in scuttling off home again; Lady Alicia Blackwood was one of the few who stayed.

She and her clergyman husband sailed for Scutari at their own expense as soon as the horrific news of the battle of Inkerman reached England in November 1854, intending to do whatever jobs were needed most. They reported to Miss Nightingale immediately they arrived and were soon in the thick of all the tragic squalor that hallmarked the whole Crimean campaign. Henry Blackwood was sent to help the army chaplains with their sick-visits and burial rotas, while Lady Alicia was shown down to the hospital sewers. Here, just discernible through the fog-like stench, were bundled nearly three hundred soldiers' wives and children. Not all of them were ill when they arrived—they might have nowhere else to go, or be in the last stages of pregnancy—but sooner or later they all succumbed to whatever disease was prevalent at the time, and no one got better. The dead were either shovelled away or, in the case of one baby Lady Alicia saw, used to stop up leaks in the sewage pipes. And nobody had the time to do anything about it until Lady Alicia arrived. Within three months she had the place cleaned out, set up partitions to give her patients some privacy, opened a subsidized 'shop' selling the odd length of muslin, bunch of ribbons, or skein of wool, and started a Sunday school for the children.

The Blackwoods stayed in the Crimea until peace was declared in March 1856, taking the odd excursion together to Therapia in Turkey or Simferopol in Russia, and only thought of telling their story when, a quarter of a century later, the horse they had brought back with them from Scutari eventually died. That suddenly brought back what had happened out there, said Lady Alicia: only now could she bear to record it.

COOPLAND, Ruth M.

A Lady's Escape from Gwalior and Life in the Fort of Agra During the Mutinies of 1857. London: Smith, Elder, 1859; pp. 316

This is not just another heart-breaking account by a widow of the Indian Mutiny, although it has all the simplicity and pathos that made Kate BARTRUM's narrative so affecting. As far as I know its author published nothing but this single book, but because she was so startlingly skilful a writer her story stands well apart from the rest. Briefly, it runs thus.

George Coopland was a chaplain to the Honourable East India Company, and was posted to Gwalior—about a hundred miles south of Agra—soon after his wedding in 1856. His young wife was much excited by the journey in a variety of bizarre vehicles from Calcutta through the sacred city of Benares and Allahabad, and when the couple arrived at Gwalior they were made both comfortable and welcome by sahibs and sepoys alike. But within weeks the astounding news began to break of European massacres at Delhi and Meerut, and by the time it was decided that the women should try to escape to safety somewhere, it was too late. All the roads were blocked by advancing mutineers: there was nothing left but prayer and resignation.

The end came on Sunday 14 June. It was the day of a child's funeral, and whilst the service was going on the sepoys first set fire to the British settlement and then waited quietly outside Coopland's church with their guns. Coopland himself was one of the first to die, shot down as he stood with his wife at the door, and the moment of utter silence that followed his slump to the dry earth was the most terrible of Mrs Coopland's life.

Most of the women were left alive, and hustled straight to the local Rajah's palace. Under his protection they were marched north to the comparative safety of the Fort at Agra where Ruth's son was born and where she spent her first wedding anniversary 'utterly stunned' in an opiate haze. After five months' squalid imprisonment at Agra it was reckoned safe enough for her to take the baby to an aunt's bungalow at Simla—a long and harrowing journey—and once there, she was able to recuperate her strength (if not her spirits) for the seven-week trek that led to Lahore, down the Indus to Karachi, and eventually to Bombay and out of India for good.

❦

DAVIES, Catherine

Eleven years' residence in the family of Murat, King of Naples. London: How and Parsons, 1841; 12mo, pp. 92

Surely Miss Davies could have chosen a better title for her book. *Eleven years' residence* sounds like an account of a quiet Italian sojourn. *The Strange and Wonderful Adventures of a Welsh Governess Abroad* might have been nearer the truth.

Catherine Davies was an Anglesey girl whose sturdy father, twice married, had thirty-three children. It was hardly surprising that Catherine should have to earn her own living as soon as possible, which she did by working 'in service' in Liverpool and London before being offered the chance to accompany an English family to Paris as governess. All went well for nearly a year, until the Peace of Amiens in 1802 and Napoleon Bonaparte's subsequent internment of the British in France. Catherine somehow came to the notice of Bonaparte's sister Caroline and her husband General Murat during her imprisonment (one of several unexplained turns of events in the book, perhaps due to the fact that the author wrote it *very* retrospectively at the age of sixty-seven); they requisitioned the Welsh girl as a fashionable adjunct to the household and Catherine settled in so well that when the order came for the deportation of all Britons, she defied it and stayed.

Bonaparte declared himself emperor in 1805, and created his brother-in-law Murat king of Naples. So the household travelled to Italy, pausing in Rome where Catherine fell violently ill; she could only follow the new royal family after six weeks' fragile convalescence. Four years after her arrival in Naples, the port was captured by the English, and Catherine vividly describes the seventeen-day siege that followed and her dramatic rescue in HMS *Tremendous*. The next few months are rather confusing, though. No sooner had the ship arrived at Trieste than Catherine found herself returning to Naples (both voyages through waters raging with men-of-war)—it seems her mysterious 'Rome illness' had returned, and she was thought safer within reach of the balmy waters of Ischia (Naples' island spa) than at sea. Soon Catherine felt *just* well enough to attempt the ghastly journey home by herself, and set out by carriage and diligence overland to Paris and at last to London and back to Beaumaris on Anglesey. And there she stayed, an exotic local celebrity, for the rest of her life.

❖

DAVIS, Elizabeth (1789–1860)

The Autobiography of Elizabeth Davis, A Balaclava Nurse, Daughter of Dafydd Cadwaladyr. Edited by Jane Williams. London: Hurst and Blackett, 1857; 2 vols., pp. 308/298, with frontispiece portrait

Quite a life, had proud and pretty Betsy Davis: it reads like a mixture of *Tom Jones* and the *Wonderful Adventures* of Mrs SEACOLE. She was born in deepest, darkest Wales and ran away from home to Chester when still a girl, from there to sail to Liverpool and join domestic service. Her first family took her to the Continent in 1814, where she was sent on strange errands in even stranger streets as far apart as Naples and Madrid. She was allowed to visit the field of Waterloo, just five days after the battle (the bodies still thrillingly littered about) and to take a turn in the Tuileries with the best of 'em. Later, she was engaged by a sea captain's wife to sail as lady's maid to the West Indies.

Having a taste for travel by now Betsy applied to another seafaring family and soon set sail for Tasmania; the next few years (she is not very good on dates) were spent with the same family, running cargo between Australia, China, India, and South America. She learned to do a little trading on her own account and eventually grew wealthy enough to speculate on some property in London. Unfortunately, her broker turned out a thief and Betsy was reduced again to service until 1854 when she volunteered as a nurse at the outbreak of the Crimean War. She ended up, after various acrimonious brushes with the authorities (who thought her too ill-bred to nurse British heroes) as a cook in the hospital at Balaclava, sworn enemy of the high-falutin' Miss NIGHTINGALE (whom she dubbed, tight-lipped, 'a petticoat'), and spirited defender of her own Welsh, rabble-rousing blood. She died back in Britain a few years after her return—a marvellous story-teller, if nothing else.

❦

DUBERLY, Mrs Henry (Frances Isabella) (1829–1903)

Journal kept during the Russian War: From the Departure of the Army from England in April, 1854, to the Fall of Sebastopol. London: Longman, Brown, Green and Longmans, 1855; pp. 311

Campaigning Experiences in Rajpootana and Central India, during the Suppression of the Mutiny, 1857–1858. London: Smith, Elder, 1859; pp. 254

It must have been one of the most incongruous sights of the Crimean War: a woman, honey-blonde and beautiful in a habit of sky-blue and gold, perched on a prancing steed and galloping into battle at the head of the Eighth Hussars: Fanny Duberly, the soldiers' darling (or the officers', more like), out for a ride.

Fanny was quite a woman. She was the only 'Lady of Quality' (as she put it) to accompany her husband all the way to the Balaclava camp. Not content with that, she managed to secure one of the few wooden huts available there, a stable for the horse she brought from England, and even a bed for husband Henry. Her greatest love was for horses: every day she would take a ride (usually to the Front, where she witnessed the 'glorious and fatal charge' of the Light Brigade) and she was the doyenne of the Balaclava Racecourse. Next she adored fine company: every available Gentleman—the higher his rank the better—was summoned on a sort of rota to escort the Regimental Paymaster's Lady on jaunts to the battlefield or rides along the Lines. A poor third came the Regimental Paymaster himself, shabby Captain Duberly, following constantly at Fanny's heels like a loyal little lap-dog.

When Sebastopol fell in September 1855, Fanny, along with the indomitable Mrs SEACOLE, was one of the first to ride into the city ('Who could keep away from a place where so many interests were at stake? Not I!'). And no sooner had the last shot been fired than her manuscript *Journal* was on its way to London: by the time she brought Henry home it was already in the booksellers'.

Fanny's second book is rather more pedestrian. Eighteen months after arriving home from the Black Sea, Henry was posted with his regiment to Bombay, to join the 'Rajputana Column' against the Indian mutineers. Of course Fanny came too, trailing over 2,000 miles in a *dhoolie* (a covered litter) contemplating the heat and the rain; it was too inclement to ride a horse (in any case the men were too tired to escort her) and there were no meaty battles—only the odd scuffle with rebels along the way. Still, it was better than sitting at home like other women: 'I have often prayed that I may *wear* out my life and not rust it out', she wrote.

The Duberlys stayed in India for another ten years after the publication of *Campaigning Experiences*; it was Fanny's last sojourn abroad and she died at home—no doubt rather worn than rusted—at the age of seventy-four.

❧

DURHAM, (Mary) Edith (1863–1944)

Through the Lands of the Serb. London: Edward Arnold, 1904; pp. 345, with 16 halftones, 5 sketches, and a map

The Burden of the Balkans [etc.]. London: Edward Arnold, 1905; pp. 331, with 9 halftones, 20 sketches, and a map

High Albania. London: Edward Arnold, 1909; pp. 352, with 11 halftones, 18 sketches, and a map

The Struggle for Scutari (Turk, Slav and Albanian). London: Edward Arnold, 1914; pp. 320, with 15 halftones

Twenty Years of the Balkan Tangle. London: George Allen and Unwin, 1920; pp. 292

The Sarajevo Crime. London: George Allen and Unwin, 1925; pp. 208

Some Tribal Origins, Laws and Customs of the Balkans. London: George Allen and Unwin, 1928; pp. 318, with 11 halftones and 23 sketches

Edith Durham was a missionary, not religious but political. She dedicated her entire travelling life to the Balkans, the 'melting-pot of Europe', in an effort to evangelize its political and cultural complexities to the mercenary West, and so save its soul.

The eastern European States now comprising Montenegro, Bosnia, Herzegovina, Serbia, Albania, and Bulgaria were relatively unfrequented by visitors when Edith made her first visit in 1900: where there were mountains they were distinctly inhospitable, their people fiercely independent and distrustful (little wonder, with Russian, Turk, Austro-Hungarian, and Briton squabbling over military and economic rights), and within the 'Balkan triangle' itself the internal conflicts and shifts of allegiance made it a dangerous place for any lone outsider to be. But Edith, a gifted artist and anthropologist who began travelling in search of health, was captivated by the land, the people, and their aspirations, and not a year went by from 1900 to the Great War (itself seated in 'her' territory) without her journeying amongst them.

She was thirty-seven when first she went, assuring herself that what she called a middle-aged woman with 'short hair, no stays, very plain and stout' and just a horse-boy and one elderly guide for company was much more likely to amuse than to threaten, and with her linguistic gifts, her practical help, and political sympathies, she soon became well known and accepted (especially in Albania) as national champion and honorary 'Queen of the Mountains'. During the revolts of 1903, 1909, and the subsequent Balkan war of 1912, she brought medical aid, food, and understanding to the 'near-savage' Albanian tribesmen whilst trying to explain to them—as well as to us, through her books—that no answer could be found to the vexed Balkan question without first appreciating the vastly different cultures of the constituent States.

After the First World War, the Albanian government offered Edith a permanent home in its country; she preferred instead to act independently as its spokesman still, and settled back in London (in poor health again by now) to arrange her rich collection of Balkan textiles and embroideries, and to write tirelessly for her cause.

◆

EATON, Charlotte (1788–1859)

The Battle of Waterloo. By A Near Observer. London: John Booth, 1815 [fourth edition; earlier editions, published the same year, not located]; pp. 116, with a hand-coloured aquatint panorama

Narrative of a Residence in Belgium during the Campaign of 1815; and of A Visit to the Field of Waterloo. By An Englishwoman. London: John Murray, 1817; pp. 351

The Days of Battle; or, Quatre Bras and Waterloo. By An Englishwoman. London: Henry Bohn, 1853 [reprinted as *Waterloo Days* by Bell in 1888]; pp. 172

Rome in the Nineteenth Century; containing A Complete Account of the Ruins of the Ancient City, the Remains of the Middle Ages, and the Monuments of Modern Times. With Remarks on the Fine Arts, on the State of Society, and on the Religious Ceremonies, Manners and Customs of the Modern Romans. In a Series of Letters written during a Residence at Rome, in the years 1817 and 1818. By An Englishwoman. Edinburgh: for Archibald Constable, 1820; 3 vols., pp. 368/433/424, with 5 engraved plans

Anyone who has read Thackeray's *Vanity Fair* will be familiar with the scene at Brussels on the eve of the Battle of Waterloo: Wellington's hastily got-up headquarters was almost as full of women as of soliders. Charlotte and Jane Waldie (later Eaton and Watts respectively) were still unmarried in the summer of 1815, and like all the regiments' mothers, sisters, wives, and sweethearts, had gladly taken the chance of an escorted tour through picturesque Bruges and Ghent whilst accompanying their men (in this case their brother) to war.

Victory (if indeed it came to that) was a foregone conclusion, they thought. But as the mood began to change from complacency to panic, those same women turned from holiday-makers into a terrified mixture of refugees, impromptu nurses, and ghoulish sightseers. Some of them actually marched with their men on to the battlefield; others fled as far away from it as possible. Many were evacuated to Antwerp as Bonaparte drew nearer—Charlotte and her sister amongst them.

Charlotte's *Narrative* was hugely popular, and rapidly ran through ten editions. It differed from most other eyewitness accounts of Waterloo in that it dealt with events behind the scenes: the desperate anxiety of trying to sift the truth from the rumours, the squalor of the makeshift hospitals in Antwerp, and the awful relief or desolation of those waiting for news of casualties. Her description of a visit to Waterloo a month later is painfully detailed: the stench is still hovering over the battlefield like a mist; there are cannon-balls lodged in tattered tree-trunks and footprints dried in the mud, and local women are greedily picking through the half-buried remains for scraps of gold or silver.

Happily, Charlotte's brother survived the battle, and a year later accompanied his two sisters and their husbands to Italy. Charlotte's masterly description of Rome, published on her return in 1820, met with immediate acclaim and remained during her lifetime the definitive guide to the city's past and present. She went on to write a brace of novels, but managed to concentrate most of her career as a writer on supervising successive editions of *Waterloo* and *Rome*.

FARMBOROUGH, Florence

Life and People in National Spain. London: Sheed & Ward, 1938; pp. 222, with 12 halftones and 2 maps

Nurse at the Russian Front: A Diary 1914–18. London: Constable, 1974; pp. 422, with 48 photographs and 4 maps

Russian Album 1908–1918. Edited by John Joliffe. [Salisbury]: M. Russell, 1979; pp. 96, with photographs throughout

Much of Florence Farmborough's work makes harrowing reading. She went out to Russia in 1908 as a governess and English teacher, living first in Kiev and then in Moscow. When war broke out, she trained there as a nurse for the Red Cross and apart from a period of convalescence from paratyphoid in the Crimea and the odd leave in Moscow, saw active service at the Russian front continuously until 1917. The nurses of her division were largely peripatetic: it was a travelling life on which she learned to rely, feeling lost and useless when the Bolsheviks forced the division to disband just before the Revolution. Dangerously, she made the journey back to Moscow alone, and when compelled to leave for England, spent six months trying to reach home via the Trans-Siberian Railway and an unhealthy assortment of steamers.

Florence witnessed another Bolshevik revolution in Spain, where she lived and worked in the mid-1930s, broadcasting the anti-Communist talks published in her first book to other English-speaking residents there. Her stunning photographs prove her as much a journalist, almost, as a nurse; they are unflinching, often shocking and always strongly evocative. She was born to an unquiet life, she says, and must record it for others as best she can.

❧

HARRIS, Mrs G.

[Anon.]: *A Lady's Diary of the Siege of Lucknow: Written for the Perusal of Friends at Home*. London: John Murray, 1858; pp. 208

The Reverend and Mrs Harris had been three years in India when, in March 1857, they were sent to Lucknow. They had just settled themselves in their new home as the Chaplain and his lady when news of the Meerut massacre of 10 May came through, and after seven ghastly weeks of rumoured horrors and real ones, the Siege of Lucknow began.

Mrs Harris was one of the very few European women involved in the Indian Mutiny who never had to part with her husband. Nor did she have any children of her own to keep alive. Perhaps this helps to make her account of the Siege of Lucknow and subsequent trek to Allahabad as comparatively calm as it is. She spent the months of the siege not in the British Residency, but in a neighbouring doctor's house, helping her husband with almost daily funerals, the odd grim christening, and the pastoral and practical care of the sick, wounded, and bereaved.

Like Lady INGLIS (whom she much admired), Mrs Harris was just as concerned for those at home as she was for her fellow-sufferers—there were enough apocryphal rumours flying around India, let alone between India and England—and her book, a simple day-to-day account of deaths, privations, and eventual relief, was published as a sort of open letter to them, to put the record straight.

HODGSON, Lady (Mary Alice)

The Siege of Kumassi [etc.]. London: C. Arthur Pearson, 1901; pp. 365, with 32 halftones and 2 maps

Lady Hodgson witnessed the Ashanti War as the wife of the Governor of the Gold Coast. When she and Sir Frederick had arrived at Kumassi (now Kumasi, in Ghana) at the end of 1899, after a long and enervating journey by hammock and canoe from Accra, there was no sign of discontent and the cheerful pair were made welcome by the various tribesmen of the Ashanti capital. But by March 1901 relations had become markedly strained, first between the different tribes disputing over whom should be their 'King Paramount', and then between the Ashanti peoples and the British. With the Hodgsons as its focal point, a concerted and bloody revolt arose in April, and was not quelled until the Crown's forcible annexation of Ashanti at the end of the year.

Lady Hodgson's diary of the months from April to July describes her imprisonment in the besieged fort of Kumassi, where she and the other resident Europeans waited for reinforcing troops from Accra and Lagos, and then chronicles her vigorous escape at the end of June, pursued through the tropical rain by rebel snipers and a kaleidoscope of fevers to the safety of the Cape coast ports, and home. And in case any other Englishwomen are thinking of following her to Kumassi (she was the first), she adds a few barbed words of advice. 'Skirts are an impediment when fleeing for your life in Ashanti-land.' It is best to learn to swim before trying white-water canoeing. And never complain about the weather: rain may make your hat achingly heavy, but it also dampens rebel gun-fire.

HORNBY, Mrs (later Lady) Edmund (Emilia Bithynia)

In and Around Stamboul. London: R. Bentley, 1858; 2 vols., pp. 320/292

Constantinople During the Crimean War [a revised edition of the above]. London: R. Bentley, 1863; pp. 500, with 4 chromolithographs

Emilia Hornby travelled out to Turkey with her husband in 1855. They sailed from Marseilles on a French troop-ship, the hold full of packages 'to Miss Nightingale at Scutari', and could hardly move in the Bosphorus for warships and cargoes of sick and wounded soldiers. Mrs Hornby was brave to come. Her husband's work (negotiating a British loan to the Turkish government) was liable to be a long-drawn-out affair; they had had to leave their baby daughter at home, and would not even have each other for company, since Edmund would be busy with the Commission. Despite all, her letters home are dutifully bright and cheerful.

The Hornbys stayed in Therapia on the Black Sea coast for the cruel summer months, and moved to a cottage on the outskirts of Istanbul for the winter. Emilia's pleasures were simple: at home she played the piano or read the copies of *Punch* and the *Illustrated London News* her family sent for her; she visited a local Greek lady to learn the language and the art of embroidery; she took her two dogs for walks, and occasionally joined the other 'British Wives' at the embassy in the city. On one of these visits she actually met Florence NIGHTINGALE, and carefully describes her appearance in every detail. There were balls to go to, the odd visit to a harem, and in summer (after peace was declared) the most delightful yacht voyages—the Crimean War might almost never have happened.

The last of Emilia's letters home is dated 1858, by which time she had been back to England to collect her daughter. Hornby's next posting was to China in 1865; sadly his wife died before she could join him there.

❦

INGLIS, Lady (Julia Selina) (1833–1904)

Letter containing Extracts from a Journal kept by Lady Inglis during The Siege of Lucknow. London: for private circulation only, 1858; pp. 31

The Siege of Lucknow: A Diary. London: James R. Osgood, 1892; pp. 240

Lady Julia was the wife of Brigadier (later Major-General) Sir John Inglis, the man responsible for the defence of Lucknow during the terrible, magnificent siege of 1857. For this reason the *Letter*, somehow smuggled out during the siege itself, held special interest for a large circle of friends at home in England, and Lady Julia's mother lost no time in printing copies for personal distribution.

The letter spoke simply and immediately of its writer's unique position as go-between during the entire siege for the dwindling community inside the Residency and those fighting outside, thanks to the privilege she had of seeing her husband for about an hour a day: it often fell to her to break the news of soldiers'—or wives' and children's—deaths to their families, and by example (however sick at heart she might be) to keep up the morale of the survivors.

Although Lady Julia was one of the first women imprisoned in Lucknow to develop smallpox, and although her three children were constantly low with fever and Sir John was wounded more than once, her family was one of the few to emerge from the siege intact. And from the escape from Lucknow afterwards, which involved a nineteen-day march via massacred Cawnpore to Allahabad, a three-week journey down river to Calcutta, and then an uncertain wait for the next available steamer home. Widows and the sick sailed first (amongst them Kate BARTRUM, whose story saddened Lady Julia more than any she had heard); when her own turn came she was alarmed to see that the decrepit vessel supposed to carry them home was already falling apart. Sure enough, it broke up in the first storm they met a few days out from Calcutta, and the exhausted passengers were forced to row the ten miles to the shore of Ceylon in lifeboats, many losing those few belongings they had cherished through the siege.

Again the Inglises survived, and so did Lady Julia's Diary, but it was thirty-four years before she felt distanced enough from the appalling events of 1857–8 to publish it.

❖

MUTER, Mrs D. D. (Elizabeth) (*c.* 1827–1914)

Travels and Adventures of An Officer's Wife in India, China, and New Zealand. London: Hurst and Blackett, 1864; 2 vols., pp. 324/314

My Recollections of the Sepoy Revolt (1857–58) [reprinted from the above and augmented]. London: John Long, 1911; pp. 266, with 16 halftones and a map

Mrs Muter was the dutiful wife of a high-ranking army officer, who travelled not by her own volition but to keep him company on his various postings around the world. Thus she found herself at Meerut when the Indian Mutiny broke out; in Delhi, where they had fled, during the siege that followed; on the troop-ship *Eastern Monarch* when it exploded outside Portsmouth harbour on the way home from India and, two years and a quick visit to Canada later, in Peking during the British Occupation, as the first Englishwoman ever to enter the imperial city. She rounded off her travelling career with a voyage to New Zealand, where Lieutenant-Colonel Dunbar Douglas Muter had been sent to report on the danger to British settlements posed by marauding Maoris. Unfortunately, so meek was Mrs Muter that she even surrendered the writing of her own recollections to her husband, admitting that he dictated them from her journals and 'used the opportunity of stating opinions he strongly held'. (She admits it only because 'he would be sorry [such opinions] should be published under the shelter of a lady's name'.) Thus her books tell us more about Muter's prejudices and grievances than about Elizabeth's remarkable travelling life: a great pity.

❖

NICOL, Martha

Ismeer, or Smyrna and its British Hospital in 1855. By a Lady. London: James Madden, 1856; pp. 350, with 2 lithotints and a wood-engraved vignette

'I was simply a British woman who had little to do at home, and having no fear of disease, was willing to be of what use I could to our poor soldiers.' Miss Nicol was one of many middle-class gentlewomen who went out to hospitals during the Crimean War not as nurses, but as Ladies. Not that the two were mutually exclusive, as Miss NIGHTINGALE proved, but there were definite distinctions between medical and lay volunteers. Miss Nicol travelled via Paris and Marseilles with a party of twenty like-minded women in March 1855, bound for the British hospital in Smyrna, now Izmir. When they arrived, they were bundled into a nine-roomed boarding-house and with only a few days' acclimatization, set to work on the cholera and typhus wards. They were certainly expected to assist the nurses in emergencies, and help with hospital administration, but their chief importance, as Miss Nicol saw it, was in 'exercising a marvellous moral influence over the soldiers'. After the fall of Sebastopol, the number of patients began to wane and those Ladies that were left (twelve had either died or gone home sick) were allowed more time for sightseeing the Greek and Turkish quarters of the town. Not until the hospital was closed at the end of 1855 and converted into a barracks did the brave little band return to England.

❧

PAGET, Mrs Leopold (Georgiana Theodosia) (*c.* 1822–1919)

Camp and Cantonment: A Journal of Life in India in 1857–1859, with some account of the way thither. To which is added a Short Narrative of the Pursuit of the Rebels in Central India, by Major Paget. London: Longman, Green, *et al.*, 1865; pp. 469, with a lithotint frontispiece

This is an Indian Mutiny account with a difference: the Pagets were not stationed in India when the mutiny broke out, but drafted there immediately afterwards from England. The major commanded a battery of the Royal Artillery and his wife, given the choice, decided to accompany him to Poona. Once there, after a four-month journey by troop-ship, railway, and forced march, the major was set to breaking-in local horses and rallying what forces he could for the field, whilst his wife spent monotonous days in desultory sightseeing and visit-paying. When the major was ordered with his men down to Dharwar, Mrs Paget followed (too frail to march) by steamer, sailing from Bombay (her premature baby was born on board) and then trekking through the leechy jungle to meet him at the camp. The pattern was repeated several times—Paget would be posted somewhere, and march off to fight whilst his wife and child followed as best they might. Terror alternated with ennui, anxiety with loneliness, until at last the major was given leave (he had cholera) and the pair sailed home to stay. Incidentally, there is a nice little footnote to Georgiana's history: some sixty years after leaving India, she distinguished herself as one of the oldest women to exercise her right to vote. She was wheeled to the polls in a Bath chair, aged ninety-six.

❧

SALE, Lady Florentia (1787–1853)

A Journal of the Disasters in Afghanistan 1841–2. London: John Murray, 1843; 12mo, pp. 451, with 2 maps

Florentia Sale was the heroine of the First Afghan War. She was dubbed the 'soldier's wife *par excellence*' by *The Times*, and 'the Grenadier in Petticoats' by her husband's fellow officers: a force to be reckoned with.

She was born in India of an old East India Company family, and educated in England. In 1808 she married 'Fighting Bob' Sale and travelled with him to Mauritius, Burma, and back to India with their five children (the only survivors of a brood of twelve) in 1823. When Sale was appointed second-in-command of the Army of the Indus, the force sent by Lord Auckland to take the Afghan capital Kabul in 1840, Florentia was only too pleased to join him. It was a prestigious post, and she thrived on prestige. By 1841 the British government considered the Afghan rebels safely quashed, and Sale was ordered to leave with his brigade for Jalalabad. His wife was preparing to follow when insurgents attacked the cantonment at the beginning of November and months of alarm and privation began.

What fighting men were left at Kabul vainly tried to frighten off the Afghans whilst their womenfolk began to perish from lack of food and warmth inside the fort. Eventually a retreat was negotiated, whereby the remaining troops and their 12,000 followers (wives, children, and servants) were promised safe conduct to Jalalabad; by now the snow had arrived, however, and, ill-equipped and weak as they were, literally thousands died each day of the retreat—killed not just by cold and starvation, but by bands of rebel snipers. Staggeringly, only one survived of the 4,500 officers and men who set out, and only those women who, with Florentia, were surrendered in January 1842 as hostages to the Afghans just thirty miles from Sale's stronghold in snow-bound Jalalabad had any chance of emerging from the ordeal alive. The sixty-three hostages were kept first in six mud-walled rooms in Buildeeabad, then moved back up towards Kabul and finally north-west to Bamian before Sale's relief expedition freed them exactly ten months and two weeks after the siege began.

Florentia described the whole ghastly episode with astonishing objectivity. Her journal (which she hid amongst her clothes and smuggled out to Sale in instalments) is full of military detail and comment (she had not much patience, she said, with absent 'Big-wigs'), and laconic references to daily horrors ('The pony Mrs. Sturt [her daughter] rode was wounded in the ear and neck. I had, fortunately, only *one* ball *in* my arm; three others passed through my poshteen [sheepskin coat] . . . without doing me any injury'). It was published in England to much acclaim shortly before the Sales arrived back to a heroes' welcome.

Florentia was back in India in 1844, a year before Sale was killed in the First Sikh War, and she remained there alone until 1853, when urged to visit South Africa for her health. She died at the Cape just three days after arriving there.

❖

SANDES, Flora (1876–1956)

An English Woman-Sergeant in the Serbian Army [etc.]. London: Hodder and Stoughton, 1916; pp. 242, with 17 halftones

The Autobiography of A Woman Soldier: A Brief Record of Adventure with the Serbian Army, 1916–1919. London: H. F. and G. Witherby, 1927; pp. 222, with 15 halftones

When Flora Sandes was growing up in her clergyman father's home in Suffolk, she used to pray that one day she might become a boy. And she did, more or less. She rode horses far faster than women should, learned to shoot, and even drove a racing car before sailing for the troubled Balkans in August 1914.

Flora was sent to Serbia by the St John Ambulance Brigade and joined the Red Cross there, attached to the Second Infantry Regiment of the Serbian army. It was not long before her staunch sympathy with the Serbian soldiers' cause and her own masculine sense of adventure led her to abandon her awkward nurse's uniform for an infantryman's, and so Private Sandes's distinguished military career began. She was genuinely accepted as one of the Serbian army's bravest fighters—not just its most eccentric—and was decorated with the prized Order of the Kara-George for her bravery after near-fatal injuries sustained in a grenade attack in 1916. She spent her brief period of sick-leave raising funds for the army at home in England before returning to the field in 1917 and remaining with her regiment—first as lieutenant and then as captain—until her marriage to her sergeant, a White Russian refugee, in 1927.

She left the army then ('Turning from a woman into a private soldier proved nothing compared with turning back again') and lived with her husband first in France and then Belgrade, where they ran a taxi service and were interned by the Germans during the Second World War. After the war Flora, a widow, came home to Suffolk and settled meekly at last—on her full Serbian army pension—in a quiet corner of England until her death.

❖

SEACOLE, Mrs Mary (1805–1881)

Wonderful Adventures of Mrs Seacole in Many Lands. Edited by W. J. S. With an Introductory Preface by W. H. Russell, Esq., the *Times* Correspondent in the Crimea. London: James Blackwood, 1857 [copy not seen]; pp. 200, with woodcut frontispiece

Punch wrote poems to her; *The Times* published letters on her; the *Illustrated London News* drew pictures of her: she was, as they say, a legend in her own lifetime. Now, a hundred years later, Mary Seacole is being rediscovered not just as one of Queen Victoria's most flamboyant subjects, but as a pioneer of feminine independence.

She was born in Jamaica during the last year of Maria NUGENT's residence there as first lady; her father was a Scottish seafarer named Grant and her mother a liberated slave, famous for her Kingston hotel and her skills as an amateur 'doctress'. Mary Grant inherited her mother's renown, taking over the hotel on her death and setting herself up as local nurse and expert in herbal remedies. She always longed to travel, she says—especially as 'an unprotected female'—and before her marriage in 1836 made two journeys to Britain, once to visit relations and once as a trader in West Indian pickles and preserves.

Edwin Horatio Seacole (named after his godfather Nelson) was already sickly and elderly when he wed Mary; when he died soon afterwards he left her enough money to start up a personal travelling fund. In 1851 she indulged herself in a dangerous journey to New Granada (now Panama) to help her brother set up a hotel on the Gold Rush road across the isthmus to California. Cholera broke out whilst she was there, and yellow fever a year later back in Jamaica: Mary relished the chance of practising her healing art and soon established herself as the local authority on tropical diseases.

In 1854 she sailed to England a third time, arriving just after news of the Battle of Alma reached London: the Crimean struggle had begun, and Mary, always susceptible to what she calls 'the pomp, pride, and circumstance of glorious war' vowed to be a part of it. But no one in London would take her seriously: Miss NIGHTINGALE's books were full of well-meaning amateur nurses; the War Office could not cope, and the Crimea Fund had more pressing calls on its resources. So Mary went out under her own steam as a 'sutler', or army provisioner, and was immediately recognized on her arrival at Scutari by soldiers who had stayed at her Kingston hotel in the past.

'Mother Seacole' soon became a legendary feature of the grisly Crimean landscape, her considerable and dusky form invariably clad in a yellow dress and blue bonnet with scarlet trimmings. She set up The British Hotel at Balaclava, offering food, accommodation, and company to the officers and men; she trundled a trolley through the front lines of the Redan, Sebastopol, and Tchernaya laden with everything from bandages to plum puddings ('and enjoyed the sight amazingly') and nursed the British, French, Turkish, and Russian wounded with equal compassion and gentleness. She left the Crimea on the declaration of peace in 1855 and returned to England a bankrupt heroine. The Seacole Fund was set up by such admiring luminaries as HRH The Prince of Wales, Lord Rokeby, and Admiral Keppel to support her in a lively dotage spent travelling between Jamaica and Britain, her adopted home.

❧

TYTLER, Harriet (1828–1907)

An Englishwoman in India: The Memoirs of Harriet Tytler 1828–1858. Edited by Anthony Sattin. Oxford: University Press, 1986; pp. 229, with 22 halftones and 2 maps

The morning of 11 May 1857 dawned in Delhi like any other. It turned out to be one of the most chilling days in the history of British India: the day the mutineers arrived. They came hot foot from Meerut, where the slaughter of the 'innocent' Europeans (chronicled by Mrs MUTER) had tolled the beginning of the Indian Mutiny the day before, and they held the old capital of Delhi in siege for four months.

Harriet Tytler was one of the lucky army wives who managed to escape from Delhi, eight months pregnant and with two children of four and two, reaching Ambala, some 120 miles distant, after days of dangerous and excruciatingly uncomfortable travel on an overcrowded bullock-cart. Unlike most of the others, however, she chose to march back to the city when the hastily got-up British force was sent in early June (Robert Tytler was its paymaster) and she was the *only* European lady to witness the Siege of Delhi from first to last, being too near the birth of her child to ride out with the rest to safety. In fact her son was born on 21 June on an old ammunition-cart (the Tytler family home) amidst the whistle of bullets and the stench of cholera and death.

Harriet's *Memoirs*, written for her family nearly fifty years after the siege and not published in book form until a generation later, describe her life in India both before Delhi and afterwards (she was born there of good army stock and lived most of her life on its plains and hills, with only occasional breaks at 'home' in England or touring the Continent) but it is the account of those four months from May to September 1857 that leaves the deepest impression. One detail more than any other seems to sum up the pathetic horror of her situation: to keep her daughter occupied during the long days they lived on the ammunition-cart, Harriet would 'scratch holes' in her own feet and tell her daughter 'she must be my doctor and stop their bleeding. This process went on daily . . . No sooner did my wounds heal when she used to make them bleed again . . . it had the desired effect of amusing her for hours.'

❧

WARD, Harriet (d. 1873)

Five Years in Kaffirland; with Sketches of the Late War in that Country, to the Conclusion of Peace. Written on the Spot. London: Colburn, 1848; 2 vols., pp. 306/352, with 2 engraved plates and a map

Harriet Ward was an army wife in the doughty mould of Lady SALE and bonny Mrs DUBERLY. When her husband set sail for the Cape in 1842 to fight the Kaffir Wars, she insisted on taking her child and following him in moral support. The family suffered both shipwreck and scarlet fever on the miserable troop-ship *Abercrombie Robinson*; but neither could deter Harriet, and after a long, unglamorous cattle-waggon trek she arrived at the Grahamstown barracks all set to terrify the Hottentots, Fingoes, and Kaffirs uprising around with her sheer Britishness.

It was a lengthy campaign, during which Harriet suffered the usual camp hardships of rumour, sickness, anxiety, and ennui, complaining at one moment of glaring political blunders (what Lady SALE would call 'Big-wig' trouble) and of niggling domestic problems the next. To fill the lonely hours she composed this history of Kaffir country, through which she threaded as objective an account of the present fighting as she could manage. It was published with exemplary expedition within months of the signing of the peace treaty at the end of 1847: so quickly, in fact, that Harriet should perhaps be regarded as a forerunner to Florence DIXIE as the first woman war correspondent in South Africa, and not a travel writer at all.

12

The Gilding Off

Many and varied are the difficulties which beset a woman, when she first exchanges her European home and its surroundings for the vicissitudes of life in the tropics. Few can realise the sacrifices they will be called upon to make in taking such a decided step; many home comforts, and the host of nameless social fascinations, so dear to a woman's heart, have to be given up, while the attractions offered by the irresistible 'day's shopping', the box at the opera, a few of our summer recreations, and nearly all our winter amusements, must be temporarily relegated to the list of past pleasures.

So begins one of the least encouraging travel manuals I have ever come across: *Tropical Trials: A Hand-book for Women in the Tropics* by Major S. Leigh Hunt (Madras Army) and Alexander S. Kenny (rather chillingly described as a 'Demonstrator of Anatomy' at King's College, London). Published in 1883, it was designed as a companion volume to the authors' previous work *On Duty Under a Tropical Sun* and was dedicated to one Anna Tatam, 'as a tribute to her late husband, by whose bedside, during his last painful illness, contracted when on service in India, she maintained an almost ceaseless vigil extending over many months'. How miserable. It is obvious that Messrs Leigh Hunt and Kenny had no high expectation of a woman ever actually *enjoying* herself in the corrosive climes of India or Africa: for the gentler sex, travel was something to be endured until time (or something less transient) rescued her.

It should be equally obvious by now that the average woman travel writer (strange thought!) was much more mettlesome than her menfolk gave her credit for. Many were far more likely to relish rather than lament the loss of 'the irresistible "day's shopping", the box at the opera' and so on: home was the endurance test for them, not travel.

There are always exceptions, though, and this chapter should perhaps be dedicated to Anna Tatam too, whoever she was. For the women here are little tragic heroines, each with their own miserable story to tell of an exile abroad. They are reluctant travellers all.

Most never wanted to travel in the first place. They were taken abroad by their husbands, planted somewhere thoroughly uncongenial, and then expected to perform their social and domestic obligations in stoical ignorance of the wilderness around. Others were forced abroad by illness, never content for wondering whether it might have been kinder after all to stay at home and die with their families rather than live a sad and solitary life elsewhere: travel can be so lonely.

Nor was the excitement of writing a book about it always guaranteed to assuage the misery of a reluctant journey. For Keturah JEFFREYS, a young widow, the exercise was purely commercial: ironically, the very wretchedness of her experiences in Madagascar between 1821 and 1825 was her only asset, and she was forced to sell the story to live. Mrs JUSTICE, on the other hand, printed her frenetic account of *A Voyage to Russia* as a sort of public affidavit against the husband whose behaviour had forced her to go there in the first place (and she needed the money, too).

Perhaps Leigh Hunt and Kenny were right after all, then? Perhaps women really are happiest at home with their 'host of nameless social fascinations'? Don't you believe it. Even here, amongst a chapter of victims, it is apparent that some of the writers are in their element. There can be something rather magnificent about disapproval— especially when it is plain to see that the disapprover relishes her role. Mrs CALDERWOOD found the annoyances of foreign travel most attractive: what could be more gratifying for a trenchant Scottish body like her than to deliver abuse where abuse was due (i.e. everywhere but Scotland)? And the American Dora HORT had censure down to a fine art, tenderly conducting her disgust around the world like a particularly precious piece of luggage. For some people, being disappointed can be a positive pleasure.

BAILLIE, Marianne (*c.*1795–1831)

First Impressions on a Tour upon the Continent in the Summer of 1818, through Parts of France, Italy, Switzerland, the Borders of Germany and a Part of French Flanders. London: John Murray, 1819; pp. 375, with 4 sepia aquatints and an engraved plate

Lisbon in 1821–2–3. London: John Murray, 1824; 2 vols., pp. 219/250, with 8 aquatints and title-page vignettes

The reader 'ought not to expect me (labouring under the two-fold disadvantage of sex and inexperience) to narrate with the accuracy and precision of a regular tourist', apologizes the poet Marianne Baillie in the introduction to her first travel book. She need not have been so modest: it was obviously thought distinguished enough to merit publication by none other than Mr Murray himself, who included the romantic demigod Byron amongst his authors.

First Impressions was one of the first accounts of the Grand Tour to be written by an Englishwoman—but its author was no born traveller. Even in Lisbon, where she lived with her husband for two and a half years, she never settled down to the foreign routine. 'Mr Baillie and I always try to enter into the society of the natives', she writes wistfully—but they never quite succeed. Both books are carefully descriptive on the surface and, after the first flush of fashionable delight in *First Impressions*, pungently homesick underneath. A mournful poem in the Lisbon book best—if somewhat melodramatically—sums up her suitably ladylike attitude to touring:

> 'Tis home alas! that I desire,
> Thither my absent spirit strays,
> While on my trembling lips expire
> Th'unconscious words of hollow praise.

❖

BARTRAM, Lady Alfred

Recollections of Seven Years Residence at The Mauritius or Isle of France. By A Lady. London: Cawthorn, 1830; pp. 208

Lady Bartram wrote this sad account for her daughters as a tribute to their father, who had been lured to Mauritius and his death, she said, under false pretences. The family had been living quite happily in the West Indies (Lady Alfred's birthplace) until 1819, when the enticing promise of 'a lucrative situation under government' persuaded them to cross the world to one of Britain's most recent (but least-known) possessions. After a pleasant voyage to Port Louis, with a stop at Cape Town on the way, they settled themselves in a house at La Grande Rivière, and waited for preferment and success.

Two years passed, and still there had been no promotion for Bartram, who was little more than a clerk: evidently he had been forgotten. So the family moved back to the comparative limelight of Port Louis where they stayed for the next five years, still hoping for the promised 'lucrative situation' to materialize, but not daring to complain for fear of losing favour altogether. They were not easy years. The French population were still smarting from the Treaty of Paris of 1814, when they lost Mauritius to the British; travel (very necessary during the warm season, when Port Louis began to stink with fever) was a cumbersome and difficult affair, there being few carriages and fewer roads; Lady Alfred was starved of congenial company (the population consisting, it seemed to her, of slaves and Frenchmen), and her overworked husband began to fall ill. He eventually died in 1826, and the widow and orphans (her word) were left with a miserable pension, to do as they pleased. Of course, they sailed home.

❧

BRADLEY, Eliza (b. 1783)

An Authentic Narrative of the Shipwreck and Sufferings of Mrs Eliza Bradley, Wife of Captain James Bradley of Liverpool, Commander of the Ship Sally *which was wrecked on the coast of Barbary in June 1818* [etc.]. Boston [Mass.]: James Wald, 1821; pp. 108 (including the publisher's 30-page History of Arabia), with 2 woodcut plates

This sensational travel account was published both in England and in America in the cheapest possible form; there must have been previous editions, but this Boston imprint is the earliest I have been able to find, and it is in the characteristic format of a flimsy popular chapbook, with crude and primitive woodcuts and a good fruity flavour. Its heroine was the only woman aboard the *Sally* when the ship set sail for Tenerife in the spring of 1818. She was just along for the ride, and thoroughly enjoyed herself until a storm blew up five weeks into the voyage and drove the hapless *Sally* on to the Barbary coast of north-west Africa.

All thirty-two on board were forced to swim for shore and found themselves spending the next five days foraging for all the food they could find (mussels and a dead seal) and searching unsuccessfully for fresh water. Captain Bradley's plan was to march inland in the hope of finding aid before the party could be discovered by hostile Arabs: a vain hope, for no sooner had they all set out than a passing caravan overtook and captured them. Eliza was allowed to ride the one available camel, upon forfeit of her intriguing bonnet, gown and shoes (the camel became quite comfortable in time, but she never got used to the sun).

The leader of the caravan, to whom she demurely refers as 'my master', proposed to sell off the men to some fellow tribesmen bound for Mogadir, who would then trade their lives for money with the British consul there.

Eliza he decided to sell direct whenever the opportunity arose, women of child-bearing age being worth twice as much as men, and fun to store in the mean time. So the Bradleys were parted, and Eliza travelled on to the leader's village where she spent the next five months.

'My master' was unexpectedly kind, letting her keep a Bible and giving her a tent to herself. She was allowed the run of the village (whose women used to come and watch her cry) and found time to note for us the character and dress of the local Arabs, until a welcome letter arrived from her husband in Mogadir. He had been redeemed by the consul and was ready to sail as soon as Eliza could join him. Payment and escort were duly sent and three months later the Bradleys were home again in Liverpool—with a most extraordinary travellers' tale to tell.

CALDERWOOD, Mrs (Margaret) of Polton (1715–1774)

Letters and Journals from England, Holland, and the Low Countries in 1756. Edited by Alexander Fergusson. Edinburgh: David Douglas, 1884; pp. 386, with an etched plate

Scotland was the destination of Thomas Cook's first 'foreign' tour in 1844—crossing the Border was even then almost as exciting and outlandish as crossing the Channel. Imagine how much more so, then, in 1756, when Margaret Calderwood journeyed south from Polton (near Edinburgh) to the piazzas and citadels of England. She was thoroughly and gleefully scathing about everything she found there—an excellent foil for Dr Johnson—and summarily dismissed fair Albion as a noisy, uncultivated, low-spirited swamp. Shaking the Sassenach dust from their slippers the Calderwood party tackled the Low Countries next (not half so low as England) and spent a splendid few weeks berating their benighted inhabitants before sailing home again to civilization in Scotland. Throughout the letters and journals collected in her book it is obvious that Polton was Margaret's touchstone for the world (and why not?)—the worse she found damn foreigners, the better pleased she was. It is one of the most uproarious travel accounts I have read.

CAMPBELL, Harriet Charlotte Beaujolois (1803–1848)

A Journey to Florence in 1817. Edited, with notes, by G. R. de Beer. London: G. Bles, 1951; pp. 174, with frontispiece and a leaf of manuscript-facsimile

A charming journal, this is, written by a young 'lady of quality' travelling with her mother and rather tiresomely attractive sister through Europe.

Harriet was only fourteen when the journey began, and gives us a uniquely fresh and unjaundiced view of the already hackneyed route to Italy. Her mother, the widowed Lady Charlotte Cameron, was Queen Caroline's favourite lady-in-waiting; such an exalted position guaranteed a smooth and aristocratic passage through the Continent to Florence—and, hoped Harriet, to certain romance. In fact Italy proved a little too romantic: it was only to be expected that her sister should acquire a husband during the trip but it is too much for Harriet when her mother does the same. The journal closes abruptly, the author vowing that she cannot write until she is happy again; and that, we are assured, will never be.

❧

CURRIE, Jessie Monteath

With Pole and Paddle Down the Shiré and Zambesi. London: Routledge, 1918; pp. 159

The Hill of Good-Bye: The Story of a Solitary White Woman's Life in Central Africa. London: Routledge, 1920; pp. 249, with 4 halftones

Jessie Monteath had only been married to the ageing Mr Currie a fortnight before he left for Africa. She was to join him a year later, he said, and they would spend their honeymoon on Mount Mulanje, up near Quelimane in Nyasaland, just a little late. Mount Mulanje—now in Malawi—means 'the Hill of Good-Bye', and could not have been better named. Currie's first words on meeting the young bride he had not seen for a year were not exactly rapturous ('You are a brick') and his welcoming gift was a loaded revolver: '"If the worst comes to the worst," he warned, "you will put the bullet in your own head."' The year Jessie spent there before she and the other Europeans on the station were driven out by looting natives was supremely lonely and ridden with illness and depression. At night she would sit in her room watching the rats and licking her quinine-papers dry; the days were dead and empty. Perhaps because these two books were written so much later (the Curries came home from Africa in 1893), Jessie could afford to be uncomfortably frank in her descriptions—what other Victorian lady traveller would speak of the stench of sweat and vomit, or dare to soliloquize on her husband's failings? Their utter lack of glamour is a salutary antidote to the brave, besotted breeziness of such as Mary HALL or May French SHELDON.

❧

DUFF GORDON, Lady Lucie (1821–1869)

'Letters From the Cape' [in *Vacation Tourists and Notes*, edited by Francis Galton]. London: Macmillan, 1864; pp. 119–222

Letters From Egypt, 1863–65. London: Macmillan, 1865; pp. 371

Last Letters From Egypt: To Which are added Letters from the Cape [etc.]. With a Memoir by her Daughter, Mrs. Ross. London: Macmillan, 1875; pp. 346, with engraved frontispiece

These *Letters* have been celebrated ever since they were first published. Lucie Austin was the daughter of high-thinking Radical parents, educated at home and in Germany, and married at eighteen to Sir Alexander Duff Gordon, with whom she kept an open house for the literary and intellectual luminaries of the day. One of them, George Meredith, described her as 'radiantly beautiful, with dark brows on a brilliant complexion'—but it was an unhealthy brilliancy, he said, indicating 'the fell disease which ultimately drove her . . . to die in exile'.

Lucie was a consumptive, forced to leave her husband and children (the youngest only three) on doctor's orders in 1861, for the healthier climate of South Africa. The letters she wrote home reveal an attitude hardly met with before in the English abroad: first of all she avoided the supposed comforts and security of Cape Town and took herself up-country to Caledon; secondly she chose the company of non-Europeans, preferring best to be with the poor Malay immigrants. The improvement in her health was lost as soon as she returned to England, and her eldest daughter's recent marriage to a bank official working in Alexandria tempted her to try Egypt in 1862. She was only well enough to return home once during the next seven years: her obvious distress at being separated from her family, and unsentimental appraisal of the seriousness of her disease, give the *Letters* something of tragedy.

After a spell in Cairo, Lucie again defied European tradition and settled into a crumbling house built amongst the ruins of Luxor. First hiring a *dahabeah* and then building one of her own, she sailed down on the flood tides each year to visit her daughter at Alexandria, but was always relieved to get home to 'my Theban palace': 'I don't like civilization so very much,' she said; 'It keeps me awake at night.' She made Luxor a real home, becoming the head of a huge and loving family of impoverished Egyptian neighbours: their doctor, teacher, and ally. She travelled amongst them as far as her health permitted, and the practical care and love she should have been spending on her real family at home fell instead to them.

Reluctant as she was to have her work printed ('I never could write a good letter'), Lucie was concerned to make her friends' situation clear. She always wrote better about people than places, and her letters are sympathetic political documents as much as anything else. The treatment of immigrant labour in the Cape disgusted her, and the Egyptian Government, urged on by the greed of Europe, was spending the country's peasantry on self-aggrandizement (more particularly on labour for the Suez Canal) and giving nothing in return.

Of course she was a travel writer too, and her *Letters from Egypt* were so successful that the Nile rapidly became *the* place to winter, and Lady Duff Gordon *the* person to visit there. She took to locking herself up as soon as she saw a steamer approach (except when it carried such esteemed visitors as the ubiquitous Marianne NORTH, Edward Lear, or the Prince and Princess of Wales). She died in 1869, and her last letters show the cumulative strain of seven years away from home. She was fully aware of her condition, and for all the love of what she called her own people and country in Egypt, it was indeed, as Meredith said, a death in exile.

❧

FANE, Isabella (1804–1886)

Miss Fane in India. Edited by John Premble. Gloucester: Alan Sutton, 1985; pp. 246, with 12 halftones and a map

General Sir Henry Fane's daughter Isabella never got on particularly well with her mother. And so when he was appointed British commander-in-chief in Bengal in 1835, he took Isabella with him as his hostess and left Mrs Fane at home (which was really just as well, as he and 'Mrs Fane' were not actually married). This wonderfully catty canon of discontent records the next few years spent in Calcutta, Simla, and touring in between, during which Isabella was wearied by countless grand and sweaty dinner parties ('I never was more bored . . . I would much rather have been hanged') and entertained one batch of silly Englishwomen after another. She it was who wrote of the EDEN sisters that 'They are both great talkers, both old, both ugly, and both s—k like polecats!' (Emily and Fanny in turn declined to comment on Miss Fane), and she variously reported on her sister memsahibs as fools, pigs, nasty creatures, looking ninety and undistinguishable from a corpse, and surrounded by hordes of pasty children like fat maggots. Then there were the mosquitoes ('I have a place on my leg which I have over-scratched, nearly a quarter of a yard long') and the cold at night, which turned one's feet to stones: what a life. Her letters appear here in their shameless natural state, completely devoid of the glamour and felicity of phrase that glosses most contemporary accounts of the upper echelons of the British Raj, and they are quite delicious.

It is sad that Isabella Fane drew no relief from leaving India in 1838. She had failed to find herself a husband there, and had no real home in England, and so after a few mean years wandering around Europe she settled in France, dull and in depressingly reduced circumstances, until her death.

❧

FOOTE, Mrs Henry Grant

Recollections of Central America and the West Coast of Africa. London: T. Cautley
Newby, 1869; pp. 221

Mrs Foote wrote these sketches to help while away her time as the wife of an
ill-starred diplomat—ill-starred by virtue of his appointment to two of 'the
most unhealthy spots on earth'. The couple spent eight years in Central
America, first at Greytown (now San Juan del Norte in Nicaragua), where
Mrs Foote was bored and disgusted by day, and sleepless (for the noise of
the insects) by night; then in Salvador. There they narrowly avoided the
devastating earthquake of April 1854 by being away on a diplomatic jaunt
which involved wading wearily towards one's destination through a morass
of heat, locusts, spiders, and sulphur fumes. Surely even West Africa, where
Mr Foote was posted in 1860, was bound to be an improvement. But Lagos
was little better. Mrs Foote was soon depressed by the city of sandy lanes
and crumbling mud walls, and tired of having to have sacks of cowrie shells
lugged around with her for change whenever she went shopping. And the
fine West African Negroes she had heard so much about were turning
rather ignoble: 'A black man in his white toga is an interesting object, but
clothed in a European coat he loses at once his native dignity and becomes
vulgarised on the spot.' Some women took naturally to the travelling life of a
diplomat's or army officer's wife—Mrs SCOTT-STEVENSON or Fanny
DUBERLY revelled in it—but Mrs Foote was only too pleased to take the
opportunity of accompanying a sick child home in 1861: Englishwomen
were meant for England, and should stay there.

❧

FORBES, Anna (d. 1922)

Insulinde: Experiences of a Naturalist's Wife in the Eastern Archipelago. Edinburgh:
Blackwood, 1887; pp. 305, with a map

Anna was the wife of prominent Victorian naturalist Henry Forbes, author
of *A Naturalist's Wanderings in the Eastern Archipelago*, a book much
respected for its sturdy scientific observations and thoroughness. His wife,
little Anna, was persuaded to publish her own account of the collecting
expedition to the islands between Sumatra and Papua New Guinea two
years after Henry's—it was to be a woman's view of travelling in the tropics,
'the interests and pleasures, drawbacks and disappointments' of the trip,
with nothing of the science and serious stuff. But there were precious few
interests or pleasures, it seems: Anna had a dreadful time. She was rarely

free of fever ('as much nervous as malarial') when accompanying Forbes on his treks into the interior, and desperately lonely and frightened when left behind with only a dwindling collection of ailing pets for company—and yet she never openly complained. *Insulinde* is a brave and sad book, more to do with duty than with travel.

◈

GRIFFITH, Mrs George (Lucinda) Darby

A Journey across The Desert, from Ceylon to Marseilles: comprising sketches of Aden, the Red Sea, Lower Egypt, Malta, Sicily, and Italy. London: Henry Colburn, 1845; 2 vols., pp. 319/318, with 2 lithotints and 18 wood-engraved vignettes

This narrative was presumably published by the Griffiths to raise money. They had been living happily together in Ceylon, where the major was stationed, until Lucinda was taken ill and advised to return to England. The major wouldn't hear of her travelling alone, and so settled for half pay and took her himself. He drew the pictures and she wrote the words. The book, like the journey, is fairly unremarkable, but it does give a good early account of the recently acquired British colony of Aden. And Mrs Griffith includes some aspects of travel that her peers tended to gloss over: the sordid details of life on board a workaday steamer, for example, choked with coal dust and swarming with rats and cockroaches, and the frustrating (still familiar) vagaries of quarantine, passport, and customs regulations.

◈

GUTHRIE, Mrs Maria (d. *c.*1797)

A Tour, Performed in the Years 1795–6, through the Taurida, or Crimea, The Antient Kingdom of Bosphorus . . . and all the other countries on the North Shore of the Euxine . . . Described in a series of Letters to her Husband, the Editor [etc.]. London: for T. Cadell, 1802; quarto, pp. 446, with 3 copper-engraved plates (some with several images), 8 wood-engraved plates, wood-engraved in-text vignettes, and 2 maps

The title-page makes this *Tour* sound like a lady's leisurely pleasure trip across picturesque Russia. Nothing could be further from the truth. The journey was really a gruelling and unremitting quest, a sort of 'field trip' that no one should have been expected to perform.

Maria Guthrie was 'formerly acting Directress of the Imperial Convent for the Education of the Female Nobility of Russia'—so she cannot have been foolhardy. It must have been something else that drove her to accept her husband's commission. He, as well as being 'Physician to the First and Second Imperial Corps of Noble Cadets . . . and Councillor of State to His

Imperial Majesty of all the Russias', was a noted antiquary; he asked Maria to report on all the ancient Greek and Roman ruins and antiquities she could find between St Petersburg and the Black (or 'Euxene') Sea. This she did, travelling not only under the normal difficulties of eighteenth-century cross-country transport but the added pressures of providing full dispatches for her husband on every stage of the journey. The substance and regularity of the letters show how conscientious she was—they are astonishingly detailed and scholarly (her own particular interest was in Euxine commerce)—just what Dr Guthrie wanted. He can hardly have been surprised when the journey began to take its toll, and Maria grew ill.

There are few hints of her illness in the letters—she teases her husband often, telling him to remember that 'I intend . . . to punish you men for your sneer at the charming disorder that must reign in the narrative of a female traveller' and dubbing herself his 'wandering spouse' and galloping hunter after antiquities. But like Lucie DUFF GORDON, she cannot help occasionally showing despair: 'That the return of my health may soon lessen the distance between me and my family is the prayer with which I finish this letter and most of my others, although not so openly expressed'. The supreme irony is that the *real* pretext for the Tour had been a search for health: Maria was ordered to leave St Petersburg for the kinder climes of the Black Sea, and the benefit of relaxed travel. Of course, she did not survive.

❖

HORT, Mrs Alfred (Dora)

Via Nicaragua: A Sketch of Travel. London: Remington, 1887; pp. 267

Tahiti: The Garden of the Pacific. London: T. Fisher Unwin, 1891; pp. 352, with photogravure frontispiece

Dora Hort was what many would call typically British: forceful, outspoken, and blessed with a sublime sense of self-importance. In fact she was an American whose travel books, for some reason, were published only in London.

The first one concentrates not so much on the journey she once made from New York to San Francisco (via Nicaragua) as with the battle it involved. That battle was with 'The Company', under whose arrangements Mrs Hort was travelling. Much to her satisfaction (I suspect), they could do nothing right. There were no interesting passengers aboard the filthy steamer; the Nicaraguans themselves were 'a hideous race' with disgusting manners, and the horse she was given to ride across the isthmus was bad tempered and ill bred. No side-saddle was provided; as riding astride was too vulgar to countenance, she was forced to borrow a pair of gentleman's 'unmentionables' (for modesty's sake) and crouch on the beast, with one leg up round its neck somewhere and the other clasping its belly. At one of The Company's hotels she was bitten by a jigger (the 'American tick' which buries its eggs to hatch under the skin); her baggage of course went astray several times, and to cap it all, when they eventually reached San Francisco their incompetent captain hove to next to a ship that was riddled with disease. She did eventually step ashore, and spent a pleasant few weeks visiting the gold- and silver-mines of California and exploring the Pacific coast. She made another (unchronicled) visit to California three years later—by then she had probably put 'The Company' out of business.

Mrs Hort's second travel book is about the islands of the Pacific where she and her husband Alfred lived for some time. He owned a farm on Morea, and an estate at beautiful Papeete. Tahiti itself was fine—Mrs Hort based most of her romantic novels on it and obviously appreciated its charms; it was when she started travelling again that the trouble began. She visited New Caledonia (nasty) and Sydney (unimpressive) and accompanied brave Alfred to San Francisco and Chile. All these journeys were marred by a bizarre series of accidents—not to her (fate would not dare), but to her beloved pet dogs. A Scotch terrier was thrown overboard by a drunken sailor; a Newfoundland died of some canine ague in Australia; a Chilean spaniel fell down a deck hatch left suspiciously open, and a Peruvian poodle had an eye torn out by an irate Tahitian cat and then was kidnapped by more sailors and returned (for a ransom) reeking of gin. Poor Mrs Hort: she was obviously not meant to travel but, being a forceful woman, tried to defy the gods. Her books should be a lesson to us all.

❧

HUGHES, Mrs T(heodore) F.

Among the Sons of Han: Notes of a Six Years' Residence in Various Parts of China and Formosa. London: Tinsley Brothers, 1881; pp. 314, with a map

Poor Mrs Hughes was a very reluctant traveller. Her book has a brave air of martyrdom about it, as she describes a succession of miserable postings in the more unglamorous cities of the Far East. She and her diplomat husband were sent first of all to Shanghai. This might have been tolerable—there was at least a European enclave there. After a few weeks, however, they were shunted up to Cheefoo (Yantai), and then inland to Foo-Chow (Fuzhou) in Kiangsi. Next followed Takon, in south Formosa (Taiwan), the worst place of all. Life there was unbearable for Mrs Hughes, who saw only one other European woman during their eleven-month stay (the wife of a passing sea captain), and grew maudlin and weak on a diet of chicken and oysters. The climate of Formosa, she noted wanly, had been likened to that of deadly

West Africa, and there was great relief on the Hughes' return to Shanghai, where they remained until Theodore's long-awaited furlough was due. They had managed to survive the lowest points of the last six years—the earthquakes and fires that followed the horrific typhoon of 1876 in Foo-Chow, and the storm that nearly wrecked the Chinese man-of-war that ferried them from Formosa to the mainland—and even enjoyed some parts. Their holiday voyage up the Yangtse River to Hankow (Wuhan) had been fun, and their visits to Nanking and the Island of Puto Shan (where they daringly took tiffin in a Bhuddist temple). But best of all, of course, was the journey home to England.

INNES, Emily
The Chersonese with the Gilding Off. London: Bentley, 1885; 2 vols., pp. 273/250

In 1883 Isabella BIRD published a book called *The Golden Chersonese*, 'chersonese' being a romantic name for any peninsula, and in this case the land between Thailand and Sumatra then known as the Malay Native States. It was a eulogistic account of her journey through tropical swamps and fragrant forests as a privileged guest of the British Protectorate: life there must be like heaven, she enthused.

Like hell, muttered Emily Innes. Emily was the long-suffering wife of a magistrate and 'collector', or local administrative official, who spent six grim years 'buried alive' in the Native States, and for whom the publication of Isabella's chirpy book was just the last straw. She wrote a two-volume rejoinder, explaining to Miss Bird and others what it is *really* like for a well-bred Englishwoman to be condemned to slow death in the company of an unambitious husband, various opportunist and unscrupulous superiors, and the disgusting inhabitants of the mouldering Chersonese. She describes their dismal residence between 1876 and 1882 (including short episodes in Singapore, Java, and Penang) in lugubrious detail, and dwells almost lovingly on all its attendant miseries, from mud and mosquitoes to mutinous coolies. Her catalogue of complaints includes the cowshed on stilts provided for their home, the inevitable fowl cooked by a repellent Chinaman for their food, and the 'tailorless, cobblerless, doctorless, bookless, milkless, postless and altogether comfortless jungle' for their entertainment. And things were no better when at last James Innes resigned his post: their long absence and the expense involved in setting up their own tea-merchant business left them 'friendless and penniless' at home. What a dreary life—almost fascinatingly so—and this is not the half of it, says Emily: 'My pages . . . are dull and gloomy, but my excuse is that my life was dull and gloomy to a degree which can hardly be conceived even from this sketch of it.' Poor old Emily Innes.

❖

JAMESON, Anna Brownell (1794–1860)

A Lady's Diary. London: H. Colburn, 1826 [first published anonymously; reprinted the same year and subsequently as *Diary of an Ennuyée*]; pp. 354

Visits and Sketches At Home and Abroad [etc.]. London: Saunders and Otley, 1834; 4 vols., pp. 301/366/305/305, with an engraved frontispiece and 9 vignettes

Winter Studies and Summer Rambles in Canada. London: Saunders and Otley, 1838; 3 vols., pp. 315/341/356

Anna Jameson is well known as 'the mother of art criticism' and a doyenne of literary finesse. She was one of the charmed circle that included Lady Byron, Elizabeth Barrett Browning, Ottilie von Goethe, and Fanny KEM-BLE: not just writers, but arbiters of the public taste. She was also an accomplished traveller, although because they are so well laced with critical discussion, her travel books are often overlooked as such. The volume best known as *Diary of an Ennuyée* was her first, written following a tour of the Continent in 1821. Anna was Miss Murphy then, the eldest of an Irish artist's five daughters, and obliged to work as a governess for her living. To make her account of the tour from France through Switzerland to Italy a little more exciting, she created a fictional heroine—a young and passionate invalid who was travelling to escape some Dark Tragedy. The Diary, says the Publisher's Preface, was found by the side of the poor 'Ennuyée' after her death from a broken heart at Autun. It was an instant success, although not acknowledged as Anna's (and a hoax) until the 1834 edition, which, with an account of a trip to Germany and some other snippets, formed the four volumes of *Visits and Sketches*.

Anna's most important travel book was her last, about a visit to her husband who was stationed in Toronto. She had married Robert Jameson in 1825, but it was an unhappy liaison and they soon separated. She only joined him now to support his bid for the vice-chancellorship of Upper Canada, and left as soon as the appointment was confirmed. Before returning to England, Anna embarked on two Grand Tours: the first via Niagara to Buffalo, Hamilton, and Detroit in the United States, and the second up Lake Huron to Mackinaw and Manitoulin Island in Chippewa Indian country. On the American journey she insisted on using 'public transport' (a stage-coach), unencumbered with solicitous servants, and she went about Lake Huron in a missionary canoe and a native 'bateau' with its five Indian oarsmen.

Anna spent the rest of her life more conventionally based in London, with occasional excursions to Germany and Italy. She regarded the Canadian journey as one of the finest things she had ever done (but not to be repeated): it was a unique, wild venture 'such as few European women of refined and civilised habits have ever risked, and none have recorded'.

◆

JEFFREYS, Keturah

The Widowed Missionary's Journal; containing Some Account of Madagascar, and also, A Narrative of the Missionary Career of the Rev. J. Jeffreys; who died on a passage from Madagascar to the Isle of France [Mauritius], *July 4th 1825, aged 31 Years.* Southampton: [by subscription] For the author, 1827; pp. 203, with a frontispiece

This is a painful book. It was obviously written for no other reason than to beg money, and one feels in reading it that one is trespassing on a very private tragedy.

John and Keturah Jeffreys were sent to Madagascar by the London Missionary Society. They left in August 1821, taking one child with them, and another was born in Mauritius on the way. The family arrived at Tamatave on the eastern coast of Madagascar in May the next year, and after a few days' rest to recover from the voyage, set off inland to the island's capital, then called Tananarive (Antananarive). It was a horrific journey: Keturah, nursing her four-month-old baby, was carried along in a suspended cot—for most of the time with her head tipped lower than her feet. It took three weeks to reach Tananarive, where they settled only long enough to set up a small school, and were then moved up-country to Ambatoumanga station. Madagascar was notorious at this time for its 'Malagasy Fever', often fatal to European visitors. The Jeffreys managed to withstand it until January 1825, when Keturah fell ill. By May it was decided that nothing would save her but a change of air, so the return trek to Tamatave was made (this time with four children, and another on the way) and a passage secured on a ship carrying cattle and rice to Port Louis, Mauritius. The Jeffreys had to share the hold with the cargo (having no money for a cabin), so there was not much hope for Keturah. But on the tenth day out from Madagascar, it was John Jeffreys and his eldest daughter who were complaining of headaches; within a few days both were dead.

So ended the missionary career of John and Keturah Jeffreys. She and the other children managed to survive the rest of the voyage to Port Louis, and were looked after there until strong enough to sail home to England and, through this book, to sell their pathos for a livelihood.

◆

JUSTICE, Elizabeth

A Voyage to Russia Describing the Laws, Manners, and Customs of that Great Empire . . .
With several Entertaining Adventures that happened in the Passage by Sea, and Land . . . To
which is added, Translated from the Spanish, A Curious Account of the Relicks, which are
exhibited in the Cathedral of Ovideo [etc.]. York: Thomas Gent, 1739; pp. 59, with 19
woodcut vignettes

This is a delightfully unprofessional account of a Wronged Woman's
journey to Russia. She was forced to go by her former husband, leaving her
children and taking up a post as governess to an English family in St
Petersburg. She explains the situation at great length in her introduction:
briefly, Mr Justice (perhaps an ironic pseudonym?) had refused to pay her
an annuity, and the lack of this money along with legal expenses incurred in
trying to obtain it left Elizabeth penniless. So on 4 July 1734 she joined the
three other passengers (all ladies, thank God) aboard a small frigate bound
for 'Crownstead-Mould' (Kronstadt), and set off.

Elizabeth begins her short account with a flourish of poesy:

> From fair Britannia's happy Land, and Coast,
> To Russia's Sands Justicia has been toss'd . . .

With that out of the way she goes on to describe the voyage via Elsinore,
during which her chief interests were reading and fishing, and reels off the
most salient points about the natives of St Petersburg (what they eat, how
they dress, where they worship, and—incidentally—what they do when a
gnat bites them).

There is an exceeding abundance of thunder and lightning in Russia, she
notes, and some of it followed her on the return voyage to England in 1737:
the crossing was uncomfortably rough. She arrived home safely, but only to
find that errant Mr Justice had still not paid his dues. The book was quickly
written to raise more money. It includes a list of those subscribers who were
kind enough to bespeak a copy, and to give them good value it is padded out
with crude 'stock' woodcuts and a strange catalogue of Catholic relics at the
end, bearing no relation to the rest of the book at all.

MARTIN, Selina

[Anon.]: *Narrative of a Three Years' Residence in Italy, 1819–1822. With illustrations of*
The Present State of Religion in that country. London: John Murray, 1828; pp. 356, with
engraved frontispiece

Selina Martin (later a successful children's writer and Irish historian) went
to Italy at the invitation of her sister, who was spending a few years touring

with her husband and children in and around Rome and Naples. In return for her holiday keep, Selina was to help look after the children and act as her sister's companion.

All went well at first: after a six-week voyage round Spain to Leghorn, and an exciting diligence journey 'with foreigners alone', Selina arrived in Rome and was immediately swept into a tireless round of sightseeing trips and social engagements. But then things began to go wrong. The sparkling holiday tour turned into a mournful vigil as first her niece, then her brother-in-law, and finally her sister fell alarmingly ill—and this in a country that might have been designed expressly to caress ailing Britons back to health. Months of anxious nursing followed (interspersed with remarkably sanguine trips to Vesuvius or Orvieto) and by August 1821 both niece and brother-in-law were dead. It remained to Selina to get her sinking sister over the Alps and home—a frightening winter journey—and to write up her experiences for the solemn edification of future visitors to Italy.

❦

MELVILLE, Mrs Elizabeth

A Residence in Sierra Leone. Described from a Journal kept on the Spot, and from Letters written to Friends at Home. By A Lady. Edited by the Hon. Mrs Norton. London: John Murray, 1849; pp. 335

Most of this account—as well it might be—is centred around the climate of West Africa. It is, says Mrs Melville, 'the most unhealthy quarter of the globe' and one's whole being, if one is a Colonial Wife out there, must be dedicated to survival. There is little European female company, since most women either die, leave fever-stricken for home, or move on with their husbands after a matter of months. Mrs Melville coped, however. She moved with her husband into a pleasant house in the hills behind Freetown, and managed to master the local patois well enough to deal with a handful of servants. She had two children to care for (one born there) and, of course, the weather to record in her Journal. Her chief pleasure came from looking out over the sea for English sails bringing the month's budget of letters. The family fell ill with predictable regularity, and in 1843, a year after his arrival in Sierra Leone, Melville was allowed sick-leave in England. But he and his wife (brave woman) were back again six months later, having left their infant children behind, and managed to hang on this time until the end of 1845, when the family was reunited back in the old country, safe and reasonably sound.

❦

POOLE, Annie Sampson

Mexicans at Home in the Interior. By A Resident. London: Chapman and Hall, 1884; pp. 183, with a halftone plate

This is a short and rather lonely account of an Englishwoman's stay at Guanajuato, some 200 miles north-east of Guadalajara, between her marriage and the birth of her first child. Her husband, who had already lived in Mexico for eight years, took her with him to take up an appointment at the mint in Guanajuato; she was given a horse to ride and a small household of servants, and left to contemplate rife and terrifying tales of native and European murders—there was some sort of local atrocity almost every day. Occasionally the couple would try a nervous holiday excursion to the hills, or a hunting trip (for birds and bandits), but they were never able to relax. 'I do so wish this was a more quiet country,' sighed Mrs Poole; 'if you could only sweep off the people it would be perfect.' When the time came for her to leave, to save the life of her new-born baby, she was not sorry to go.

❖

RADCLIFFE, Ann (1764–1823)

A Journey made in the summer of 1794, through Holland and the Western Frontier of Germany, with a Return down the Rhine: to which are added Observations during a tour to the Lakes of Lancashire, Westmoreland and Cumberland. London: for G. G. and J. Robinson, 1795; quarto, pp. 500

Gentle Mrs Radcliffe, one-time resident of Bath and now of London, overwhelmed everyone (including herself) by producing *The Romance of the Forest* in 1791. The book's instant success, and her subsequent plunge into the literary limelight was all too much for the shy authoress; as soon as the even more popular Gothic novel *The Mysteries of Udolpho* was published three years later, Mrs Radcliffe fled the country. She and her husband William planned a holiday tour of the Continent (no matter that the whole of it seemed to be at war: as long as one avoided the battlefields, one was as safe as any other traveller in a foreign land). They had hoped to get as far as Switzerland; in fact they only reached Fribourg on the German border before being turned back as suspected spies.

On reading Ann's account (with its prolix political asides by William), one cannot help feeling the whole trip was a bit of an anticlimax. Holland, for example, although its *trechtschuyts* (or canal-boats) were fun and its cities clean and neat, was undeniably dull. 'They must come from very wretched countries who can find pleasure in a residence at Cologne'; and Koblenz was no better, with its rows of houses standing up 'like tombs'. Even though the Rhine proved satisfyingly picturesque—terror-inspiring, shadowed in 'gloomy grandeur and venerable ruins'—none of what the writer saw matched the brilliant imagined landscapes of her novels and poetry: in future she contented herself with these and just the occasional voyage abroad: to the Isle of Wight.

❧

ROBERTSON, Janet

Lights and Shades on a Traveller's Path: Or, Scenes in Foreign Lands. London: Hope and Co., 1851; pp. 356

Castles Near Kreuznach. London: Williams and Norgate, 1856; pp. 134

In 1826 Anna JAMESON first published her famous fictional travel account the *Diary of an Ennuyée.* No one could be quite as desolate and deliciously miserable as its heroine, one thinks—until one reads Miss Robertson's book. It begins promisingly: 'for I went alone to a strange land, to a house of sickness and gloom—a gloom that presaged death'. Janet Robertson, we gather from the *Shades on the Traveller's Path*, was an orphaned Scottish spinster, ousted from her brother's affections by his overweening wife and suffering from a general unstringing of the nerves. She travelled to Italy 'poor and homeless' to stay with friends, and for a few months wandered broodily amongst the tombstones of Florence before repairing to the brighter air of Lucca for the summer. She nearly enjoyed herself there—were it not for the 'tuft-hunting' British tourists and their hypocritical gossip and snobbery, she might almost have been happy. After a depressing visit home to Scotland a few years later (how much better it would have been for her to die young, she sighs) she ventured over to France. There, in 1847, she was trapped in Paris by the Revolution, eventually escaping to Boulogne in 1848 and home again soon afterwards, floating 'on the winds and waves of the world, equally indifferent to its sunshine and its storms'. In England she prepared a two-volume collection of lugubrious romances—'sketches from life'—published anonymously as *Affinities of Foreigners*, and the *Lights and Shades* book. Then something happy must have happened, for when we next hear of her, cruising with a friend along the Rhine in 1855, she shocks us by being as merry as can be. And after that, there is only blissful silence.

❧

SHEIL, Lady (Mary Leonora)

Glimpses of Life and Manners in Persia. With Notes [by her husband] on Russian Koords, Toorkomans, Nestorians, Khiva and Persia. London: John Murray, 1856; pp. 402, with 7 wood-engraved plates

It is interesting to compare the narrative of Lady Sheil's four years in Persia (from 1849 to 1853) with that of Ella SYKES, who spent three years there some half-century later. Ella loved the journey to Kerman, revelled in the country's strangeness and hardships, and travelled the inhospitable uplands as much as she could. For poor Lady Sheil, though, the long haul to Tehran (where her husband was British Envoy and Minister at the Shah's court) was 'tedious . . . miserable . . . dreary' and Persia itself 'cheerless . . . squalid . . . dull . . . monotonous': everything bad. When we hear how the Lady had to travel (in an uncompromising box on wheels, with her poor maids 'one on each side of a mule . . . where, compressed into the minutest of dimensions, they balanced each other and sought consolation in mutual commiseration of their forlorn fate in this barbarian land') and how, once arrived in Tehran, she was forced to keep indoors for fear of offending her Muslim hosts, her reaction is hardly surprising. In winter it was almost too cold to move and in summer, even up in the hills, the ground literally reeked with heat; time dragged terribly. In fact there was nothing else to do but write a book; that done, and safely home again, Lady Sheil could comfortably forget about the whole dismal experience.

❦

TRANT, Clarissa (1800–1844)

The Journal of Clarissa Trant 1800–1832. Edited by C. G. Luard. London: John Lane, 1925; pp. 335, with 2 coloured and 7 halftone plates

Clarissa Trant was the exhaustingly well-travelled daughter of an Irish army officer whose career spanned the best and worst of the Napoleonic wars. Clarissa first saw the light of day in Lisbon, and from the ages of five (when her mother died) to thirty-two (when she married) she lived a fragmented and lonely life in a succession of European billets from Portugal to Italy and most points in between. There was no novelty in travelling for Clarissa; the act of getting from one place to another—'dismal' Brussels to 'unwhole-some' Paris, perhaps—was uncomfortable and tedious, and once one had arrived there was little to do but unpack, write one's journal, and then pack up again. Even for a comely young lady at the dawn of the empire's halcyon days, travel could be a distinctly unglamorous business.

13

Life in the Bush

MOST of the authors I have talked about in this guide to women travellers have been just that: travellers. People who make journeys, who leave home and go away and then, God willing, come home again. But this final chapter is different. The women here left home sure enough, and went away—but there they stayed.

Some of them were straightforward emigrants, like Rebecca BURLEND and Harriett KING, driven abroad by hardship and incentive to make a new life in a new country. Following the supposed success of the convict deportations of the late eighteenth century (witnessed by Mary PARKER in Chapter I above), the British government invested from the 1830s onwards in a series of mass emigration schemes, designed to relieve the mother country of her less profitable sons and daughters and to provide the underpopulated expanses of North America (or British Africa or Australia and so on) with much-needed labour and stock. Free or 'assisted' passages to specified destinations were offered in appallingly overcrowded steamers and once arrived, the bewildered volunteers would be dispatched by cart or railway truck to their allotted tract of wilderness. The land cost nothing (or near enough), but the working of it all too often bankrupted the emigrants' means, their strength, and sometimes even their lives.

'I would not live in such a place for worlds', wrote Mrs HALL after experiencing first-hand the life of a settler in Manitoba in 1883. She was lucky: she did not have to. She was one of the 'temporary emigrants' of the chapter: women who either went out to visit friends or relatives abroad, wrote books about their lives there, and then scuttled thankfully back to civilization, or who could afford to treat emigration as a speculative enterprise: if things went wrong, they simply gave up and came home.

Of course, not all emigrants were miserable. Some relished the challenge of a strange new life and grew to love their adopted countries as their own. Elegant Lady BARKER delighted in the rough novelty of her role as a New Zealand sheep-farmer's wife, and Catherine TRAILL felt herself a born backwoodswoman, even writing a handbook to encourage other female emigrants to try their mettle on the wilds of Canada as she had.

The taste of 'Life in the Bush' (bitter or sweet as it might be) was not exclusively emigrant fare. All sorts of expatriate residents found themselves homes from home abroad, from the ostrich-farmer in South Africa (married to Annie MARTIN) to the plantation-slave owner in the abundant West Indies (where Mrs CARMICHAEL lived between 1820 and 1826); for them, and for their womenfolk represented here, the novelty of their bizarre circumstances never quite had the time to wear off.

It was the same with Karen BLIXEN, author of one of the most sublime travel books ever written. She was more of a true 'traveller' than many of the women here. Her first journey was made in 1914, when she left her native Denmark for the Ngong Hills behind Nairobi in Kenya. During the seventeen years she spent there as a coffee-farmer, Karen got to know every glowing stone of the tribal lands around her through her own explorations and safaris, and, being a child of her time, from the air. She never lost the sense of wonder that greeted her first arrival in Africa. Another aerial traveller, and in the same country, was Karen's contemporary Beryl MARKHAM, who had been taken out from England by her parents at the age of three. She became one of the pioneers of the skies, crowning a colourful career by crossing the Atlantic solo in 1936.

For Karen and Beryl, their new homeland was irresistibly alluring, drawing them not only to discover its local secrets, but beyond. Which is something all these exiles have in common to a greater or lesser extent, emigrants and residents alike. More than any other kind of traveller leaving their old countries for a new one, if they were to have any chance of surviving happily at all, they had to leave their old selves behind too and start afresh.

So perhaps this final chapter is not really about travelling at all. It is about arriving.

BARKER, Lady Mary Anne (afterwards Lady BROOME) (1831–1911)

Station Life in New Zealand. London: Macmillan, 1870; pp. 238, with a lithographed plate

Travelling About Over New and Old Ground. London: Routledge, 1872; pp. 353, with 6 woodcuts, a title-page vignette, and 5 maps

Station Amusements in New Zealand. London: Wm. Hunt, 1873; pp. 278, with a wood-engraved plate and a map

A Year's Housekeeping in South Africa. London: Macmillan, 1877; pp. 335, with a wood-engraved frontispiece and 8 lithographs

Colonial Memories. London: Smith, Elder, 1904; pp. 301

There is an engaging mixture of the pompous and the down-to-earth in Lady Barker. Life started conventionally enough; at twenty-one Mary Stewart married a promising young artillery officer who was knighted in 1859 and with whom she travelled to India, where he was stationed, the following year. Even then it was an unremarkable journey—hundreds of women of her generation and upbringing did the same (although far fewer came home again—but that is another story). Just a year later Barker was dead, and to help support herself and her two sons his widow returned to England and began to write. She had established a mild literary reputation when in 1865 she met and married Frederick Napier Broome, and her real colonial adventures began. He was a farmer eleven years her junior, who took his new bride—aged thirty-five and pregnant with their first child—out to the vast sheep station he was setting up near Christchurch, New Zealand. The next three years were spent trying to keep the station going, but snow, flood, and Broome's unbusinesslike tendency to feyness and financial confusion eventually defeated them. They returned to a sedentary life in England, where Lady Barker (for so she remained, only changing her professional name when Broome was knighted in 1884) churned out books and articles whilst schooling her husband in the useful art of Establishment respectability. He was duly appointed colonial secretary in Natal, then Mauritius, governor of Western Australia and finally of Trinidad in 1891.

Lady Barker/Broome's travel books are delightfully eager-spirited and humorous, and were highly regarded on their publication as the work of a thoroughly admirable and 'modern' woman. Had she stayed at home, I suspect she would have settled into a frustrated and overbearing old age; devoting her considerable energies to travel, however, and its accompaniments of sheep-shearing, pig-sticking, bone-setting, or bush-trekking, she became a colourful and tenacious spokeswoman for what her contemporaries liked to call the Spirit of the Empire.

❖

BLIXEN, Baroness Karen (pseud. Isak DINESEN) (1885–1962)

Out of Africa. London: Putnam, 1937; pp. 416

Shadows on the Grass. London: Michael Joseph, 1960; pp. 106, with 5 halftones

Karen Blixen made only two great journeys from her native Denmark: one to Africa in 1914, and the other to America in 1957, and yet she is the author of one of the best-known travel books there is. As Isak Dinesen she published six collections of curiously crafted short stories—in fact it was as a literary doyenne that she travelled to the United States, on her election to the eclectic American Academy. But it is for *Out of Africa* that she is really remembered.

Karen Blixen first saw the dramatic, flame-skied expanse she writes about so well when she arrived in Kenya with her cousin and husband, Baron Bror Blixen-Finecke, to start up a coffee farm 6,000 feet up in the Ngong Hills. She ran the farm herself until drought, disease, and falling prices forced her away in 1931. It was never easy—especially after her divorce from Blixen in 1921—but Karen always claimed that 'to my mind the life of a farmer in the East African highlands is near to an ideal existence.' She was a proud and independent woman whom the Africans named 'Her Honourable Lioness'; an accomplished artist and sportswoman, and the local doctor, champion, and Chief Friend. It was her special relationship with the Masai and Kikuyu tribespeople living around her that set Karen Blixen apart as a European traveller in Africa: she accepted their culture and way of life unconditionally and where others only sought to learn, she *lived.*

❖

BURLEND, Rebecca (1791–1872)

[Anon.]: *A True Picture of Emigration: or Fourteen Years in the Interior of North America; being a full and impartial account of the various difficulties and ultimate success of an English family who emigrated from Barwick-in-Elmet near Leeds in the year 1831.* London: G. Berger [1848]; pp. 62

Rebecca Burlend never wanted to go to America. She and her husband had lived all their lives in west Yorkshire, making a living (albeit meagre) on their farm there. However, as the family began to grow, the price of corn began to fall, and they found themselves in trouble.

Following an emigrant friend's advice to join him in prosperity, they decided to take advantage of a grant of land in Pike County, Illinois, and start life again. They left behind the eldest two (of seven) children, travelled to Liverpool—a new world in itself to Rebecca, who had never strayed more than forty miles away from home—and there secured a steerage passage to New Orleans. Most women travellers would have found the conditions below deck quite insupportable—but Rebecca accepted them merely as tedious and depressing. Nor was her first sight of America any consolation: 'a sort of melancholy came creeping over me as I gazed upon it', she says.

As instructed, the Burlend family took a steam-packet from New Orleans up the Mississippi, and twelve days later, on a sharp December night, were put off at what they were assured was their final destination. There was nothing there. No house, no cabin, no welcome, nothing. Rebecca's first night was spent huddled in the open with her children on one of the two beds they had brought with them (along with two boxes and some pots and pans) whilst her husband went in search of life.

In her *True Picture of Emigration* Rebecca describes how the family managed to survive; it is written, says Rebecca, so that others might learn by their mistakes. She explains what to do when the milk freezes almost before it is out of the udder in winter, how to cope with the mosquitoes in summer, how to thatch a hut, make soap from suet, and cure 'Illinois mange' with sulphur. She explains the (corrupt) system of land purchase in Pike County and records every item of expense incurred, including the journey out from Barwick (£30).

After fourteen years, the Burlends began to support themselves and make a small profit, enabling them to buy and stock more land and pay for Rebecca's first visit to the children she left behind. She dictated this narrative to her son (the educated one of the family) during that visit; he added a postscript to say that his sister and her family accompanied Rebecca back to Illinois, and four other local families soon followed. Their descendants, of course, are still there.

❦

CARMICHAEL, Mrs (A. C.)

Domestic Manners and Social Conditions of the White, Coloured, and Negro Population of the West Indies. London: Whittaker and Treacher, 1833; 2 vols., pp. 336/338

Mrs Carmichael—author of the popular children's book *Tales of a Grand-mother*—spent the years between 1820 and 1826 in St Vincent and Trinidad as the wife of a plantation owner. During the whole of that time, she says, she never had a single day to devote 'to the gratification of mere curiosity': instead, every afternoon would be spent walking or riding over the Carmichael and neighbouring estates, to find out as much as she could about the people working there. In this remarkably unsentimental account, published a year after Earl Grey's Reform Bill officially abolished slavery throughout the British Empire, she reports on what she found.

Unlike Fanny KEMBLE or Harriet MARTINEAU, Mrs Carmichael was no Abolitionist. She rarely saw a slave-owner or manager use physical cruelty, she says, and contrary to popular belief she never saw a slave-driver wield a whip. The coloured and Negro slaves she spoke to and visited (including first-generation Africans) were usually satisfied with the way of life to which they had been set, and had a far more congenial and relaxed life than servants and labourers at home in England. Their native culture and customs were valuable and colourful: they were *different* from white men, and suited to slave work, but certainly not *inferior*. 'It is our bounden duty, as Christians, to instruct the negroes in religion and help them forward in civilisation,' said Mrs Carmichael (she founded schools both in St Vincent and Trinidad); 'but if by civilisation it is intended to make them live in the same manner as Europeans, I would say that the negroes would not submit to such an arrangement, and beyond a doubt it would make them most uncomfortable and unhealthy.'

Sadly, Mrs Carmichael's observations were as naïve as they were sympathetic: 'civilization' had already taken hold and by 1826 a series of frenzied uprisings in Trinidad made it no longer safe for white planters to stay on their own estates. In 'failing health and spirits' the Carmichaels sailed for England, as sad for the slaves they left behind as for themselves.

❖

CHISHOLM, Caroline (1808–1877)

Mackenzie, K.: *The Emigrant's Guide to Australia with a Memoir of Mrs Chisholm* [and several chapters by her]. London: Clarke, Beeton [1853]; pp. 187, with engraved portrait and half-title

The A.B.C. of Colonization [etc.]. London: John Ollivier, 1850; pp. 42

Caroline Chisholm, called variously 'a second Moses in bonnet and shawl', 'the guardian angel of her helpless sex' and, most simply, 'the emigrants' friend', was a true philanthropist. She was educated at home in Northamptonshire, and brought up by her mother to believe strongly in the moral influence of her sex and the virtue of practical charity.

Caroline's philanthropic travels began in 1832 when her husband (of the

East India Company) was posted to Madras; there she founded a school for soldiers' daughters which soon became famous as the 'Female School for Industry'. When the Chisholms moved to Australia in search of health (and found it) in 1838 Caroline turned her generous attention to the piteous state of women convicts and 'voluntary' immigrants recently transported from England, many of whom were dangerously left to drift around Sydney in search of whatever livelihood they could find. She established a free registration centre and agency for working women and a hostel in which they could live until a 'position' was found: then she set off along the coast of New South Wales as far as Brisbane, occasionally branching inland to set up a network of twelve similar 'depots'. She took parties of immigrant women with her on her repeated journeys, rather like Annie MACPHERSON in Canada, all loaded on to bullock drays and deposited wherever there was room for them.

Thoroughly fired by now, Caroline returned to England in 1846 to tackle the problem of provision for emigrants at its source by chivvying Earl Grey into instigating a Select Committee in the House of Lords, and publishing several meticulously researched pamphlets and petitions for her cause. In 1849 she realized a long-held ambition by helping found the Family Colonization Loan Society to encourage whole families to emigrate together, and five years later returned to New South Wales where she stayed until ill health and her own children's educational welfare forced her finally home to England in 1866.

❖

GODLEY, Charlotte (1821–1907)

Letters from Early New Zealand [etc.]. With an Introduction by Prof. A. P. Newton. [Plymouth]: for Private Circulation only, 1936; pp. 377, with 5 halftones and a map

Charlotte Godley was a sort of temporary emigrant. She and her husband John Robert sailed out to Dunedin in 1850 to supervise the founding of the Canterbury Settlement, a non-denominational, mixed-class emigrant colony in South Island. It was John Robert's own brain-child, designed to establish 'a complete segment of English Society' on the other side of the world. The Godleys lived for two years in New Zealand, sailing between Dunedin, Wellington, and Christchurch, during which time Charlotte wrote these lively letters home.

Charlotte described life as surprisingly easy (no doubt to calm her family who had tried to dissuade her from accompanying Godley for fear she might spoil her hands) and even travelling never involved 'roughing it', she claimed, except for one trip inland to Ricarton. The country was beautiful and life was good.

The letters are unique as a record of the early British colonization of New Zealand—which only began in 1838—from a slightly rarefied point of view. As Godley had hoped, Charlotte confirmed that the Canterbury Settlement was indeed a microcosm of home life, with necessary limitations ('it is not quite the thing to be away here, and not see the [1851 Great] Exhibition'), but it is obvious that it came complete with all the social niceties and peccadilloes of life at home too. It is interesting to compare the account Charlotte gives of emigrant life with that of Rebecca BURLEND or Mrs Harriett KING, for example: perhaps the difference between them is just the evidence of one more legacy from the Old Country.

❖

HALL, Mrs Cecil (M. G. C.)

A Lady's Life on a Farm in Manitoba. London: W. H. Allen, 1884; pp. 170, with 3 line-drawings

Although she wrote this book for the use of future colonists in Manitoba and Colorado, Mrs Cecil Hall was no pioneer herself. Her brother had emigrated in 1882 and bought 480 acres of Canadian prairieland; 'Life on a Farm' for her meant a three-month holiday visit.

Mrs Hall sailed out from Liverpool with her sister on an elderly steamer loaded with sixty first-class passengers, *twelve hundred* emigrants, and a hold full of potatoes. They took the train from New York to Washington, fitted in a reception at the White House there, and then travelled up through Chicago and St Paul to Winnipeg, the nearest city to their brother's home. During their stay they made a lengthy expedition 'up West' towards Rapid City and the Oak River. Where there were no trains they went by wagon, and where there were no hotels they camped. On the sisters' way home they took in visits to Denver, Colorado, and to a friend's ranch in Uncompaghre Park in the Rocky Mountains.

Mrs Hall's brief taste of a colonist's life did not suit her: the land was desolate ('its vastness, dreariness and loneliness is appalling'), the weather was harsh, and the work was endless. Three months of rising at six and doing all the chores a servant should have done at home was more than enough for her: 'My advice to all emigrants is to leave their pride to the care of their families at home before they start . . . I would not live in such a place for worlds.'

❖

HARGRAVE, Letitia (1813–1854)

The Letters of Letitia Hargrave. Edited, with introduction and notes by Margaret Arnett Macleod. Toronto: The Champlain Society, 1947 [limited edition of 550 copies]; pp. cliv + 310, with 8 halftones and 2 maps (including 1 on endpapers)

Letitia Mactavish was one of the brave band of pioneers—mostly Scots—who went out to Canada in the flourishing days of the Hudson's Bay Company. Her father was in the fur trade and her brothers did their apprenticeship at the company's York Factory settlement, six miles from the mouth of the Hayes River in northern Manitoba. Their supervisor there was one James Hargrave, also a Scot, who visited the Mactavish household whilst on a wife-finding mission to Britain in 1837. Two years later, Letitia was installed at York Factory—as bleak and unforgiving a spot as she could imagine—and there she stayed (with one visit home) for the next eleven years.

Five children were born to the Hargraves (although not all survived); Letitia's time was taken up with feeding, clothing, and amusing them, with playing her elevated part in the social life of the settlement (with its complicated hierarchy of half-castes and Indian mistresses), and longing for each August, or 'ship-time', when news from Scotland came. By the end of 1851, his family ill, exhausted, and melancholy, Hargrave was at last promoted. Sault-Ste-Marie, further south on the shores of Lake Huron, was considered quite healthy and comfortable, and Letitia looked forward to a new life in this 'solitary land'. She and the children sailed home to recover their health and spirits in Scotland whilst Hargrave made the move, and they joined him in the summer of 1852. So much for a new life: just two years later, worn out to 'a perfect skeleton', Letitia died. She was forty-one.

❧

KING, Harriett Barbara

Letters from Muskoka. By An Emigrant Lady. London: Richard Bentley, 1878; pp. 289

This tells the sad story of a widow who lived comfortably with her family in France until the outbreak of the Franco-Prussian War in 1870. Various circumstances then led to the loss of her home and most of her money, and *en masse* the family decided to try life afresh in Canada. Emigration was still strongly encouraged from Europe to this big, new country: each settler was entitled to his own grant of land, and, given hard work and patience, it was said that one could even find one's fortune there.

What no one told the Kings was that one needed a little capital to start off with. Harriett arrived at her miserable log cabin in the bush by Lake Muskoka (Ontario) with only debts to her name, and immediately fell into a long and paralysing depression. She suffered two years of unsuccessful 'pioneering' before eventually giving up the cabin and retiring with those members of the family that were left to the nearby township of Bracebridge. And there she stayed to write this cautionary tale, closing it with a miserable warning: 'I went into the Bush of Muskoka strong and healthy, full of life and energy . . . I am now a helpless invalid, entirely confined by the doctor's orders to my bed and sofa, with not the remotest chance of ever leaving them'.

❖

LADY, a.

My Experiences in Australia: Being a Recollection of a Visit to the Australian Colonies in 1856–7. London: J. F. Hope, 1860; pp. 367, with 6 lithotint plates

This Scots author sailed for Australia with her husband and daughter in 1856, to visit the family sheep station at Keera, some three hundred miles from Sydney. They planned to stay for a year and a half: like Mrs Cecil HALL in Canada, she was just a temporary emigrant, and very grateful for it. Life in Australia, she said, held precious few attractions. In Sydney, one had to deal with 'the small affectations and fine-lady airs of the wives of . . . minor colonial officials'. Then the privations of bush travel were almost overwhelming, even if, with good Scots canniness, one had brought one's own cart from home with a tent, iron bedsteads, and cork mattresses. And when one finally arrived at one's ramshackle destination, miles and perhaps days from the nearest neighbour, the sheer slog of survival took away all one's health and leisure. Still, she realized that some poor women might have to spend their lives in Australia, and so she livens up her *Recollection* with descriptions of the towns and bush-stations she visited, including any gold-diggings or aboriginal encampments *en route*, and for the benefit of future *emigrées* lists the most common bush diseases and hazards, from snake bites to ophthalmia, and one or two handy recipes.

❖

MARKHAM, Beryl (1902–1986)

West with the Night. Boston: Houghton Mifflin, 1942 [London: Harrap, 1943]; pp. 294

When Ernest Hemingway read *West with the Night* he was stunned. 'She has written so well, and marvellously well,' he wrote, 'that I was completely ashamed of myself as a writer . . . it is really a bloody wonderful book.' Other reviewers ranked it with Karen BLIXEN's *Out of Africa*: it was warm, poetic, and richly sensitive—in fact so unlike Beryl Markham herself, they said, that she probably had not written it at all.

Beryl was taken to what was then British East Africa by her father in 1905, and as she grew up amongst the hills outside Nairobi she acquired a variety of rather startling skills: 'hunting wild pig with the Nandi, later training race-horses for a living, and still later scouring Tanganyika and the waterless bush country between the Tana and Athi Rivers, by aeroplane, for elephants'. She was well known amongst the tight-knit and aristocratically decadent white community in Kenya for her cool nerves and beauty, and picked her way consummately between marriages and lovers both there, in England, and in America. Beryl took up flying in the early 1930s, soon after receiving her racehorse trainer's licence; she became one of Africa's first commercial pilots, and would give friends lifts to Paris or London—the freedom of the air, she said, was intoxicating. In 1936 she made her most famous solo flight 'west with the night' across the Atlantic; after her arrival in America (nose-down in a Nova-Scotian bog) she stayed there, on and off, for thirteen years. She wrote the book in Hollywood—just the place for the scandal that broke almost as soon as it was published. Perhaps Beryl's husband and 'collaborator' Raoul Schumacher had written it, said her critics—it was far too fine for a glamorous opportunist like her.

Beryl died back in Nairobi, where she returned at the end of the Second World War to train her horses. By the time she had reached her eighties she was almost forgotten, an eccentric recluse wrapped in rumours of a colourful past and an all-too-present vapour of vodka fumes, and even now no one quite knows whether she wrote that 'bloody wonderful book' or not.

❧

MARTIN, Annie

Home Life on an Ostrich Farm. London: George Philip, 1890; pp. 288, with 10 halftones

Home life on a *what*? Annie Martin did not take the idea seriously at first, either. The only ostrich farm she had seen was included on a sightseeing trip down the Nile as a quaint Egyptian novelty. But then she met a man who confessed to having earned a tidy living as an ostrich-farmer in South Africa, and who suggested they found another farm together as husband and wife. The idea was irresistibly romantic. In 1881 the Martins sailed to Port Elizabeth and, after a few weeks' prospecting on the Karroo, managed to find twelve thousand acres and forty-nine ostriches: enough, they hoped, to make their fortune. Annie's book tells of the 'several busy and most enjoyable years of ostrich-farming life' they spent there, with a varied household of angora goats, merino sheep, a collie dog, Jacob the secretary-bird (who swallowed kittens whole), Bobby the crow, the odd visiting leopard, and, of course, a thoroughly grumpy collection of ostriches. Sometimes the Martins would take a holiday to the nearby diamond-fields, or an excursion (ostensibly to round up some stray birds) on horseback through Hottentot dust and thunderstorms; it was a 'delightful, rough South-African life', said Annie, and she missed it very much.

❧

MEREDITH, Louisa Anne (Mrs Charles) (1812–1895)

Notes and Sketches of New South Wales, during a Residence in that Colony from 1839 to 1844. London: John Murray, 1844 (in their Home and Colonial Library series); pp. 164

My Home in Tasmania, During a Residence of Nine Years. London: John Murray, 1852; 2 vols., pp. 274/275, with 18 wood engravings

Some of My Bush Friends in Tasmania: Native Flowers, Berries, and Insects Drawn from Life, Illustrated in Verse and Briefly Described. London: Day and Son, 1860; folio, pp. 106 (including 2 chromolithographed leaves), with 13 chromolithographed plates

Over the Straits: A Visit to Victoria. London: Chapman and Hall, 1861; pp. 284, with 16 wood-engraved vignettes

Our Island Home: A Tasmanian Sketch Book. Hobart: J. Walch, London: Marcus Ward, 1879; folio, ff. 45 (printed on recto only), with 12 autotype plates

Tasmanian Friends and Foes Feathered, Furred, and Finned: A Family Chronicle of Country Life, Natural History, and Veritable Adventure. London: Marcus Ward, 1880; pp. 259, with 8 chromolithographed plates, and wood-engraved vignettes throughout

When Louisa Meredith sailed to Sydney on a honeymoon voyage in 1839, she was already well known (under her maiden name of Twamley) as an author and illustrator of poems and prose. Her subject then had been the local countryside: so it continued for the rest of her life (only the locality changed), with a string of novels, sketches, and travel accounts about the other side of the world.

Charles Meredith was Louisa's cousin; he had emigrated to Tasmania when a child and married her on his first return to England. At first they planned to settle in New South Wales, where Charles had invested his modest fortune in a number of dubious enterprises. But Meredith proved a singularly inept business man and just as bad a sheep-farmer, so after only four years he and Louisa retreated to his family home just across the Straits. Louisa had hated Sydney—hated the climate and the company; Tasmania's gentle, lusher landscape suited her talents as artist and sentimentalist far better. Soon the family were able to build their first home near Swansea on the east coast. They moved house thereafter with exhausting regularity, building a new one each time, but never strayed very far from their original starting-place.

Louisa's accounts of life on the island, bringing up children and grandchildren there, and managing a series of homesteads and farms were regarded by her public as colourful celebrations of Tasmania by its most illustrious adopted daughter. Privately, however, she never considered herself as an emigrant, but as a traveller and an exile. She visited England— 'home'—once, in 1889 when she was seventy-seven, and said then that her whole life had been a struggle against an unwelcome fate: she was never meant to be a traveller at all.

❧

POOLE, Mrs Sophia (1804–1891)

The Englishwoman in Egypt: Letters from Cairo, written during a Residence there in 1842, 3, and 4, with E. W. Lane Esq. By his sister [preface signed 'Sophia Poole']. London: Chas. Knight, 1844 (in their Weekly Volume series); 2 vols., 12mo, pp. 232/240, with 19 wood-engraved plates, 3 plans, and vignettes throughout

The Englishwoman in Egypt: Letters from Cairo, written during a residence there in 1845– 46 [etc.]. Second Series. London: [as above], 1846; 12mo, pp. 249

Egypt, Sinai, and Jerusalem: A Series of Twenty Photographic Views by Francis Frith. With Descriptions by Mrs Poole and Reginald Stuart Poole [her son]. London: Virtue [1860]; elephant folio, each photograph with a leaf of letterpress

Cairo, Sinai, Jerusalem, and The Pyramids of Egypt: A Series of Sixty Photographic Views by Francis Frith. With Descriptions [as above]. London: Virtue [1860–1]; folio, published in 20 parts, each photograph with a leaf of letterpress

At the suggestion of her brother, the eminent Arabic scholar Edward Lane, Sophia Poole first travelled out to Cairo in the summer of 1842. She took her two young sons with her and together they stayed for seven years. When her highly popular accounts of a lady's life in Egypt were published back in London, they caused a mild sensation. It might be permissible for a learned chap like Lane to immerse himself in the exotic culture of the East—but an *Englishwoman*? A Christian wife and mother dressing herself up in Turkish 'trowsers' and visiting the city's harems? Living in what she insists is a haunted house, and witnessing barbarous murders almost on her own doorstep? And, worst of all, taking *Turkish baths* with the natives? Sophia tempered the sensationalism with a serious study—to complement Lane's *Manners and Customs of the Modern Egyptians*—of the habits and customs of harem life in Cairo (the first in the English language based on personal observation) and qualified herself admirably to write a definitive text to Frith's stupendous photographs of Egypt in the 1850s.

❧

RESIDENT, a Lady

The Englishwoman in India: Containing Information for the use of Ladies Proceeding to, or Residing in, the East Indies, on . . . Outfit, Furniture, Housekeeping, the Rearing of Children, Duties and Wages of Servants . . . and Arrangements for Travelling. To which are added Receipts for Indian Cookery. London: Smith, Elder, 1864; pp. 211

This book is exactly what it says: a collection of helpful hints for the ladies of the Raj, and a modest forerunner to Flora STEEL's epic *Complete Indian Housekeeper and Cook.* Its author had seven years' experience of the vagaries of Indian housekeeping and travel ('I once marched about 1,000 miles alone with three very young children, all in delicate health'), and her advice is full of common sense. Take especially pretty night-dresses with you, for instance, so that when the fever comes, you need not be embarrassed. Learn to cook a good curry in emergencies (when the cook has just died, perhaps) and try to keep your amah, or wet-nurse, away from hot chillies, or else your baby is liable to turn rather fierce. Half the book is given up to recipes and menus for everything from Scotch broth to syllabub.

❧

THOMSON, Mrs

Life in the Bush. By A Lady. [London: Chamber's Miscellany of Useful and Entertaining Tracts, 1845.] pp. 32, with 2 wood-engraved vignettes

This is a short and cautionary tale for emigrants to the Australian bush. Mrs Thomson went there in 1838 with her husband, child, two brothers, a shepherd, his wife, and two female servants, to farm a sheep and dairy station in Victoria. The ship from England had taken them as far as Hobart Town, Tasmania: there the party temporarily split up, the men sailing on to buy the stock and build the homestead (about a hundred miles inland from Port Philip), and the women staying to learn dairying on a farm near Launceston. Soon the station was ready, and the Thomsons spent a hard-working two years getting it established. During these two years Mrs Thomson trekked to Melbourne, several days away, in a tarpauline-covered bullock-dray to have another baby; she learned how to catch and cook a kangaroo, and—her *pièce de résistance*—how to bake a parrot pie. She was thoroughly happy. In 1840, however, the younger brothers decided that they should run the station on their own, and Mr and Mrs Thomson were banished from the bush to Melbourne. Soon after arriving, after a broiling summer spent under canvas for want of a house, Mrs Thomson became dangerously ill, and the tale ends with the family again split up: the husband clinging on to a hard-won office job in Melbourne, and the wife and children sailing home to England, sick, exhausted, and disillusioned.

⟡

TRAILL, Catherine Parr (1802–1899)

[Anon.]: *The Backwoods of Canada; being Letters from the Wife of an Emigrant Officer, Illustrative of the Domestic Economy of British America.* London: Chas. Knight (Library of Entertaining Knowledge series), 1836; 12 mo, pp. 351, with 19 wood engravings and a map

'Canada Fever' was at its height during the late 1820s and early 1830s, when tens of thousands of hopeful emigrants left the United Kingdom for the brave new world of 'British America', often with little more to their names than a government grant of uncleared land in the backwoods of what is now Ontario. Amongst them were three members of the talented Strickland family, Samuel, Catherine (Mrs Traill), and Susanna (MOODIE), all three of whom wrote books about their new homes.

Catherine's account of life around Cobourg, Peterborough, and the Rice Lake plains was perhaps the most practical. The Traill family had more than its fair share of crop failures, fire damage, and financial crises, and like her sister, Catherine was forced to write for the little money her work brought in. Unlike Susanna, however, and even though she was welcomed to its shores by cholera and utterly exhausted by the tasks of life and travel in this untamed land, Catherine loved the country from the start. She wrote improving children's books about it, romantic studies of its natural history, and, in 1854, the *Female Emigrant's Guide*, a sensible and thorough handbook 'for all classes, and more particularly for the wives and daughters of the small farmers' (reprinted several times as the *Canadian Settler's Guide*). Her autobiographical account, listed here, was not as popular as Susanna's classic *Roughing it in the Bush*, but still found a niche in emigrant literature as a lively book suitable (she said) for 'the wives and daughters of ... the higher class'.

Catherine obviously thrived on a rough life: she outlived Sam, Susanna, and several of her own children to die, in slightly gentler poverty than she had been used to, at the age of ninety-seven.

Maps

1. Europe ('the Continent')

and Asia Minor

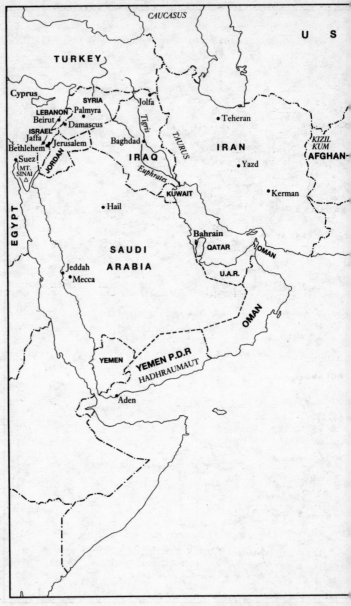

2. The Middle East, Arabia,

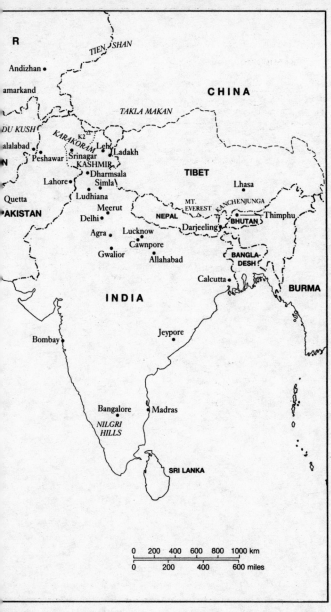

Central Asia and India

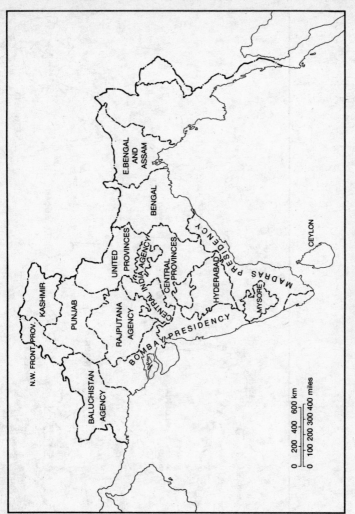

2a. Historical map: India in 1907

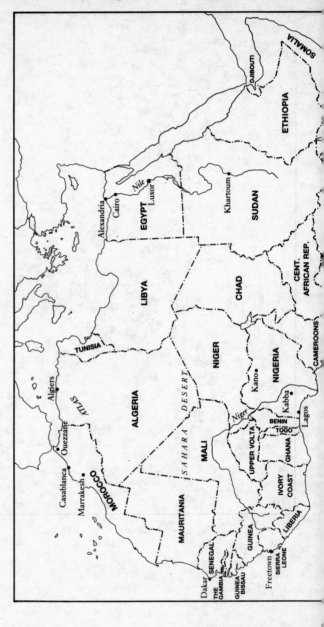

3. Africa

3a. Historical map: Africa in 1910

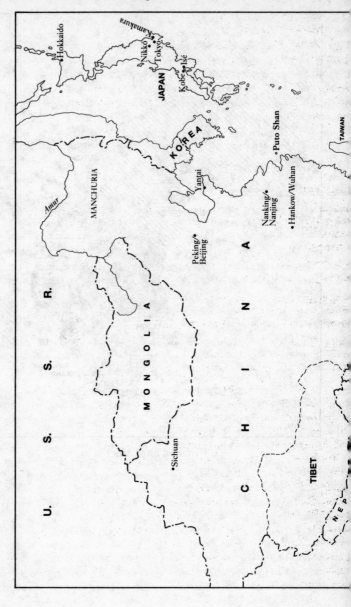

Hokkaido

Kamakura

Nikko
Tokyo
JAPAN
Kobe Ise

KOREA

Puto Shan

MANCHURIA

Yantai

Nanking/
Nanjing

Hankow/Wuhan

TAIWAN

Amur

Peking/
Beijing

C
H
I
N
A

R.

MONGOLIA

S.

S.

Sichuan

TIBET

U.

NEP

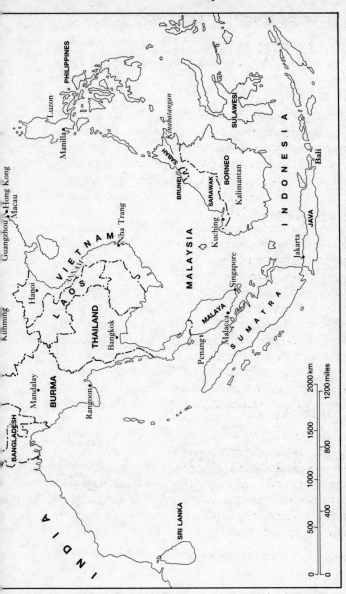

4. China, Japan, and South-East Asia

INDONESIA

PAPUA
NEW
GUINEA

Sepik R.

Fly R.

Solomon
Islands

Tuvalu
Funafuti

Western
Samoa

Torres Strait

Timor

Darwin

Vanalu
Levu

Fiji

New
Hebrides

Viti Levu

Suva

Tonga

New
Caledonia

Brisbane

AUSTRALIA

NEW
ZEALAND

Swansea

Paramatta

Sydney

Auckland

Napier

Botany Bay

Wellington

Bendigo

Melbourne

Canterbury

Christ Church

Great Australian Bight

Adelaide

Launceston

Dunedin

Tasmania

Hobart Town

5. Australasia and

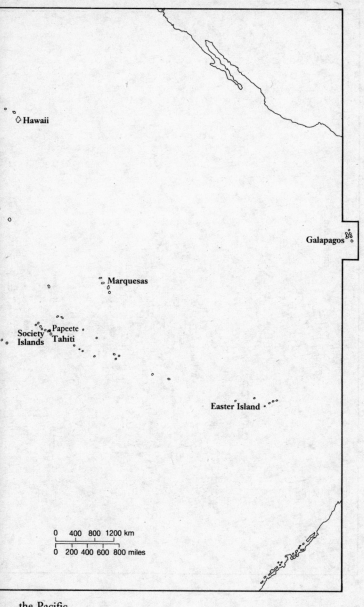

Hawaii

Galapagos

Marquesas

Society Islands Papeete Tahiti

Easter Island

0   400   800   1200 km
0   200 400 600 800 miles

the Pacific

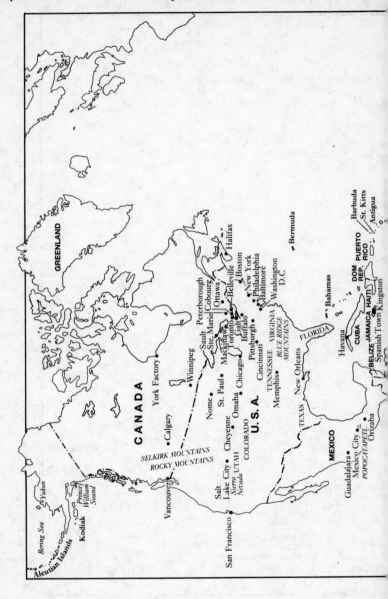

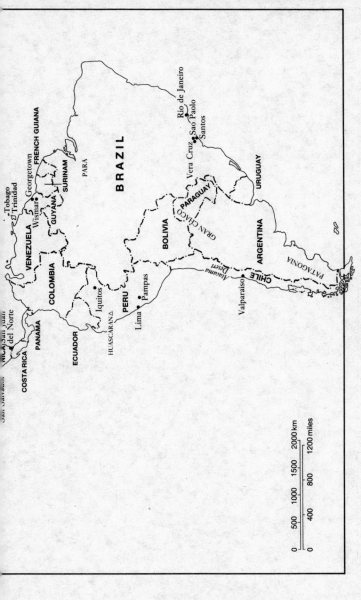

6. The Americas

Useful Reference Books

ADAMS, William H. Davenport. *Celebrated Women Travellers of the Nineteenth Century.* London: Swan Sonnenschein, 1883.

AITKEN, Maria. *A Girdle Round the Earth: Adventuresses Abroad.* London: Constable, 1987.

ALLEN, Alexandra. *Travelling Ladies: Victorian Adventuresses* [Daisy Bates; Isabella Bird; Mildred Cable, Evangeline and Francesca French; Alexandra David-Neel; Jane Digby El Mezrab; Kate Marsden; Marianne North; May French Sheldon]. London: Jupiter, 1983.

BARR, Pat. *The Memsahibs: The Women of Victorian India.* London: Secker and Warburg, 1976.

BIRKETT, Dea. *Spinsters Abroad: Victorian Lady Explorers.* Oxford: Basil Blackwell, 1989.

BLANCH, Lesley. *The Wilder Shores of Love* [Isabel Burton; Jane Digby El Mezrab; Aimée Dubucq du Rivery; Isabelle Eberhardt]. London: John Murray, 1954.

GREENHILL, B., and GIFFORD, A. *Women Under Sail.* London: David and Charles, 1970.

KEAY, Julia. *With Passport and Parasol: The Adventures of Seven Victorian Ladies* [Emily Eden, Anna Leonowens, Amelia Edwards, Kate Marsden, Gertrude Bell, Daisy Bates, and Alexandra David-Neel]. London: BBC, 1989.

LOMAX, Judy. *Women of the Air.* London: John Murray, 1986.

MACMILLAN, Margaret. *Women of the Raj.* London: Thames and Hudson, 1988.

MIDDLETON, Dorothy. *Victorian Lady Travellers* [Isabella Bird; Marianne North; Fanny Bullock Workman; May French Sheldon; Annie Taylor; Kate Marsden; Mary Kingsley]. London: Routledge and Kegan Paul, 1965.

MILLER, Luree. *On Top of the World: Five Women Explorers in Tibet* [Nina Mazuchelli; Annie Taylor; Isabella Bird; Fanny Bullock Workman; Alexandra David-Neel]. London: Paddington Press, 1976.

RUSSELL, Mary. *The Blessings of a Good Thick Skirt: Women Travellers and their World.* London: Collins, 1986.

TILTMAN, Marjorie Hessell. *Women in Modern Adventure* [eighteen contemporary travellers]. London: Harrap, 1935.

TROLLOPE, Joanna. *Britannia's Daughters: Women of the British Empire.* London: Hutchinson, 1983.

WILLIAMS, Cicely. *Women on the Rope: The Feminine Share in Mountain Adventure.* London: G. Allen and Unwin, 1973.

Geographical Index

This index is arranged to correspond to the maps, beginning with Europe and moving eastwards round the world.

Europe ('The Continent')

France and Corsica

Belgium and the Netherlands

Scandinavia and Polar Regions

USSR

Eastern Europe and the Balkans

Greece and Turkey

Iran and Iraq (Persia and Mesopotamia)

Arabia and Middle East

Egypt

Africa

Africa *cont*.

Sheldon, May French 27
Speed, Maud 74
Speedy, Mrs Charles 75
Strickland, Diana 75
'Tully', Miss 248

Vassal, Gabrielle 248
Ward, Harriet 272
Wazan, Emily, Shareefa of 250
Workman, Fanny Bullock 30
Wortley Montagu, Lady Mary 32

Madagascar, Mauritius, Ascension, and St Helena

Bartram, Lady Alfred 275
Chapman, Olive Murray 85
Colvile, Lady Zélie 207
Gill, Mrs 237
Graham, Maria 44

Hare, Rosalie 239
Jeffreys, Keturah 287
Murphy, Dervla 97
Pender, Rose 219

India, Pakistan, and Sri Lanka

Ali, Mrs Meer Hassan 228
Aynsley, Harriet 80
Baillie, Mrs W. W. 64
Barker, Lady Mary 295
Bartrum, Katherine 254
Becher, Augusta 203
Benn, Edith 229
Brassey, Lady Anna 203
Bridges, F. D. 84
Burton, Lady Isabel 232
Cameron, Charlotte 176
Canning, Charlotte 204
Chisholm, Caroline 298
Clerk, Mrs Godfrey 109
Coopland, R. M. 256
David-Neel, Alexandra 9
Davis, Elizabeth 258
Dibble, L. Grace 88
Donaldson, Lady Florence 208
Duberly, Mrs Henry 259
Dufferin, Lady Harriot 209
Eden, Hon. Emily and Frances 209
Elwood, Mrs Colonel 14
Falkland, Amelia, Viscountess 212
Fane, Isabella 280
Fay, Eliza 43
Forbes, Rosita 91
Fountaine, Margaret 132
Gardner, Mrs Alan 68

Gordon Cumming, Constance 93
Graham, Maria 44
Graham Bower, Ursula 46
Griffith, Mrs George 282
Harris, Audrey 95
Harris, Mrs G. 263
Hill, Rosamond and Florence 134
Hobson, Sarah 48
Inglis, Lady 265
Jenkins, Lady 70
Kindersley, Mrs 242
Leonowens, Anna 138
Lloyd, Sarah 97
Lushington, Mrs Charles 214
Mackenzie, Mrs Colin 215
Maillart, Ella 51
Maitland, Julia 215
Marryat, Florence 185
Montez, Lola 55
Murphy, Dervla 97
Muter, Mrs D. 266
Nugent, Lady Maria 217
Paget, Mrs Leopold 267
Parkes, Fanny 218
Postans, Mrs 220
Resident, A Lady 306
Roberts, Emma 190
Sale, Lady Florentia 268
Savory, Isabel 74

Central Asia and Himalaya

China

Japan

Bird, Isabella 81
Bridges, F. D. 84
Chesterton, Mrs Cecil 39
D'A[lmeida], Anna 110
Fountaine, Margaret 132
Gordon Cumming, Constance 93

Harris, Audrey 95
Howard, Ethel 135
Hutchison, Dr Isobel 135
Mannin, Ethel 183
Moss, Lady F. N. 118
Stopes, Marie 150

South-East Asia

Behn, Aphra 128
Bigland, Eileen 175
Bird, Isabella 81
Bridges, F. D. 84
Brooke, Lady Margaret 230
Brooke, Sylvia 231
Clifton, Violet 233
D'A[lmeida], Anna 110
David-Neel, Alexandra 9
Ellis, Beth 211
Forbes, Anna 281
Fountaine, Margaret 132
Gordon Cumming, Constance 93

Hanbury Tenison, Marika 238
Hare, Rosalie 239
Harris, Audrey 95
Innes, Emily 285
Johnson, Osa 50
Le Blond, Mrs Aubrey 20
Leonowens, Anna 138
McDougall, Harriette 165
Mannin, Ethel 183
Moss, Lady F. N. 118
Vassal, Gabrielle 248
Workman, Fanny Bullock 30

Australia, Tasmania, and New Zealand

Anley, Charlotte 126
Barker, Lady Mary 295
Bates, Daisy 127
Brassey, Lady Anna 203
Broad, Lucy 154
Cable, Mildred 155
Chisholm, Caroline 298
Clacy, Mrs Charles 206
Clerk, Mrs Godfrey 109
Davis, Elizabeth 258
Du Faur, Freda 11
Fountaine, Margaret 132
Franklin, Lady Jane 92
French, E. and F. 155
Godley, Charlotte, 299

Gordon Cumming, Constance 93
Hall, Mary 15
Hare, Rosalie 239
Hill, Rosamond and Florence 134
Hort, Mrs Alfred 283
Lady, a 302
Leonowens, Anna 138
Meredith, Louisa 304
Miles, Beryl 185
Montez, Lola 55
Muter, Mrs D. D. 266
Parker, Mary 23
Shaw, Flora 193
Stevenson, Margaret 246
Thomson, Mrs 306

Pacific Islands

Bird, Isabella 81
Brassey, Lady Anna 203
Broad, Lucy 154
Cameron, Charlotte 176
Cheesman, Evelyn 130
David, Mrs Edgeworth 234
Dodwell, Christina 89
Forbes, Anna 281
Forbes, Rosita 91
Fountaine, Margaret 132

Gordon Cumming, Constance 93
Grimshaw, Beatrice 182
Hall, Mary 15
Hort, Mrs Alfred 283
Johnson, Osa 50
King, Agnes 213
Paton, Maggie 168
Routledge, Katherine 147
Stevenson, Margaret 246

Canada

Bosanquet, Mary 84
Dibble, L. Grace 88
Duffus Hardy, Lady 111
Hall, Mrs Cecil 300
Hargrave, Letitia 300
Hasell, Eva 161
Hubbard, Mrs Leonidas 16
Jameson, Anna 286
King, Harriett 301

Macpherson, Annie 166
Marryat, Florence 185
Moodie, Susanna 22
Murray, Hon. Amelia 145
Muter, Mrs D. D. 266
Pilley, Dorothy 73
Ratcliffe, Dorothy 99
Simcoe, Elizabeth 224
Traill, Catherine 307

North America

Bird, Isabella 81
Bosanquet, Mary 84
Bromley, Mrs 108
Burlend, Rebecca 296
Cameron, Charlotte 176
Clerk, Mrs Godfrey 109
Davison, Ann 10
Duffus Hardy, Lady 111
Fay, Eliza 43
Franklin, Lady Jane 92
Gordon Cumming, Constance 93
Hall, Mrs Cecil 300
Herbert, Agnes 69
Hort, Mrs Alfred 283
Howard, Lady Winefred 115
Jameson, Anna 286
Kemble, Fanny 240

Mannin, Ethel 183
Marryat, Florence 185
Martineau, Harriet 143
Miller, Christian 54
Moffat, Gwen 71
Montez, Lola 55
Morris, Jan 188
Murray, Hon. Amelia 145
Pender, Rose 219
Pilley, Dorothy 73
Schaw, Janet 222
Stenhouse, Fanny 171
Stevenson, Margaret 246
Stuart Wortley, Lady Emmeline 121
Stuart Wortley, Victoria 121
Townley, Lady Susan 247
Trollope, Frances 197

Caribbean

Barker, Lady Mary 295
Bromley, Mrs 108
Brown, Lady Richmond 129
Carmichael, Mrs 297
Falconbridge, Anna 211
Forbes, Rosita, 91
Fountaine, Margaret 132
Franklin, Lady Jane 92
Gaunt, Mary 181

Kindersley, Jemima 242
Layard, Mrs Granville 116
Lloyd, Susette 214
Murray, Hon. Amelia 145
Nugent, Lady Maria 217
Riddell, Maria 221
Schaw, Janet 222
Seacole, Mary 270
Stuart Wortley, Lady Emmeline 121

Central and South America

Bromley, Mrs 108
Burton, Lady Isabel 232
Calderon de la Barca, Frances 233
Cameron, Charlotte 176
Clementi, Mrs Cecil 207
Clerk, Mrs Godfrey 109
Cressy-Marcks, Violet 42
Davis, Elizabeth 258
Dixie, Florence 65
Foote, Mrs Henry 281
Forbes, Rosita 91
Fountaine, Margaret 132
Graham, Maria 44
Hanbury Tenison, Marika 238
Hobson, Sarah, 48
Hort, Mrs Alfred 283

Howard, Lady Winefred 115
Kindersley, Jemima 242
Lester, Mary 140
Miles, Beryl 185
Mills, Lady Dorothy 187
Murphy, Dervla 97
Peck, Annie 24
Poole, Annie 290
Richardson, Gwen 56
Seacole, Mary 270
Simpson, Myrtle 101
Stuart Wortley, Lady Emmeline 121
Stuart Wortley, Victoria 121
Swale, Rosie 58
Townley, Lady Susan 247
Tullis, Julie 60

Islands

Barkly, Fanny (Seychelles,
  Heligoland) 201
Brassey, Lady Anna (Cyprus) 203
Clifton, Violet (Andaman, Nicobar) 233
Evans, Katharine (Malta) 159
Fountaine, Margaret (various) 132

Gordon Cumming, Constance
  (various) 93
Hutchison, Dr Isobel (Pribilof,
  Aleutian) 135
Montefiore, Lady Judith (Malta) 168
Scott-Stevenson, Esmé (Cyprus) 223

World Tourists

Bird, Isabella 81
Bridges, F. D. 84
Broad, Lucy 154
Cameron, Charlotte 176
Clerk, Mrs Godfrey 109
Forbes, Rosita 91

Fountaine, Margaret 132
Franklin, Lady Jane 92
Gordon Cumming, Constance 93
Mannin, Ethel 183
Mills, Lady Dorothy 187
Morris, Jan 188

Index of Authors

This index includes the titles of books by unacknowledged or unknown authors, with cross-references where necessary.

OXFORD

MORE OXFORD PAPERBACKS

This book is just one of nearly 1000 Oxford Paperbacks currently in print. If you would like details of other Oxford Paperbacks, including titles in the World's Classics, Oxford Reference, Oxford Books, OPUS, Past Masters, Oxford Authors, and Oxford Shakespeare series, please write to:

UK and Europe: Oxford Paperbacks Publicity Manager, Arts and Reference Publicity Department, Oxford University Press, Walton Street, Oxford OX2 6DP.

Customers in UK and Europe will find Oxford Paperbacks available in all good bookshops. But in case of difficulty please send orders to the Cash-with-Order Department, Oxford University Press Distribution Services, Saxon Way West, Corby, Northants NN18 9ES. Tel: 0536 741519; Fax: 0536 746337. Please send a cheque for the total cost of the books, plus £1.75 postage and packing for orders under £20; £2.75 for orders over £20. Customers outside the UK should add 10% of the cost of the books for postage and packing.

USA: Oxford Paperbacks Marketing Manager, Oxford University Press, Inc., 200 Madison Avenue, New York, N.Y. 10016.

Canada: Trade Department, Oxford University Press, 70 Wynford Drive, Don Mills, Ontario M3C 1J9.

Australia: Trade Marketing Manager, Oxford University Press, G.P.O. Box 2784Y, Melbourne 3001, Victoria.

South Africa: Oxford University Press, P.O. Box 1141, Cape Town 8000.

OXFORD LIVES

STANLEY

Volume I: The Making of an African Explorer
Volume II: Sorceror's Apprentice

Frank McLynn

Sir Henry Morton Stanley was one of the most fascinating late-Victorian adventurers. His historic meeting with Livingstone at Ujiji in 1871 was the journalistic scoop of the century. Yet behind the public man lay the complex and deeply disturbed personality who is the subject of Frank McLynn's masterly study.

In his later years, Stanley's achievements exacted a high human cost, both for the man himself and for those who came into contact with him. His foundation of the Congo Free State on behalf of Leopold II of Belgium, and the Emin Pasha Relief Expedition were both dubious enterprises which tarnished his reputation. They also revealed the complex—and often troubling—relationship that Stanley has with Africa.

'excellent . . . entertaining, well researched and scrupulously annotated' *Spectator*

'another biography of Stanley will not only be unnecessary, but almost impossible, for years to come' *Sunday Telegraph*

OXFORD LIVES

'SUBTLE IS THE LORD'

The Science and the Life of Albert Einstein

Abraham Pais

Abraham Pais, an award-winning physicist who knew Einstein personally during the last nine years of his life, presents a guide to the life and the thought of the most famous scientist of our century. Using previously unpublished papers and personal recollections from their years of acquaintance, the narrative illuminates the man through his work with both liveliness and precision, making this *the* authoritative scientific biography of Einstein.

'The definitive life of Einstein.' Brian Pippard, *Times Literary Supplement*

'By far the most important study of both the man and the scientist.' Paul Davies, *New Scientist*

'An outstanding biography of Albert Einstein that one finds oneself reading with sheer pleasure.' *Physics Today*

OXFORD LETTERS AND MEMOIRS
RICHARD HOGGART

A Local Habitation
Life and Times: 1918–1940

With characteristic candour and compassion, Richard Hoggart evokes the Leeds of his boyhood, where as an orphan, he grew up with his grandmother, two aunts, an uncle, and a cousin in a small terraced back-to-back.

'brilliant . . . a joy as well as an education' Roy Hattersley

'a model of scrupulous autobiography' Edward Blishen, *Listener*

A Sort of Clowning
Life and Times: 1940–1950

Opening with his wartime exploits in North Africa and Italy, this sequel to *A Local Habitation* recalls his teaching career in North-East England, and charts his rise in the literary world following the publication of *The Uses of Literacy*.

'one of the classic autobiographies of our time' Anthony Howard, *Independent on Sunday*

'Hoggart [is] the ideal autobiographer' Beryl Bainbridge, *New Statesman and Society*

PAST MASTERS

General Editor: Keith Thomas

SHAKESPEARE

Germaine Greer

'At the core of a coherent social structure as he viewed it lay marriage, which for Shakespeare is no mere comic convention but a crucial and complex ideal. He rejected the stereotype of the passive, sexless, unresponsive female and its inevitable concommitant, the misogynist conviction that all women were whores at heart. Instead he created a series of female characters who were both passionate and pure, who gave their hearts spontaneously into the keeping of the men they loved and remained true to the bargain in the face of tremendous odds.'

Germaine Greer's short book on Shakespeare brings a completely new eye to a subject about whom more has been written than on any other English figure. She is especially concerned with discovering why Shakespeare 'was and is a popular artist', who remains a central figure in English cultural life four centuries after his death.

'eminently trenchant and sensible . . . a genuine exploration in its own right' John Bayley, *Listener*

'the clearest and simplest explanation of Shakespeare's thought I have yet read' Auberon Waugh, *Daily Mail*

PAST MASTERS

General Editor: Keith Thomas

KIERKEGAARD

Patrick Gardiner

Søren Kierkegaard (1813–55), one of the most original thinkers of the nineteenth century, wrote widely on religious, philosophical, and literary themes. But his idiosyncratic manner of presenting some of his leading ideas initially obscured their fundamental import.

This book shows how Kierkegaard developed his views in emphatic opposition to prevailing opinions, including certain metaphysical claims about the relation of thought to existence. It describes his reaction to the ethical and religious theories of Kant and Hegel, and it also contrasts his position with doctrines currently being advanced by men like Feuerbach and Marx. Kierkegaard's seminal diagnosis of the human condition, which emphasizes the significance of individual choice, has arguably been his most striking philosophical legacy, particularly for the growth of existentialism. Both that and his arresting but paradoxical conception of religious belief are critically discussed, Patrick Gardiner concluding this lucid introduction by indicating salient ways in which they have impinged on contemporary thought.

HISTORY IN OXFORD PAPERBACKS
TUDOR ENGLAND
John Guy

Tudor England is a compelling account of political and religious developments from the advent of the Tudors in the 1460s to the death of Elizabeth I in 1603.

Following Henry VII's capture of the Crown at Bosworth in 1485, Tudor England witnessed far-reaching changes in government and the Reformation of the Church under Henry VIII, Edward VI, Mary, and Elizabeth; that story is enriched here with character studies of the monarchs and politicians that bring to life their personalities as well as their policies.

Authoritative, clearly argued, and crisply written, this comprehensive book will be indispensable to anyone interested in the Tudor Age.

'lucid, scholarly, remarkably accomplished . . . an excellent overview' *Sunday Times*

'the first comprehensive history of Tudor England for more than thirty years' Patrick Collinson, *Observer*

HISTORY IN OXFORD PAPERBACKS

THE STRUGGLE FOR THE MASTERY OF EUROPE 1848–1918

A. J. P. Taylor

The fall of Metternich in the revolutions of 1848 heralded an era of unprecedented nationalism in Europe, culminating in the collapse of the Hapsburg, Romanov, and Hohenzollern dynasties at the end of the First World War. In the intervening seventy years the boundaries of Europe changed dramatically from those established at Vienna in 1815. Cavour championed the cause of *Risorgimento* in Italy; Bismarck's three wars brought about the unification of Germany; Serbia and Bulgaria gained their independence courtesy of the decline of Turkey—'the sick man of Europe'; while the great powers scrambled for places in the sun in Africa. However, with America's entry into the war and President Wilson's adherence to idealistic internationalist principles, Europe ceased to be the centre of the world, although its problems, still primarily revolving around nationalist aspirations, were to smash the Treaty of Versailles and plunge the world into war once more.

A. J. P. Taylor has drawn the material for his account of this turbulent period from the many volumes of diplomatic documents which have been published in the five major European languages. By using vivid language and forceful characterization, he has produced a book that is as much a work of literature as a contribution to scientific history.

'One of the glories of twentieth-century writing.'
Observer